VLSI DESIGN OF NEURAL NETWORKS

THE KLUWER INTERNATIONAL SERIES
IN ENGINEERING AND COMPUTER SCIENCE

VLSI, COMPUTER ARCHITECTURE AND
DIGITAL SIGNAL PROCESSING
Consulting Editor
Jonathan Allen

VLSI DESIGN OF NEURAL NETWORKS

edited

by

Ulrich Ramacher

Corporate Research & Development, SIEMENS AG

and

Ulrich Rückert

University of Dortmund

SPRINGER-SCIENCE+BUSINESS MEDIA, B.V.

Library of Congress Cataloging-in-Publication Data

VLSI design of neural networks /   edited by Ulrich Ramacher and Ulrich Rückert.
    p. cm. — (The Kluwer international series in engineering and computer science. VLSI, computer architecture, and digital signal processing)
  Includes bibliographical references and index.
  ISBN 978-1-4613-6785-7   ISBN 978-1-4615-3994-0 (eBook)
  DOI 10.1007/978-1-4615-3994-0
  1. Integrated circuits—Very large scale integration—Design and construction. 2. Neural networks (Computer science)—Design and construction. I. Ramacher, Ulrich. II. Rückert, Ulrich.
III. Series.
TK7874.V559 1991
621.39′5–dc20   90–48757
                                       CIP

Contents

vi

CONTRIBUTING AUTHORS

P. Jespers
M. Verleysen
Univ. Cathol. de Louvain, Belgium

Ch. Nielsen
J. Staunstrup
Univ. of Lyngby, Denmark

J. Ouali
G. Saucier
INPG/CSI, Grenoble, France

P. Bessiere
LGI/IMAG, Grenoble, France

A. Chams
A. Guerin
J. Herault
C. Jutten
J.C. Lawson
LTIRF/INPG, Grenoble, France

J. Trilhe
SGS-THOMSON, Grenoble, France

P.Y. Alla
J.D. Gascuel
J. Roman
M. Weinfeld
Ecole Polytechnique,
Palaiseau, France

G. Dreyfus
A. Johannet
L. Personnaz
Ecole Supérieure, Paris, France

S. J. Prange
H. Klar
Techn. Univ. of Berlin, FRG

U. Rückert
Univ. of Dortmund, FRG

M. Kespert
P. Richert
M. Schwarz
FhG-IMS Duisburg, FRG

J. Nijhuis
A. Siggelkow
L. Spaanenburg
IMS Stuttgart, FRG

J. Anlauf
J. Beichter
N. Brüls
U. Hachmann
W. Raab
U. Ramacher
B. Schürmann
M. Wesseling
SIEMENS, Corp. R&D, FRG

J. Hoekstra
Delft University, The Netherlands

M. Chevroulet
E. Dijkstra
M.A. Maher
O. Nys
H. Oguey
E. Vittoz
CSEM, Neuchâtel, Switzerland

D. J. Baxter
S. Churcher
A. Hamilton
A. F. Murray
H. M. Reekie
Univ. of Edinburgh, U.K.

M. Brownlow
L. Tarassenko
Univ. of Oxford, U.K.

S. Jones
K. Sammut
Univ. of Nottingham, U.K.

J. A. Vlontzos
Siemens Corp. Research,
Princeton, USA

S. Y. Kung
Princeton Univ., USA

FOREWORD

The early era of neural network hardware design (starting at 1985) was mainly technology driven. Designers used almost exclusively analog signal processing concepts for the recall mode. Learning was deemed not to cause a problem because the number of implementable synapses was still so low that the determination of weights and thresholds could be left to conventional computers.

Instead, designers tried to directly map neural parallelity into hardware. The architectural concepts were accordingly simple and produced the so-called interconnection problem which, in turn, made many engineers believe it could be solved by optical implementation in adequate fashion only. Furthermore, the inherent fault-tolerance and limited computation accuracy of neural networks were claimed to justify that little effort is to be spend on careful design, but most effort be put on technology issues.

As a result, it was almost impossible to predict whether an electronic neural network would function in the way it was simulated to do. This limited the use of the first neuro-chips for further experimentation, not to mention that real-world applications called for much more synapses than could be implemented on a single chip at that time.

Meanwhile matters have matured. It is recognized that isolated definition of the effort of analog multiplication, for instance, would be just as inappropriate on the part of the chip designer as determination of the weights by simulation, without allowing for the computing accuracy that can be achieved, on the part of the user.

Hardware design of neural networks is now more frequently preceded by the investigation of the constraints introduced by the application and system environment, the performance required and the comparison of various technologies. Designers strive for architectural solutions of the interconnection problem, trying to establish to what extent it is possible to deviate from the massively parallel networking and yet to satisfy an application-specific real-time constraint.

The papers presented in this volume reflect this new sight and mark the entry to a more sophisticated era of VLSI design of neural networks.

Ulrich Ramacher

Ulrich Rückert

PREFACE

The book originated from a workshop on MICROELECTRONICS FOR NEURAL NETWORKS which was held at the University of Dortmund, W. Germany, in June 1990. It presents the approach to VLSI design of Neural Networks developed by leading experts from Europe over the last couple of years.

The intention was to provide a collection of revised and extended papers which an interested VLSI-designer or student of Electrical Engineeringor Computer Science can use as an introduction to research and as a reference for further work by his own.

In Chapter 1 some important considerations preceding VLSI implementation of neural nets are recalled.

Chapter 2 gives a review of CCD technology and focusses particularly on the neural net potential provided by Junction-CCDs. Chapter 3 presents a full account of the possibilities for analog storage of adjustable synapses and discloses results recently obtained. The computational precision of analog design is dealt with in Chapter 4; a specific method is presented to reduce the errors due to mismatching of components in a VLSI neuron.

Chapters 5 and 6 cope with VLSI architectures which follow the biological models. Digital, analog and pulse stream techniques are discussed.

Chapters 7 to 10 are devoted to the presentation of various types of application specific neural chips. Pulse density modulated neural nets, adaptive associative memories and arbitrarily structured neural nets are supported by special silicon. Digital as well as analog design by means of a neural silicon compiler, gate arrays or full custom design are described.

Chapters 11 to 14 deal with the design of neurocomputers. Chapter 11 presents the fundamental ring systolic architecture and discusses system architectures using off-the-shelf DSPs. Architectural and implementation issues concerning the processor granularity of ring systolic systems are presented in Chapter 12. Chapter 13 discloses a detailed study of neural algorithms which results in a unified description of neural models by 3 equations. A related fine-grain neurocomputer architecture based on a VLSI Signal Processor is described in Chapter 14.

The book concludes with chapter 15 which contains the computer scientist's view on neurocomputer design. Software design and its implication on hardware are particularly emphasized.

VLSI DESIGN OF NEURAL NETWORKS

GUIDE LINES TO VLSI DESIGN OF NEURAL NETS

U. RAMACHER

INTRODUCTION

The response and the characteristics of present models of artificial neural nets are primarily investigated by simulation on vector computers, workstations, special coprocessors or transputer arrays. The fundamental drawback of such simulators is that the spatio-temporal parallelism in the processing of information that is inherent to the neural net is lost entirely or partly and that the computing time of the simulated net especially for large associations of neurons (tailored to application-relevant tasks) grows to such orders of magnitude that a speedy acquisition of "neural" know-how is hindered or made impossible.

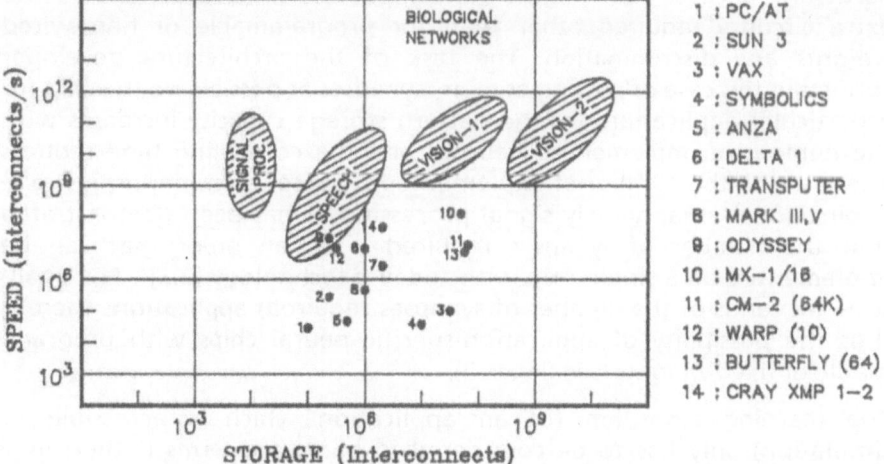

Figure 1   Computational capabilities of neural network simulators versus computational requirements of some applications

Figure 1 shows the performance obtainable with commercially available simulators [1] in terms of weights and weights per second. This must be confronted with the applicational needs. It becomes obvious that today´s hardware capabilities are limiting the development of neural network research.

An appreciable reduction in computing time and thus the handling of largish tasks or those that are to be executed in realtime become possible with specially designed neural hardware. Apart from the shortest possible computing time, neural hardware offers a very much smaller structural volume than can be implemented with simulators for the same task. This aspect is especially important when neural hardware is to be incorporated in terminals for man-machine communication or mobile robotics.

VLSI OPPORTUNITIES FOR NEURAL NETS

Depending on the application under consideration, the user will tell whether his problem is accessible by simulation on conventional computers or not.

If yes, the applicational task under consideration can be well defined and the user will specify the kind of data format and the degree of weight resolution, the size and type of the network and the processing speed for the recall mode. If the real-time requirement cannot be satisfied by software implementation, it makes sense to think about designing special hardware. Because the weights are computable in advance, there is no extra circuitry required other than for programmable or hard-wired weights and discrimination. The task of the architecture developer consists in this case of putting as many synapses as possible on the chip for a particular application, for the pattern storage capacity increases with the number of implemented neurons and the computing time reduces linearly with the number of implemented synapses. Considering just one applicational area, namely signal processing, it has been demonstrated that the number of synapses required is of an order that can be implemented on a single chip with today's technology [2,3] . For small-scale (in terms of the number of synapses required) applications there is thus the possibility of application-specific neural chips with programmable or fixed weights (see figure 2).

The learning algorithm (of an application which is accessible to simulation) only has to be considered in hardware terms if there is a relevant real-time requirement. The latter is imaginable, for instance, if the learnt knowledge is valid for a short time and new learning is repeatedly necessary. Obviously, supporting the learning will be at the expense of the number of synapses implemented, and it has to be checked whether single-chip integration will be possible at all. Wafer Scale Integration may turn out to provide the integration potential for computing and storing synaptic weights and intermediate results produced by the learning algorithm [4,5]. An alternative popular proposal is to distribute the implementation of a neural net plus learning algorithm over several chips and cascade these.

Neural nets for applications like vision or speech, on the other hand, overtax the single-chip integration potential of present technology as well as that of future 0.3-μm technology by whole orders of magnitude

3

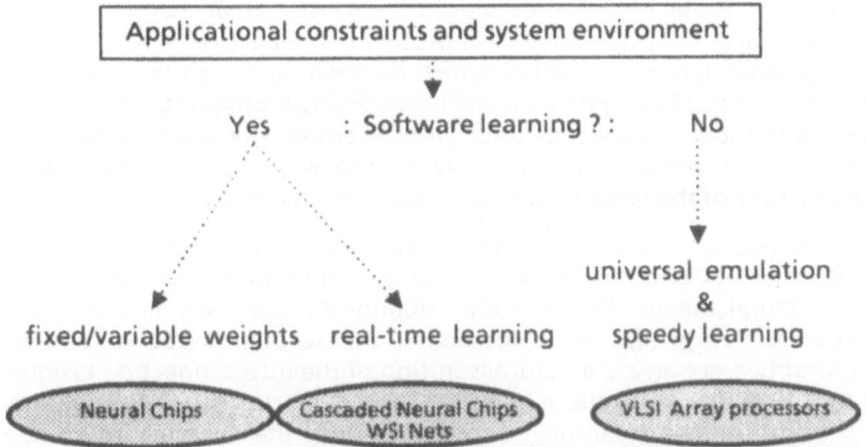

Figure 2   Applicational dimensions and VLSI opportunities

(see figure 1). Particularly the weights will have to be stored off-chip. The size of such a net will not permit simulation (especially of the learning phase) within a reasonable period of time and therefore the weights cannot be determined by simulation. This means that little or no engineering know-how has been accumulated for these large applications. Consequently, the VLSI architecture must be designed for universal emulation of neural network structures and speedy learning. VLSI architectures of this type will obviously look different from those considered in small-scale applications.

In the sequel the two opportunities and tries to present a guide line to neural VLSI design with special focus on the requirements caused by applicational needs, technological constraints and system environment.

SINGLE-CHIP INTEGRATION

Single-chip integration of a neural net means on-chip storage of the weights (not more than a few ten thousands of synapses today [2]). Consequently the amount of information a neural chip can process will be limited in size. This means in turn that the application must match the implementation potential of the technology. Fortunately, the learning for neural nets that can still be integrated on a chip can be performed on conventional computers with reasonable time expenditure, so the applicational task under consideration can be well defined and the hardware matched optimally to the task.

On the one hand, on-chip storage of weights offers an easy way to achieve real-time action by neural networks, since there exists no pad-bandwidth problem for the weights. On the other hand, the VLSI designer faces the interconnection problem, irrespective of which design style (analog, binary, digital) or which technology is used [6,7]. If several ten thousands of weights were to be connected physically to a neuron and some thousand neurons to be implemented, the wiring area would grow to such orders that the delay on the wires tends to exceed the latency time of the functional block representing a neuron.

In principle, there are two ways to fight the interconnection problem: firstly, by reducing the technological structure size, and secondly, by architectural means. The first way dominated the early era of neural hardware design [8,9,10] . Nowadays, the second course is followed preferably, since an architectural solution of the interconnection problem is the cheapest. As a rule, a designer should check first for the required processing time (realtime conditions) of the application under consideration and then think out to what extent it is possible to deviate from the ideal massive parallel networking of a neural net. Together with the decision as to whether analog or digital signal processing concepts are to be applied and the selection of a technology, there results an initial architecture draft. Instead of reviewing architectures the following sections show what influence the technology and signal processing concept exert on the further design of an architecture whatever it may look like.

Analog Design Issues

Particularly for analog realization, the emphasis of circuit design is on the exploitation of the functional properties inherent in basic circuit elements. Generally the architecture draft needed for the analog implementation of a neural net will concentrate on representing the synaptic weighting, the neural ignition response (discriminator function) and the controlling of data input and output. An important requirement is the compactness of the connection element, because the cell size mainly determines the overall area of the network.

An advantage of analog circuits in comparison to digital circuits is that they can "process more than 1 bit per transistor". If this benefit is to be made use of, however, the following problems have to be mastered:

-- Nonvolatile storage of analog weights provides very high synaptic density, but may be not sufficiently often programmable [11].

-- The design of a synapse, the size of a neural network and the degree of analog resolution are dependent on each other [12].

-- A major design problem with analog circuits is the relation of accuracy to chip area. The more precisely one wishes to control the matching of the analog components, the more chip area is needed. An analog depth of not more than 8 bit is recommendable. Crosstalk and susceptibility to coupled-in interference make special precautions necessary for analog signal processing.

-- The minimal chip area is also influenced by factors like noise and current consumption. Low current consumption calls for high-valued resistors in resistor-capacitor circuitry. Low noise creates certain limits for minimal transistor Q surfaces and capacitors; this applies to switch-capacitor as well as to resistor-capacitor technique.

-- The temperature dependence, clock feed-through and process-parameter dependence can be reduced in analog circuits so that they no longer interfere, but this is done at the expense of circuit complexity and has an effect on the chip area.

-- With future 0.3-μm transistor channel length, a lower supply voltage than 5 V must be expected, so questions of low-voltage design have to be considered. Accurate transistor modelling, innovative circuit techniques and design cleverness will be significant here like in the cases above.

The limited precision in grading the weights (realised, for example, in the form of ohmic resistances or switched capacitors) means, on the one hand, limited computing accuracy for an analog implementation and therefore influences the number and complexity of the patterns that can be reliably processed with an analog net. This applies equally to the selection of the discriminator: the computing accuracy of the entire analog chip has to be considered.

The limited precision in processing information by the analog neural net means, on the other hand, that the information must be encoded in a redundant or fuzzy fashion, i.e. only as sharply defined as is necessary for secure recognition.

If a learning algorithm (with its multiple iterations) is to be implemented in analog circuitry, it is necessary to ensure a fortiori whether the intended application can at all be learnt.

Therefore it is the task of the user concerned with problem analysis and modelling to characterize those tasks of pattern processing in which deviations of the actual weight values from the pre-computed ones can be tolerated and where it is possible to make do, for example, with ternary weighting of binary input/output signals. For, the latter poses the least problems to the analog designer.

In the analog implementation of neural nets it is consequently a matter of bringing together the application-oriented problem analysis and the circuit architecture; this is the only way to determine the application spectrum that can be implemented with analog hardware. Isolated definition of the effort for analog multiplication, for instance, would be just as inappropriate on the part of the chip architect as determination of the weights by simulation, without allowing for the computing accuracy that can be achieved, on the part of the user.

Digital Design Issues

What is of clear advantage here is that the application-oriented problem analysis and modelling of net characteristics can be made independently of circuit design. For by simulation one can determine the computation accuracy (word width) for each function block before implementing it. The technology and the circuit architecture only determine how many neurons can be integrated on a chip and what processing time can be achieved.

It is the task of the chip designer then to implement the neural algorithms in accordance to the state-of-the-art in Digital Signal Processing. A point to be tackled, for instance, is the problem of the so-called limit cycles, i.e. parasitic oscillations. These can appear in recursive digital structures (in the learning phase or in feedback nets for example) as a result of amplitude quantization (in analog recursive structures this problem consequently does not occur). With the neurons the word limiting, for algorithmic reasons, has to be implemented as a saturation characteristic anyway, so there is no danger of large limit cycles. The small limit cycles, which can be produced by bit manipulations (rounding, truncation), are dependent on the word lengths used and the structure of the feedback net. The problem of limit cycles can be eliminated however by appropriate effort in computing accuracy and numeric range (floating point). If this effort becomes unsupportable, it will be necessary to check whether such limit cycles appear and whether they can have a disturbing effect. One should also investigate whether limit cycles can be suppressed by selecting appropriate algorithms and measures.

The finer resolution of the weights, which is easy to implement in digital architecture, allows more individual weighting of the input signals than in analog processing. Furthermore, digital design can supply much easier the word width necessary to implement a learning algorithm. Thus it can be concluded that digital neural chips take on a field of application in their own right.

Analog versus Digital

Figure 3 displays the various ways how analog, binary or digital data can be processed. As discussed above, if the information must be processed with high precision (say with not less than 8 bits) or learning is to be supported on-chip, digital circuitry is the right candidate for implementing a neural net; conversely, for applications which do not need hardware support for learning and for less severe requirements in computation precision, analog design seems to dominate.

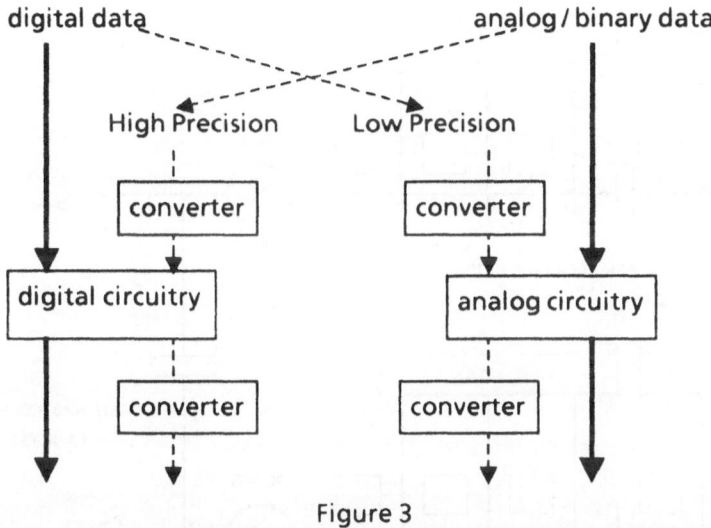

Figure 3

This picture changes if the costs for interfacing the data with the neural circuitry are to be taken into account. For instance, video images with, say, 8 bit pixels could be operated on by analog circuitry. But this would require fast D/A converters which were to be provided at the input and output ports. The costs of these could speak in favour of an all digital design [2].

Also, if the requirements for computing accuracy are not too high, various coding and transmission methods can be used that offer high noise immunity, e.g. pulse-density modulation or pulse-width modulation. But then various converters are indispensable. The effort for such complex circuit blocks and for pads can be kept within supportable limits through the use of multiplexers. However the use of time-division multiplex with continuous-time circuits (i.e. RC circuits) is not meaningful if the resistance of the switches affects the function. In this case, time-division multiplex is much easier to implement with discrete-time circuits , i.e. analog switch-capacitor circuits or digital circuits.

Meanwhile, a great number of technologies have been tried for direct implementation of neural nets, and a dozen chips built [13] . To get an idea of what processing speed and net size can be achieved by single-chip implementation some recent examples are reviewed.

An example for mostly digital design is presented by the configurable CMOS neural network chip described by H. P. Graf [2]. The chip provides 256 blocks each of which consists of 128 binary synapses. The ternary weights are stored in a 6-transistor cell and the multipliers are realized as inverted XORs. The multipliers' outputs are accumulated in analog

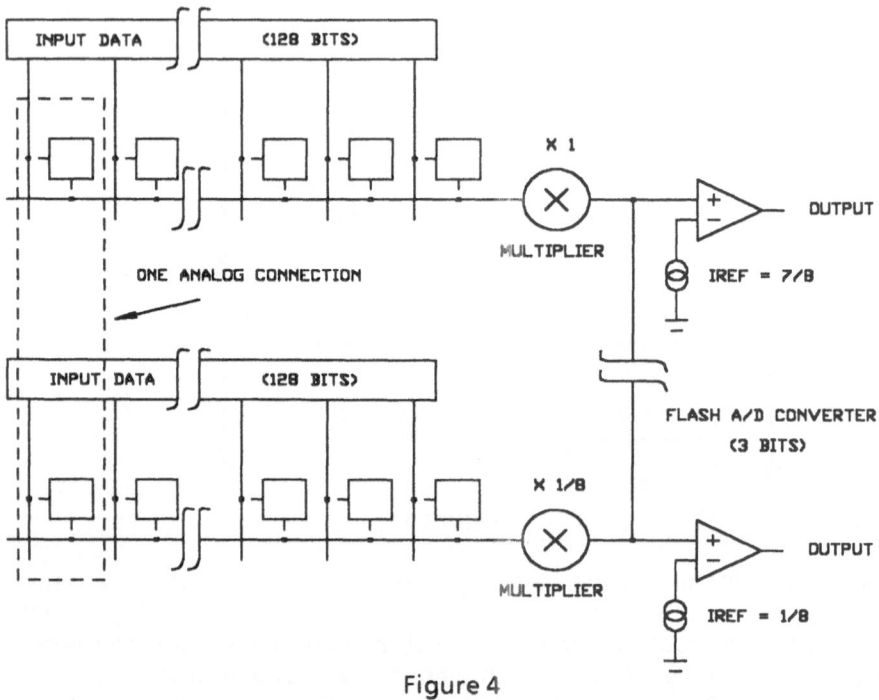

Figure 4

fashion, and the resulting current is presented via a subsequent multiplier
to a comparator. The additional multiplier is set to one of the values 1,
1/2, 1/4 or 1/8, and the comparator can be bypassed (see explanation
below). Beside the wire that connects the outputs of the synapses, the
comparator is the only analog functional unit, all other components are
designed digital. Configuration of the chip is possible with regard to the
number B (≤ 8) of synaptic blocks attributed to a neuron, and the word
width W (1to 4 bit) of the input data (see figure 4). For instance, with
B = 4 and W = 1 a neuron posseses 4x128 = 512 binary synapses and there
are 256/4 = 64 such neurons. Since the chip's bus width is 128 bit, 4 clock
cycles are necessary to present the 512 binary signals to the 64 neurons. In
case that a configuration with B = 4 and W = 4 has been chosen, a neuron
has 512 synapses each of which consists of 4 1-bit multipliers.
Consequently the number of neurons is cut to 64/4 = 16. After having
served the most significant bit of 512 4bit-words in 4 clock cycles, in the
next 4 clock cycles the next lower bit of these words are transferred to
associated 1-bit synapses, and so on. In total it takes 4x4 = 16 clock cycles
to present the 512 4-bit words to the 16 neurons. Note that every
neuron's additional multiplier (mentionned above) is preset in

accordance to the bit valence. The outputs of these multipliers are summed up again in analog fashion. The resulting value is presented to 4 comparators the threshold of which is preset according to the bit valence or any other number representation.

In terms of the coordinates used in figure 1 and a clock frequency of 10 MHz, the chip´s performance ranges from (32768,320G) for 256 neurons, each with 128 binary inputs, to (8192,10G) for 8 neurons, each with 1032 4bit-inputs. Consider then a video environment with 50 pictures recorded per second, each carrying 512x512 4bit-pixels. Depending on the selected configuration the chip would run 100 to 3 times faster than needed. Thus the chip´s computational performance is well balanced with respect to the applicational requirements and system environment met in low-level image recognition, and leaves a sufficient reserve for pictures with 10^6 pixels (25 to 0.8).

Two other chips have recently been realized. The one uses CCD and floating-gate technology, allowing for analog input which is weighed digitally (6 bits) by a multiplying D/A converter. This chip has 2016 programmable weights and performs about 3 billion weights per second [6]. The other chip developed by the group of M. Holler is fully analog in storage and multiplication of weights [7]. It uses EEPROM and NMOS technology, all functional blocks are realized in analog design with about 4bit analog depth of resolution. The number of synapses amounts to maximally 10240 (half of them reserved for feedback connections) and the performance is expected to reach 10 billion weights per second with a 64 bit wide bus. In contrast, Graf´s CMOS chip, if configured for 4 bit resolution, performs 80 billion weights per second (see table 1) . The lower speed of the analog chip [7] is mainly due to the floating gate device used for storage of weights.

APPROACH	SYNAPTIC DENSITY	4BIT CONNECTIONS PER SECOND
[2] digital 0.9μm	8 000 x 4bit at 0.15cm^2	80 billion (128 bit bus)
future 0.3 μm	ca.500000 x 4bit per cm^2	
[7] analog 1μm	10 000 x 4bit at 0.2 cm^2	10 billion (64 bit bus)
future 0.3μm	?	

Table 1   Comparison of the analog and digital approach

Comparing the synaptic density obtained either with digital VLSI design and 0.9 μm CMOS process [2] or with analog VLSI design and 1 μm NMOS technology [7] , it turns out that both approaches can achieve about 50 thousand 4bit-weights per cm^2. Note that the area specified for the analog design is based on the area of a single synapse; the overhead for interwiring and power supply etc. couldn´t be included because of lack of data [7]. In contrast, the area specified for the digital design is the chip´s area. It appears that even more aggressive design can not achieve a clear lead of analog processing over digital one.

Moreover, future 0.3 μm CMOS processes will allow for chip areas of about 4 cm^2, giving a "digital" upper limit of about 2 million 4bit-weights. As analog design does not scale linearly with structure size, digital design seems to be going to dominate neural chip design.

It must be remembered, however, that analog VLSI design of neural networks will need to be pursued because the integration of analog sensors and neural pre-processing is of outmost importance to many applications [14] and brings about performance advantages at the system level (like reduced costs for interfaces).

VLSI CIRCUITS FOR NEUROCOMPUTERS

Neural hardware can be tailored
--   for an application asking for a single neural network fixed in size and topology,
--   for an application which calls for a set of different neural networks which are programmable as to size and concatenation,
--   or be designed to serve general-purpose neurocomputing.

The first applicational species amounts mainly to single-chip integration as described above; the second overtaxes in most cases the single-chip integration potential of present technology as well as that of future 0.3-μm technology by whole orders of magnitude, and a fortiori does the third species. Applicational domains like speech or vision can not be implemented on a single chip. So one has the choice (see figure 5) either

--   to reload repeatedly a new set of weights into the memory provided by the neural chip(s),
--   to distribute the weights over a sufficiently large number of neural chips
--   or to distribute the weights off-chip in DRAMs and the neural processing over dedicated signal processors.

Reloading is paradoxical, because the time for reloading the weights will exceed the computation time of the chip by orders of magnitude. For instance, to reload the chip of H.P. Graf considered above takes 256 clock cycles whereas the multiplication of the stored weights is performed in 1.

If the number of neural chips is increased to an order that some 10 millions of weights could be distributedly stored on-chip (see figure 1, vision1: static pictures), the question arises if such a multi-chip system is not too costly and voluminous compared to a system which has the weights stored off-chip in cheap DRAMs and the chip area reserved for signal processing only (see figure 5).

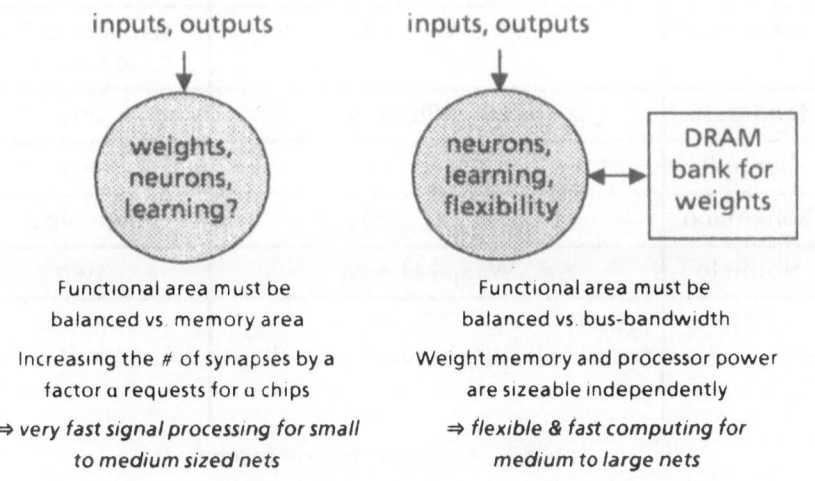

Functional area must be balanced vs. memory area

Increasing the # of synapses by a factor a requests for a chips

⇒ *very fast signal processing for small to medium sized nets*

Functional area must be balanced vs. bus-bandwidth

Weight memory and processor power are sizeable independently

⇒ *flexible & fast computing for medium to large nets*

Figure 5   On-chip versus off-chip storage of weights

In the latter case, the chip architect will use the saved area for emulation of learning algorithms and a large variety of neural networks. This is urgently needed, since the size of large nets will not permit simulation (especially of the learning phase) within a reasonable period of time and therefore the weights cannot be determined by simulation.

In consideration of the resulting lack of neural engineering know-how for applications as vision or speech, it is thus a matter of designing a system and circuit architecture that

--   produces a sufficient measure of flexibility and expansion capacity for coping with the known paradigms and those to come,

--   and supports in optimal fashion the massive parallel networking.

The ultimate purpose of neurocomputer design is to avail the user of sufficient power and flexibility so that any application-specific study can be undertaken and an application-specific neural system architecture be worked out in terms of dedicated software and hardware.

Table 2 shows some of the more common paradigms, table 3 corresponding learning rules. As can be seen from the tables, the functional complexity for either search or learning is very low, scalar product operation
is the fundamental and compute-bound operation which is shared by all
models.

MODEL	WEIGHTING	DISCRIMINATION
linear	$z_i = \Sigma W_{ij} x_j + \Theta_i$	identity
non-linear	$z_i = \Sigma W_{ij} x_j + \Theta_i$	step, ramp, sig-moidal, tanh, ...
feed-back	$z_i = \alpha \cdot y_{i,old} + \beta \cdot \Sigma W_{ij} x_j$	any
Sigma-Pi	$z_i = \Sigma W_{ij} \cdot (\Pi x_k)_j$	any
Boltzmann	$z_i = (\Sigma W_{ij} x_j + \Theta_i) \cdot \lambda_i$	sigmoidal
Hopfield	$z_i = \Sigma W_{ij} x_j + \Theta_i + w_i \cdot l_i$	step
PAN	$y_i(t+1) =$ $\min\{1, G_i \cdot [\delta_i \cdot y_i(t) + \max(0, s_i(t) - \tau_i)]\}$	$s_i(t) = \Sigma$ $(W_{ij} + \Delta w_{ij}(t)) \cdot y_j(t)$

Table 2   Common neural models

RULE	AUTHOR
$\Delta w_{ij} = \eta \cdot \Sigma_p y_{i,p} \cdot x_{j,p}$	Hebb
$\Delta w_{ij} = \eta \cdot \Sigma_p (d_{i,p} - y_{i,p}) \cdot x_{j,p}$	Widrow-Hoff, Rosenblatt
$\Delta w_{ij} = \eta \cdot \Sigma_p y_{i,p} \cdot (x_{j,p} - w_{ij})$	Großberg
$\Delta w_{ij} = \eta \cdot (Z_{i,new} - Z_{i,old}) \cdot (y_{i,new} - y_{i,old}) \cdot x_j$	Barto, Sutton
$W_{ij} = \Sigma_p (2x_i^{(p)} - 1) \cdot (2x_j^{(p)} - 1)$	Hopfield
$\Delta w_{ij} = \eta/T \cdot (P_1(s_i s_j) - P_0(s_i s_j))$	Hinton, Seynowski
$\Delta w_{ij}(t+1) = \sigma_{ij} \cdot \Delta w_{ij}(t) + \lambda_{ij} \cdot y_i(t+1) \cdot y_j(t)$	Cruz-Young et. al.

Table 3   Common learning rules

Generally, a versatile neurocomputer architecture must be built up on the
elementary compute-intensive algorithmic strings shared by all known

neural paradigms. In addition, a set of universal substrings has to be derived from those algorithmic strings which are not common to all the models. Elementary strings and substrings would compose the functional blocks, which have to be implemented in silicon, whereas concatenation of strings or substrings would be initiated by instructions. As the variety of models can be created to a large extent by dedicated sequences of such instructions, the design of a neurocomputer must be preceeded by a careful analysis of the neural network models in use.

To give an example of the compute-bound algorithmic kernel of a fairly large class of neural networks consider these three equations [15] :

$$y_{ip} = f_i \left(\sum_{j=-1}^{N} W_{ij} \cdot y_{jp} \right) \tag{1}$$

$$f'_{np} \cdot \left\{ \frac{\partial E}{\partial y_{np}} \cdot \chi(n) + \sum_{i=1}^{N} W_{in} \cdot \Delta_{ip} \right\} = \Delta_{np} \tag{2}$$

$$W_{ij}(t+1) = W_{ij}(t) - a \cdot \left[\sum_{p \in P} \Delta_{ip} \cdot y_{jp} + \left(\frac{\partial E}{\partial W_{ij}} \right)_{explicit} \right] \tag{3}$$

The first equation represents the algorithm used for the search mode of an arbitrarily structured neural net for time-independent pattern recognition. The second and third equation constitute the algorithms used for the learning phase. The dynamics of learning is controlled by a very general objective function

$$E = \sum_{k,p} E_y(y_{k\,p}) + \sum_{k,l} E_W(W_{kl}) + \sum_{k,l,m,p} E_{yW}(y_{k\,p}, W_{ml})$$

$$+ \sum_{k,l,m,n} E_{WW}(W_{kl}, W_{mn}) + \sum_{k,l,p,q} E_y(y_{k\,p}, y_{lq}) + higher\,order\,terms \dots \ ,$$

which covers supervised learning as well as unsupervised learning. Included are all models which specify a cost function, a Ljapunov function or any other kind of objective function. The Boltzmann net , the family of associative nets, the Multi-layer Perceptron etc. are all contained. One can as well derive hybrid learning rules by devising hybrid objective functions, containing for example Hopfield-type and error functions [16] .

The compute-intensive operations explicitly and implicitly present in these equations are the multiplication of matrices and scalar product operations. Consequently, this set of compute-intensive operations has to be implemented by specially designed VLSI circuitry.

It has been shown in [17,18] that systolic arrays present a favourable solution to the architectural problem of mapping neural algorithms into silicon. The basic module of the array is a VLSI Array Processor that integrates these elementary strings as hard-wired functional blocks. By adding hardware from the shelf for low-throughput operations (like reading a discriminator table or adding "noise" to an updated weight) and designing a controller and communication architecture that fits the peculiarities of fine-grain systolic data flow at chip and board level, a system architecture of its own comes into being.

As the use of analog concepts is detracted from, besides others, by the low computing accuracy of only a few percent, and application-oriented problem analysis and modelling of neural network characteristics cannot be made independently of analog circuit design, digital design appears to be best suited for neurocomputers.

The neurocomputer concept outlined in figure 6 is sizeable to the application domain in terms of processing power, memory and flexibility [19]. Throughput rates at the chip site of the order of 500 MC/sec (1 Connection = 16 bit) can be realized with 1µm CMOS technology. By systolic extension to the board level a system performance of 20 GC/sec can be achieved with 40 of these chips, which is sufficient for vision 1 (see figure 1). Note that such a system configuration is equivalent to one chip of H.P. Graf in terms of computing power. However, in terms of network size and support for various neural nets and learning a neurocomputer concept of this kind gets to new dimensions. It will enable the user to access real-world applications in reasonable time and study in breadth and depth the potential of neural networks.

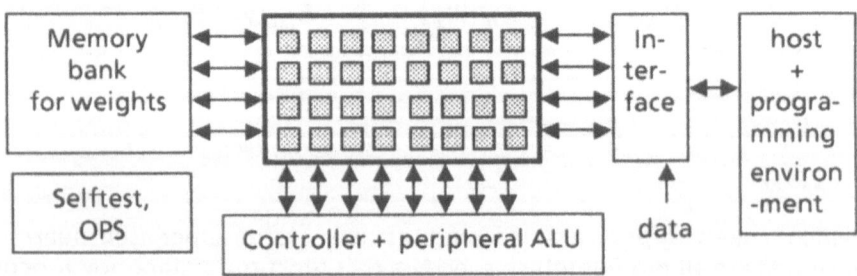

Goal: VLSI for fast & flexible emulation of neural networks for real-world applications
Features: Recall & learning in real-time, flexibility
Solution: Systolic synthesis of elementary compute-intensive algorithmic strings
Chip: VLSI Array Processor for execution of elementary algorithmic kernel
Board: 2-dimensional array of VLSI Array Processors, dedicated control & communication

Figure 6   System Concept of a Neurocomputer

CONCLUSIONS

In consideration of the applicational needs and the performance limits of convential computers indicated in figure 1, two important opportunities for dedicated VLSI can be envisaged:
-- single-chip integration of a complete neural net for small applications like signal processing;
-- implementation of compute-bound neural algorithms on dedicated VLSI array processors for large applications like speech or vision.

The question of whether analog or digital design style is preferable can be answered according to the size of the application. With single-chip implementation, the costs for interfacing a neural chip to the system environment will be the decisive factor. However, if an application calls for on-chip support of learning, digital design seems to be inevitable.

For larger applications, support for flexibility and learning is indispensible and on-chip storage of the weights is no longer manageable. The prevailing concept then is digital, since analog design fails to provide sufficient computation accuracy for the learning algorithms.

Since MOS technology comes up against limits in terms of on-chip storage potential of weights and processing power, there is a request for competing technologies. CCD technology, for instance, is reported to provide 10^6 synapses with analog dynamic range equivalent to 8 bits accuracy, and to perform 10^{10} weights per second, using a 2μm process [20]. Unfortunately, to take advantage of this potential one needs neurocomputers which do the learning. Since optics is far from being a mature technology for the construction of neurocomputers, in the near future CMOS technology seems to be favoured for the construction of neurocomputers.

ACKNOWLEDGEMENT

I like to express my thanks to Prof. Hosticka for the many discussions on analog design issues.

REFERENCES

[1] DARPA Neural Network Study, pp.34 (figure 2.14-15), pp.330 (figure 28.5), AFCEA International Press, Nov. 1988

[2] H.P. Graf, "A Reconfigurable CMOS Neural Network", Digest of Technical Papers of the Int. Solid State Circuits Conf., vol.33, pp.144, San Francisco, Febr. 1990

[3] J.B. Theeten et al., "The L-Neuro-Chip: A Digital VLSI With On-Chip Learning Mechanism", Proc. Int. Neural Network Conf., pp. 593-596, Kluwer Academic Publishers, July 1990

[4] D. Hammerstrom, E. Means, "System Design for A Second Generation Neurocomputer", Proc. IJCNN-90-Wash-DC, vol. II, pp. 80-83, LEA Publishers, Jan. 1990

[5] M. Yasunaga et al., "Design, Fabrication And Evaluation Of A 5-Inch Wafer Scale Neural Network LSI Composed of 576 Digital Neurons", Proceedings of the IJCNN-90, vol. II, pp. 527-535, San Diego, June 1990

[6] A. Chiang et al., "A Programmable CCD Signal Processor", Digest of Technical Papers of the Int. Solid State Circuits Conf., vol.33, pp.146, San Francisco, Febr. 1990

[7] M. Holler et al., "An Electrically Trainable Artificial Neural Network (ETANN) with 10240 Floating Gate Synapses", Proceedings of the IJCNN-89, pp. II-191, Washington DC, June 1989

[8] H.P. Graf et al.,"VLSI-Implementation of a Neural Network Memory with Several Hundreds of Neurons", AIP Cof. Proc. Neural Networks for Computing, Snowbird, Utah,1986

[9] C.A. Mead et al.,"VLSI-Architectures for Implementation of Neural Networks", California Institute of Technology, Pasadena, USA. AIP Cof. Proc. Neural Networks for Computing, nowbird, Utah, 1986

[10] A.P.Thakoor et al.,"Electronic Neural Network with Optically Modulated Variable Strength. Resistive a-Si:H Interconnects for Analog Computation", MRS Spring Meeting, paper E 7.5 , Reno, Nevada, 1988

[11] E. Vittoz et al., "Analog Storage of Adjustable Synaptic Weights", 3rd chapter of this book

[12] M. Verleysen, P. Jespers, "Precision Of Computations In Analog Neural Networks, 4th chapter of this book

[13] DARPA Neural Network Study, pp.372 (table 33.1-5), AFCEA International Press, Nov. 1988

[14] J. Harris et al., "Resistive Fuses: Analog Hardware for Detecting Discontinuities in Early Vision", pp. 27-56, in Analog VLSI Implementation of Neural Systems, edited by C. Mead and M. Ismail, Kluwer Academic Publishers 1989

[15] U. Ramacher, B. Schürmann, "Unified Description of Neural Algorithms For Time-independent Pattern Recognition", 13th chapter of this book

[16] B. Schürmann, J. Hollatz, U. Ramacher, "Adaptive Recurrent Neural Networks and Dynamic Stability", in: L. Carrido (ed.), Proc. XI Sitges Conf. on Neural Networks, 1990, Springer, Heidelberg, to appear

[17] S. Y. Kung, J. N. Hwang, "Parallel Architectures for Artificial Neural Nets", vol.2, IEEE Int. Conf. on Neural Networks, San Diego, July 1988

[18] U. Ramacher, "Systolic Architectures for Fast Emulation of Artificial Neural Networks", poster paper, presented at IEEE Int. Conf. on Neural Networks, San Diego, July 1988

published in: Proceedings of the Int. Conf. on Systolic Arrays, Killarney, Ireland, Prentice Hall 1989

[19] U. Ramacher et al., "Design of a 1st Generation Neurocomputer", 14th chapter of this book

[20] A.J. Agranat et al., "A CCD Based Neural Network Integrated Circuit With 64K Analog Programmable Synapses", Proc. IJCNN-90, vol. II, pp. 551-555, June 1990

(JUNCTION) CHARGE-COUPLED DEVICE TECHNOLOGY FOR ARTIFICIAL NEURAL NETWORKS

J. HOEKSTRA

INTRODUCTION

In this chapter the use of charge-coupled device technologies for the implementation of artificial neural networks (ANNs) is discussed and the design principles for these networks with junction charge-coupled devices are described. The implementation of neural networks with junction charge-coupled devices is illustrated by a simple test chip[1], on which a McCulloch-Pitts neuron is fabricated and successfully tested at a clock frequency of 40 MHz.

Artificial neural networks are interconnected networks of simple processing elements (neurons) and their hierarchical organizations, which functionally interact with the objects of the real world in the same way as biological neural networks do. The basic computation performed by a processing element (PE) is the summation of weighted inputs, which are outputs of other PEs. Subsequently, the PE produces an output value not-equal to zero if the summation exceeds a threshold value. In most neural networks a learning rule adapts the weights, as a result of training. The weights can be both positive and negative.

The biological neural network is very good at processing patterns. Even a complex task such as association, the involuntary connection of related images, is easily done by a small child. Computers, on the contrary, are poor at tasks such as association. In the brains, ten thousands of neurons are functioning in a parallel manner on performing a single task. On the contrary, however, most applications for computer systems that consist of tens of processors are very hard to program. If computers are organized more like the brain, will it be possible to solve these contradictions? Answering this question is one of the main goals of research on artificial neural networks.

The artificial neural networks designed in charge-coupled device technologies incorporate the following significant features: (1) analog storage of weights, (2) parallel computation of the weighted sum of inputs, and (3) reduction of wiring by using transport of charges through silicon device channels. In addition, the bipolar junction charge-coupled device technology adds to these features: (4) short-term memory, (5) on-

---

1   The integrated circuit was fabricated at Philips Elcoma, Nijmegen, The Netherlands

chip learning, and (6) suitability to support pulse-coded neural network algorithms.

Besides the organization of this chapter,the use of analog VLSI circuits for the implementation of artificial neural networks and the use of charge-coupled device technology, in particular, are discussed in this section.

Why analog VLSI circuits?

Many of the practical advantages that are being claimed for artificial neural networks [18] can only be realized if the networks are implemented on high performance systems. Without these systems many advanced neural network applications cannot be realized within a reasonable time period. Fast PEs on integrated circuits and integrated circuits holding many interconnected PEs are then key aspects.

In contrast to digital circuits, which are important since they take advantage of standard technologies that make the highest density in devices available, analog circuits can take advantage of the modest precision that is required in neurobiological network models, resulting in small and fast devices. In addition, integrated analog circuits for artificial neurons can be very simple, as is illustrated by the artificial neuron presented in this contribution. Other advantages are the computation of the sum of weighted inputs by analog currents or charge packets and the use of nonlinear or parasitic effects of the devices to realize exponential (or sigmoid) functions [8]. The favorable properties compensate the much longer design time necessary for non-standard analog circuits, the efforts mastering good theoretical knowledge about transistor physics and transistor modeling, and the various experimental test circuits that have to be made.

Why charge-coupled device technology?

In charge-coupled device (CCD) technology charge packets are stored and transported. These charge packets represent analog information. In essence, the CCD is a delay line or analog shift register. Consequently, if the device is designed in a loop, a memory cell is obtained. By using charge restoring techniques in which the amount of charge can be adapted, the CCD loop is used as an adaptive weight. In addition, by applying charge input techniques and charge sensing or charge output techniques, currents are used in combination with these charge packets.

The possibilities to implement large artificial neural networks are often limited by the fact that each PE has to be connected by conductors to all (or many) others. Charge-coupled devices deal with this limitation in two ways. First,in some places charge packets can be used instead of currents; second, the serial character of the delay line can be used to multiplex (in time) PEs and connections. This last mechanism can also reduce the number of pins to the chip.

The last advantage of CCDs mentioned here is a threshold-like function, which is an intrinsic property of a CCD if the amount of charge that cannot be contained in the maximum charge packet can be properly detected. Using this property, the CCD produces the neurons activation (output) function.

Organization of this chapter

The next section, Section 2, briefly describes the basic principles of charge-coupled devices. Section 3 gives an overview of recent research on CCD implementations of artificial neural networks. Section 4 describes junction charge-coupled device (JCCD) technology, and Section 5 the JCCD implementation of artificial neurons. Section 6 discusses the implementation of learning rules. Section 7 outlines the realization of a McCulloch-Pitts neuron and presents experimental results. The conclusions are formulated in Section 8.

BASIC PRINCIPLES OF CHARGE-COUPLED DEVICES

Charge-coupled devices are dynamic analog shift registers in which charge packets represent sampled analog data. Charges are introduced electrically or optically, and can be stored, transferred, and detected. The detection can be destructive or non-destructive. Under the application of a proper sequence of clock pulses these devices move "potential wells" filled with quantities of electrical charge in a controlled manner across a semiconductor substrate.

Introduction

Figure 1 shows some stages of a MOS charge-coupled device. The gates on top of the insulator are periodically interconnected and divided into three sets. By applying similar voltages (clock voltages) to these sets, which are mutually phase-shifted, a traveling potential well is induced in the semiconductor. A voltage at the input (source) is transformed into an analog charge quantity and is introduced into the potential wells every clock period. The traveling potential wells carry these quantities towards the output, where they are detected (sensed) and drained. In operation, every clock period all charge packets are shifted one stage, here three gates; at the output, the oldest quantity of charge is detected first.

The CCD behaves as an analog delay line where the product of the number of stages (N) and the clock period (T) gives the total delay time. Of importance are [6]:

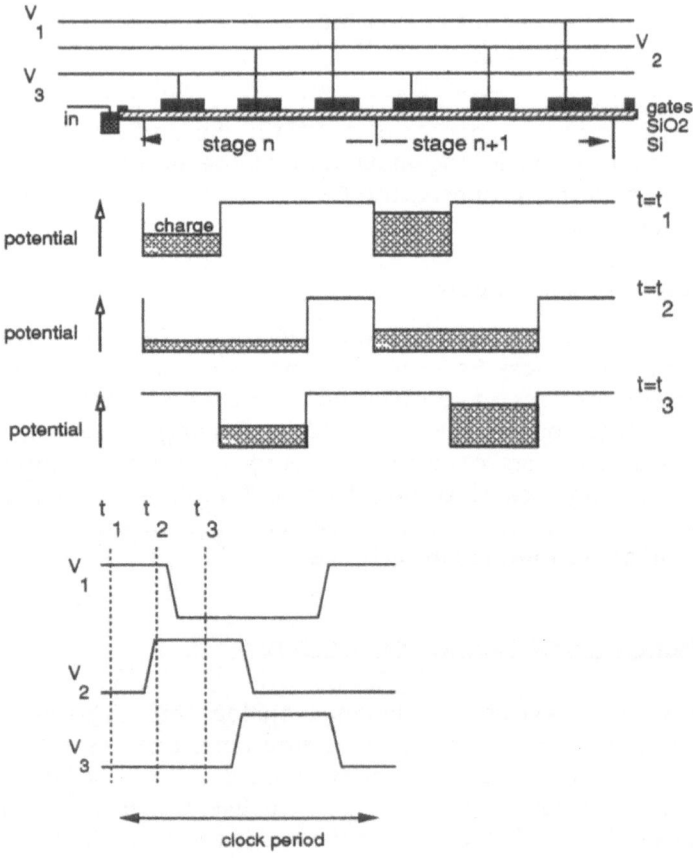

Figure 1   A simplified cross-section of two stages of a 3-phaseMOS n-channel charge-coupled device and its internal charge distribution

<u>An efficient transfer.</u> The relative amount of charge transferred per step is denoted by η and the relative amount of charge staying behind per transfer is denoted by ε. Thus $\eta + \varepsilon = 1$. In practice ε lies between 10^{-6} and 10^{-3}.

<u>A high speed transfer.</u> The minimum time required to obtain an efficient transfer determines the maximum transfer rate. The maximum clock frequencies for CCDs in which charges are transported in the top of the semiconductor and those in which the charges are transported in the bulk of the semiconductor are approximately 50MHz and 250MHz, respectively; however, when these CCD circuits are used in logic circuitry or artificial neural networks the maximum clock frequencies obtained

with the above mentioned CCD types are 10MHz and 40MHz, respectively.

A high charge handling capability. The maximum amount of charge that can be transferred. It determines in conjunction with the noise level the dynamic range of the device.

A long storage time. The CCD operates under nonthermal equilibrium conditions. Under these circumstances net generation of electron-hole pairs occurs. These dark currents fill the potential wells of a CCD and may therefore obliterate the signal charges. This limits the storage time. A typical (JCCD) dark current of 1 nA/cm^2 allows a storage time, without restoring, of about 1 second if a maximal filling of 10 % of a full packet is tolerated.

Types of CCDs

In this section the different types of CCDs and some of their characteristics are summarized briefly.

SCCD. A surface charge-coupled device (SCCD) is a MOS CCD. Inversion charge packets are transferred at the $Si-SiO_2$ interface. Surface states will interact with the charges by capture and emission. Captured charges may be emitted later on and be added to a following charge packet. This strongly limits the transfer efficiency. Because of a low drift field, which is a consequence of the transport at the surface, the transfer speed will be low.

BCCD. To eliminate the influence of the surface states on the transfer efficiency, the buried channel CCD (BCCD) was introduced. The BCCD consists of a lightly p-type doped substrate on top of which is a thin layer (typically a few microns thick) more heavily doped with n-type dopant. The surface of the semiconductor is covered with an insulator and metallic electrodes as in the SCCD. Transfer and storage of the charges inside the n-layer gives only interaction with bulk traps. Their concentration can be better controlled than that of the surface states. Moreover, for a thick layer device, high drift fields, directly induced by the externally applied gate potentials, result in high transfer speeds. The charge handling capability, however, will be one order of magnitude lower than that for the SCCD.

P^2CCD. A large charge handling capability and the elimination of surface states are obtained in the profiled peristaltic CCD (P^2CCD), in which the charges are transferred by two channels: one near the surface and one in the bulk. In the P^2CCD a combination of a lightly doped n-layer with a heavy but shallow ion implantation (profiled doping profile) is used. Small packet are thus propagated at high speed through the n-layer relatively deep in the semiconductor, while large packets are carried in the more highly doped region close to the semiconductor surface.

JCCD. The gate structure of the MOS CCD can be substituted by a pn-junction. To increase the charge handling capability the JCCD also uses a profiled layer [16, 30]. The JCCD technology and properties are outlined in Section 4.

Figure 2 shows a comparison of the various CCDs in terms of their schematic doping profiles.

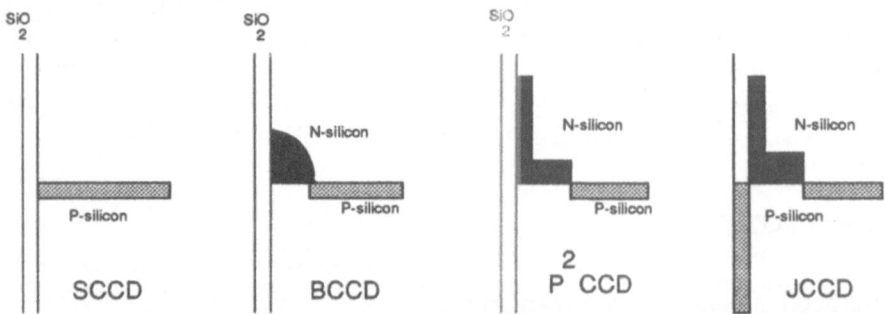

Figure 2   Schematic comparison of the main types of CCDs in terms of doping profiles

Potential analysis, input and output structures, delay line and memory

Potential analysis. The charge-coupled devices can be quantitatively described by solving Poisson's equation:

$$\nabla^2 \Phi = - \frac{\rho}{\varepsilon_0} \tag{1}$$

Due to the three-dimensional character of the charge packets, a three-dimensional solution of the Poisson equation is necessary. In general, this solution can only be obtained using time consuming computer simulations. In case of the JCCD, in which during the process of charge storage the pn junction can become forward biased, even time has to be included in the program using the Continuity equations. The amount of charge, Q_s, that can be transported in a single clock cycle per unit of gate area is:

$$Q_s = -q \int N dx dy \tag{2}$$

where N is the number of doping atoms/m^3.

Input techniques. There are basically two different approaches to the electrical inputting of information into a CCD: in the first one, the charge packet size is controlled in such a way that, ideally, it is proportional to the input voltage; in the second one, the charge packet size depends on a continuous input current and on the clock period. Figure 3 shows, as an illustration of the first approach, the potential equilibration technique, or fill and spill method, for putting charge into a CCD channel. The gate adjacent to the source diffusion is held at a fixed potential, while the input channel is applied to the second electrode. The source is pulsed to a low potential for a short period during which the storage well under the second electrode fills with charges (holes). The source is then returned to a relative high potential, which causes excess charge from the storage well to spill back into the source. The size of the resultant charge packet in the well beneath the second gate varies linearly with the voltage on this gate.

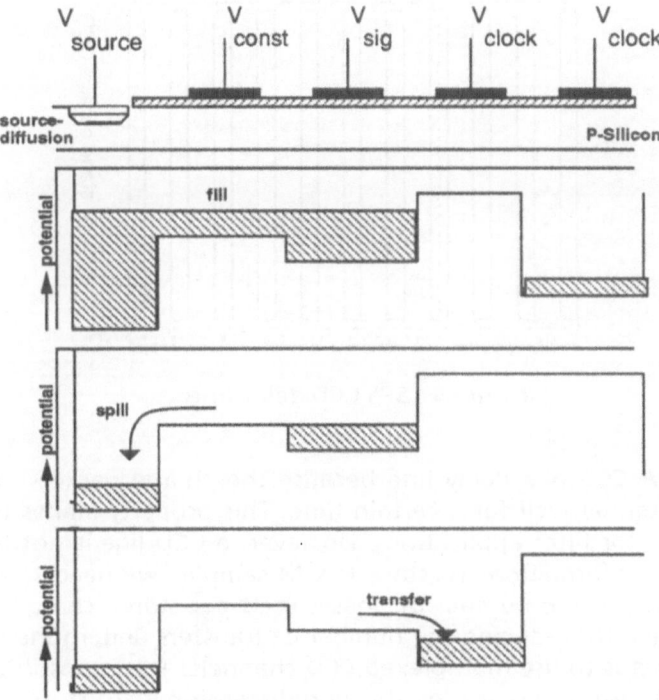

Figure 3   Putting charge into a CCD using the fill-and-spill method

As an optical imaging device the CCD may be used to collect and read out optically generated signals. An optical system is used to focus a single light spot on the front face of the CCD. Photons entering the silicon substrate, either through or between the electrodes, generate electron-hole pairs by virtue of the photoelectric effect. The charges can be collected in the potential well.

Output techniques. There also exist two different approaches to the outputting of information out of the CCD: destructive or non-destructive.

Non-destructive outputs make use of a floating gate output, or in case of a JCCD, a forward biased pn junction. In the non-destructive floating-gate sensing technique, the charge transfer in the plane above the substrate is effected in the normal way, except that the CCD electrode above the floating-gate is held at a fixed bias rather than being clocked. When a charge packet in the CCD is brought under the floating gate, it produces, by capacitive coupling, a change in potential of the floating gate; this change in potential is sensed by an on-chip MOST. Destructive outputs can make use of floating diffusions (drains).

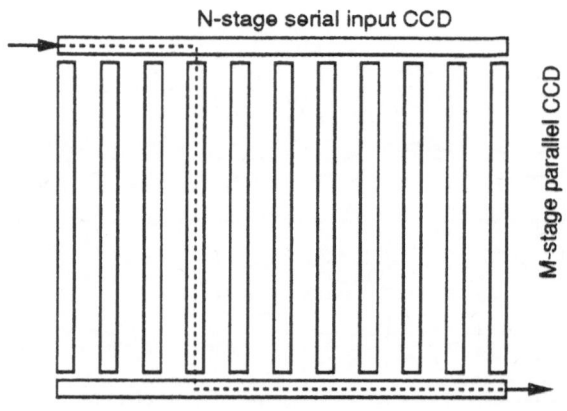

Figure 4   SPS CCD delay line

Delay line. A CCD is a delay line because the charge packets are kept within a potential well for a certain time. This property makes the CCD very suitable for filter applications. However, a CCD line is not the best way to store information. To store N x M samples we need N x M x p gates, where p is the number of phases used per stored charge packet. One technique for reducing the number of transfers undergone by each charge packet is to use multiplexed CCD channels. This is possible at the device level by making use of the two-dimensional structure of CCDs which are constrained to surface rather than to a line. The best known form is the serial-parallel-serial (SPS) structure. The scheme is shown in Figure 4. The structure stores N x M samples at the cost of only (N + M)p transfers of any particular charge packet. Furthermore, all the charge packets in the parallel array are only moved at a rate of (clock frequency of the serial line)/N. The dotted line represents a typical path that a charge packet traverses.

Memory. A memory cell with a CCD is obtained by a CCD loop. Charge is injected into the CCD and at the end the charge is sensed and regenerated in the first cell. Figure 5 illustrates the loop.

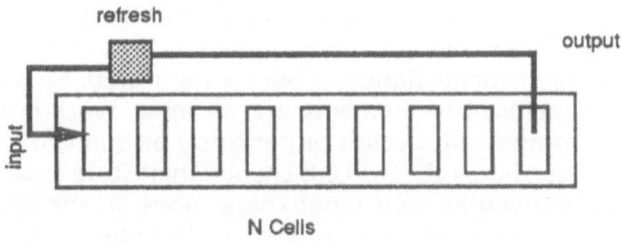

refresh

output

input

N Cells

Figure 5   Memory structure for N/p charge packets

More detailed information on the principles of CCDs can be found in various textbooks on this subject [4,7,14,27].

RECENT RESEARCH ON CCDS AND NEURAL NETWORKS

The first direct connection between charge-coupled devices and artificial neural networks was described in a, almost classical, paper of Sage, Thompson, and Withers from the Lincoln Laboratory at MIT [26], which was presented at the conference on Neural Networks for Computing at Snowbird in 1986.

However, the description of their ideas and more recent research efforts are preceded by some remarks on digital logic functions designed with CCDs. Remarks that indicate an indirect connection between CCDs and artificial neural networks: in neural network terminology these logic functions can be, in many cases, called McCulloch-Pitts cells (McCulloch-Pitts neurons).

CCD logic devices and McCulloch-Pitts cells

Consider a CCD-channel in which two charge packets are brought together in a potential well that can only contain one packet. If the surplus charge that cannot be contained in the well is detected, a threshold function is realized. In addition, if this threshold function is extended with an absolute inhibition, that is a single (inhibitory) signal that can block the response of the function, then a McCulloch-Pitts cell [20] is obtained.

Figure 6 shows a few typical McCulloch-Pitts cells. Each cell is represented by a circular figure. From each cell leads a single line or wire, called the output fiber of the cell. This output fiber may branch out after leaving the cell; each branch must ultimately terminate as an input connection to another cell, or to the same cell (feedback), or as an output connection of the total network.Two types of termination are allowed. One is called an excitatory input and is represented by a little black arrow. The other is called an inhibitory input and is represented by a little black circle. Any number of input connections to a cell are permitted. Each cell operates in discrete moments. The moments are assumed synchronous among all cells. At each moment a cell is either firing or quiet; these are the two possible states of the cell. The cells change their state as a consequence of the pulses received at their input connections. In the open part of the circle representing a cell there is written a number, called the threshold of that cell. This threshold determines the state-transition properties of a cell C in the following manner. If, at any moment t, one or more cells that have an inhibitory connection to C are firing, then C will not fire at time t + 1. Otherwise, if the number of cells that have excitatory connections to C and are firing is equal or greater than the threshold of C, then C will fire at time t + 1.

A simple artificial neural network using this threshold logic system can be written as:

$$x_i(t+1) = sgn\,[\,\Sigma A_{ij} x_j(t) - T_i\,]]_i] \qquad (3)$$

where sgn(w) = + 1 if w $\geq$ 0 and 0 if w < 0; T_i denotes the threshold value of node i; if A_{ij} = 1 then there is a connection from neuron j to neuron i, if 0 then there is no connection. The variables x_i in the equation are often called short term memory (STM) traces, or activations [9]. In most artificial neural networks the weights A_{ij} are adaptable; these adaptive weights are called long term memory (LTM) traces.

Figure 6 also shows the equivalent cells of the McCulloch-Pitts examples realized in logic junction CCD technology [12].

The analog JCCD output is used as a digital one. Because the CCD is clocked the cells fire at time t + 1 if the inputs are offered as charge packets, which are transported towards the well performing the threshold function.

The artificial neural networks with McCulloch-Pitts cells only use binary input and output values. Because CCDs are analog building blocks more powerful networks can be designed. In general, logic circuits using CCDs can easily be used for implementing activation and output functions of artificial neurons. One of the main problems of most logic CCD circuits are the limited dynamic range of the sensing output, that results in binary output functions, and the inability to pass output signal over long distances (compared with the device dimensions), due to the capacitance of the wiring. Junction CCDs do not have these limitations, they are therefore very suitable for implementing artificial neurons.

The possibilities to use logic CCD technologies for implementing artificial neural networks mainly depend on the availability of circuitry to process

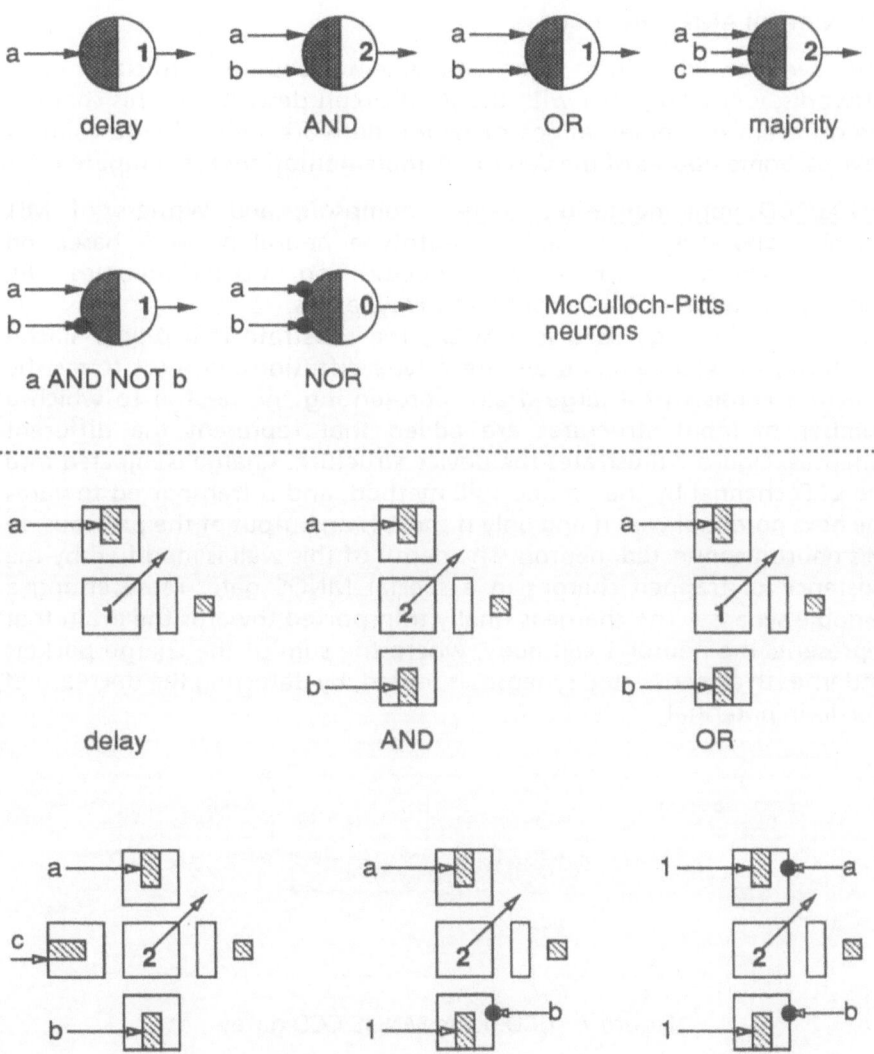

Figure 6   McCulloch-Pitts cells and equivalent Junction CCD cells

the node output function and the weight adaptation on-chip. References to the various CCD logic circuits are [3,11,15,17,21,23,31].

Overview of ANNs with CCDs

This overview describes three implementations of artificial neural networks, which together with the JCCD circuit described in this chapter, are different implementations of neural networks with charge-coupled devices. Some figures of the described implementations are compared.

MNOS/CCD implementation. Sage, Thompson, and Withers of MIT Lincoln Laboratory have built a prototype neural network based on MNOS (metal nitride oxide semiconductor) and CCD technologies [26]. The chip implements 13 neurons and 169 synapses.
The CCD technology used is a SCCD, the substrate is a p-type silicon substrate, the source and drain are n-type diffusions. In short terms the structure consists of a large drain representing the neuron to which a number of input structures are added that represent the different synapses. Figure 7 illustrates the device structure. Charge is injected into the CCD channel by the fill and spill method, and is transported towards the next potential well if and only if there is an output of the previous, to this neuron connected, neuron. The depth of this well is modified by the existence of trapped charges in a special MNOS gate, representing a variable synapse. The charge is finally transported towards the drain that represents the neuron's cell body, where the sum of the charge packets underneath all connected synapses is sensed, by detecting the decrease of the drain potential.

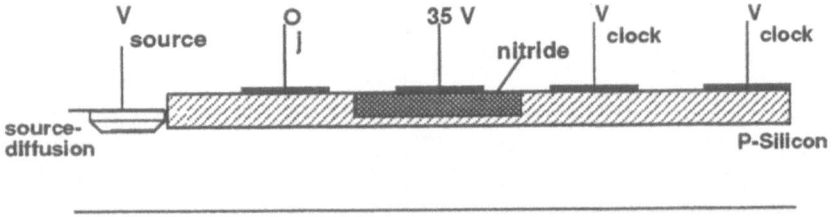

Figure 7   CCD using MNOS/CCD gates

The MNOS gate is similar to the normal MOS gate except that (1) the main gate insulator is silicon nitride and (2) the silicon oxide insulating layer is so thin (25 Angstrom) that, at gate voltages of approximately +35V or -35V, electrons and holes can move by quantum mechanical tunneling between the underlying silicon and long-life traps in the nitride layer. When the gate voltage is kept below 10V virtually no tunneling occurs, and the trapped charge is essentially permanent. The effect of the trapped charge is to make the voltage on the gate to appear to be higher or lower than its actual value, thus controlling the size of the potential well.
In principle the tunneling of holes can be controlled during the network

31

operation, thus providing on-chip learning. The binary neuron states are, however, sensed on-chip but processed off-chip.

CCD Neural Network Processor. The use of CCD technology can produce a fast on-chip analog "multiply and add" operation, for the summation in charge domain can be performed in a single transfer. To use this CCD technology as a multiply and add accelerator, the input signal and weights are preloaded in the IC and immediately after the operation the signal is taken from the IC. The multiply and add operation can be used to implement a neural network in which the output of the neurons from the chip is just the weighted sum of all inputs.

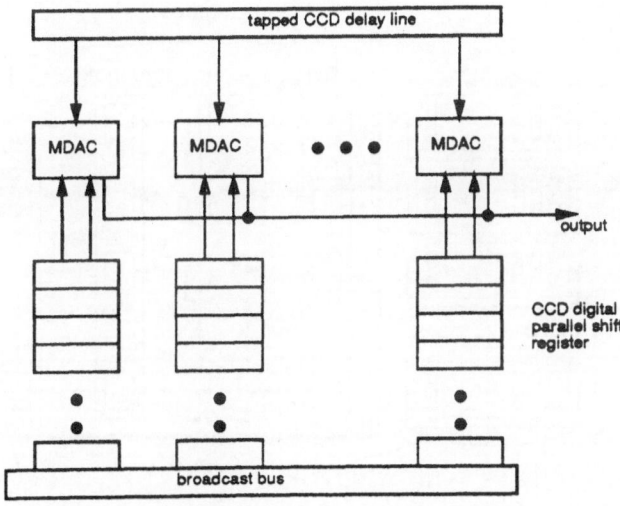

Figure 8   Block diagram of a part of a CCD signal processor

Chiang, et al. of MIT Lincoln Laboratory have built the weighted sum operation with CCD technology on a single chip; it is called the programmable neural network processor [5]. It implements 2016 programmable interconnections between 144 input nodes and 14 output nodes. The basic CCD structure is a serial-parallel structure: the input values are serially stored in a tapped delay line, and parallel CCD structures couple by a floating gate to the analog input port of corresponding multiplying D/A converters (MDACs). Other inputs to the MDACs are 14 stage 6-bit CCD digital memories. The output of the MDAC is a charge packet proportional to the product of the analog input and the digital weight. The outputs of the MDACs are summed together in the charge domain. The weights are serially preloaded over a broadcast bus.

Figure 8 shows a part of the block diagram. In operation, weights are preloaded into the 14x144 digital memory locations, and a set of 144

32

analog inputs is read in. Then, each of the 14, 144-word bit-parallel, CCD digital memory vectors is sequentially applied to the MDACs. After the first clock period the total summed charge from the first output node is computed, after the second clock period the total summed charge of the second output node is computed, etc.

CCD Hopfield network. Agranat and Yariv of the California Institute of Technology described a Hopfield network using CCD technology [1]. In this implementation an array of N (number of PEs) rows of N stage CCD memories, each row containing all the weights belonging to one PE, is preloaded.

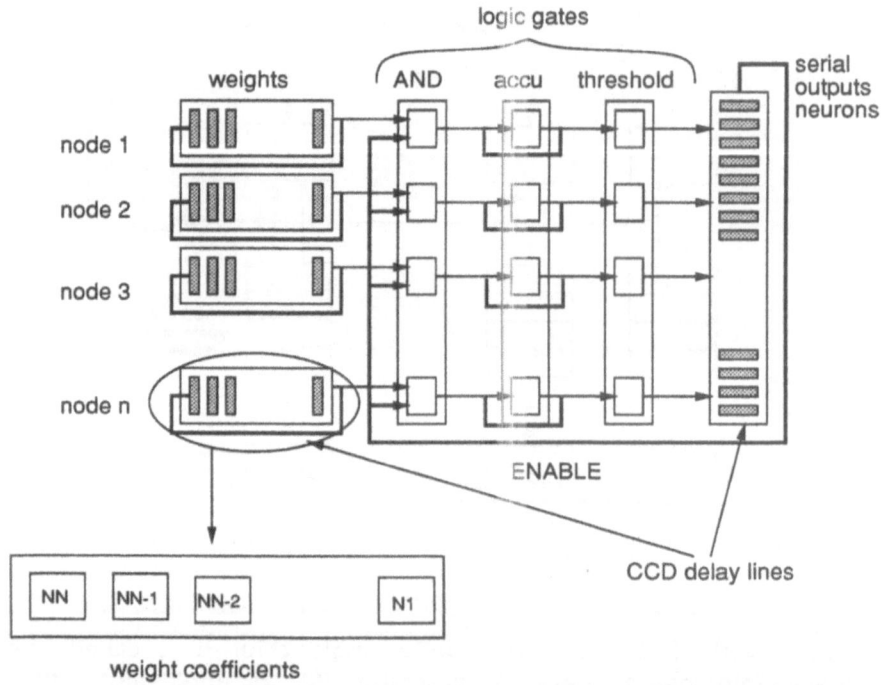

Figure 9   Block diagram of a part of the Hopfield implementation

Figure 9 shows the layout and the sequence in which the weights appear (at a specific clock pulse). In a parallel manner the rightmost weights of these rows are sensed and fed into an accumulator if and only if there is a signal that represents the output 1 of the corresponding PE. At the next clock pulse the next weight is fed into the accumulator if and only if again the corresponding PE has output 1. After the complete row is offered to the accumulator, the output of the PE is 1 if the sum of weighted inputs exceeds a threshold value, and is stored in an element of a binary N stage CCD delay line. This delay line is serially read out at the top element. In this way the outputs of the PEs are serially used to enable the weights to all the accumulators. The system continues until a stable state is reached. Possible future developments will include replacing the CCD shift registers by CCD detector arrays which will enable optical loading of weights and PE output values.

At the Washington IJCNN-90 a late paper was presented in which a working network consisting of 256 artificial neurons, fully interconnected (Hopfield network) having 65536 weights, operating at 10 MHz clock frequency was shown.

Some characteristics of chip implementations. In this subsection some of the characteristics of the different implementations are compared. In addition to the three above described implementations the characteristics of the JCCD McCulloch-Pitts neuron, which is described in Section 7, and the expected characteristics of an artificial neural network with JCCDs are given. The shown characteristics indicate the status of the built networks and an extrapolation indicating the expected capabilities of a JCCD artificial neural network. The connectivity is full if all processing elements are connected to all the other processing elements. In case that only all N inputs are connected (by uni-directional links) to a layer of M output processing elements by N x M weights a feedforward network is obtained. To obtain a network with more than one layer, a cascade of ICs is necessary. Indicated is the on-chip or off-chip processing of the output and of the processing elements, and the on-chip learning capabilities. On-chip and off-chip learning is discussed in Section 6. All inputs to the CCD circuits can be parallel or serial, in the last case making use of a serial-parallel structure to save pins. The neuron and synapse state can be analog or digital. The number of neurons indicates the number of processing elements that compute the weighted sum of its inputs. The last row shows the operating clock frequencies.

JUNCTION CHARGE-COUPLED TECHNOLOGY

The design presented in Section 7 is based on the concept of junction charge-coupled devices, which is a bipolar version of the more conventional metal-oxide-semiconductor (MOS) charge-coupled device (CCD). The essential differences between MOS-CCDs and our bipolar CCD are (1) the use of the capacitance of a reverse-biased pn-diode instead of the capacitance of the gate-oxide, and (2) the possibility to have vertical charge transport, that is, transport of charge perpendicular to the charge transport in the plane above the silicon substrate.

In junction charge coupled device (JCCD) technology charge packets are stored and transported. The charge packets represent analog information, and are transmitted to the end of the delay line, where the charge packets can be converted into a current representing an analog

Implemen-tation	MNOS/CCD	Network Processor	Hopfield Network	McCulloch-Pitts	JCCD Networks
Status	built	built	built	built	expected
Connectivity	full	feed-forward	full	feed-forward	full
Neuron output	off-chip	off-chip	on-chip	on-chip	on-chip
Learning	off-chip	off-chip	off-chip	off-chip	on-chip
Parallelism	full	semi	semi	full	semi/full
Synapse state	analog	discrete	analog	analog	analog
Neuron state	binary	analog	binary	binary	analog
Number of neurons	13	144	256	5	-
Number of synapses	169	2016	64K	7	-
Clock	10 MHz	10 MHz	10 MHz	40 MHz	> 10 MHz

Table 1   Main characteristics of the CCD implementations

output value. JCCD technology is adequately developed. It is used for solid-state imagers, filters [22], and logic circuits [10]. The technology is developed at the Delft University of Technology.

Basically, a JCCD consists of a p-type substrate and a n-type layer with diffused p-gates. If appropriate clock voltages are applied to gates, moving potential wells are created, in which charges (electrons) can be transported. Charge packets can be moved to each other, and stored in a single potential well. Figure 10 shows the combination of two charge packets creating a current outwards the collecting well and the injection at another place on the IC.

If well more electrons are offered to this well than can be contained in it, a current results at the gate's contact. This current, which results if more charges are offered then a certain threshold value, can be injected elsewhere on the chip. This 'threshold' gate produces an artificial neuron's output. To obtain a structure that is capable to act as an artificial neuron, the input potential wells must be varied in size. This can be obtained by applying a changing voltage to the gate, varying the depth of the well, or by pre-charging the well itself. If a logic function is

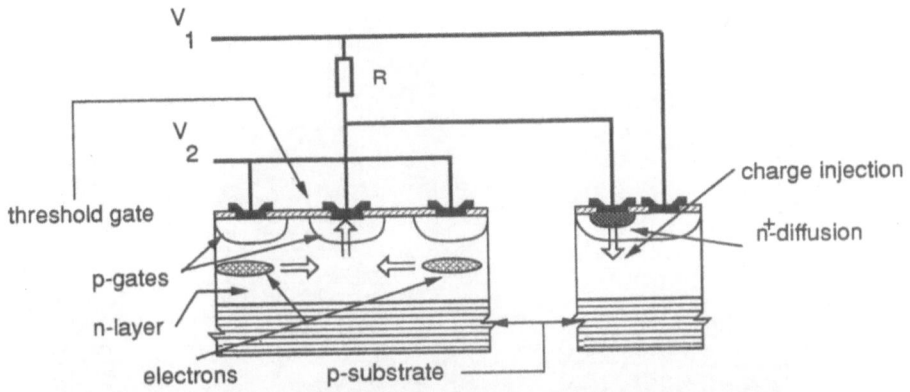

Figure 10   The JCCD structure for vertical charge transport

built using these design principles, it is a McCulloch-Pitts cell. However, a JCCD neuron can be more if we make use of the analog inputs and output.

In more detail, a JCCD consists of a lightly doped p-type substrate and a n-type layer with diffused p-gates. The drain consists of a n+-diffusion. For a given JCCD structure all electrons are removed from the n-layer by applying a sufficiently large positive voltage to the drain, while the gates and the substrate are kept at ground potential. In this state, the depletion layers extending from the gate-(n-layer) interface and from the substrate-(n-layer) interface touch. At this point the electrical potential in the semiconductor reaches a maximum. This is a global maximum; for an ideal JCCD it is constant throughout the n-layer. The maximum potential in the device is referred to as the channel potential.

In operation, a local potential maximum in the n-layer, underneath a gate, is created by clocking a gate to a positive voltage. This local maximum serves as a well for electrons. Figure 11 shows a perspective view of the two-dimensional potential distribution. The potential distribution is a result of the sequence of gate voltages: gate A = 2V, gate C = 7V, and gate B = 0V.

The current fabrication process of JCCD is an adapted standard process for fabricating bipolar circuitry. The ideal JCCD technology meets the following requirements:

-- Smooth potential distribution, there are no distortions in the potential of the channel when all gates are at ground potential and the n-layer is fully depleted.

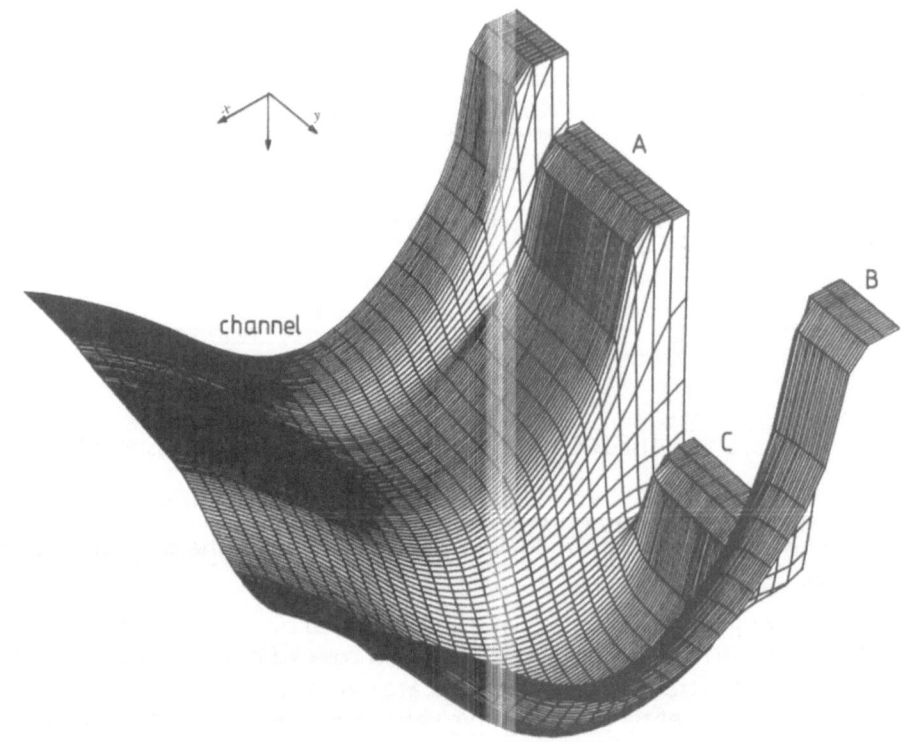

Figure 11   The potential well in a JCCD

-- No parasitic barriers, the formation of parasitic barriers underneath the gates must be avoided, because they are hardly affected by the voltage differences between the gates.

-- only small parasitic wells, in a technology using p-type islands for lateral confinement, parasitic wells will be present. Their size must be as small as possible.

-- High reach-through voltage, the terminal voltage difference between the p-gates at which reach-through occurs should be well above the channel potential.

-- optimal properties of pnp-substrate transistor, the maximum value of the collector transport current of the substrate transistor should be high, and the decay time should be short.

-- Compatible with standard bipolar circuitry, a technology in which peripheral circuitry can be realized is necessary.

Of course, if some requirements cannot be simultaneously fulfilled a compromise must be made. The concentration profile of charges through a p-gate resulting from the fabrication processis shown in Figure 12.

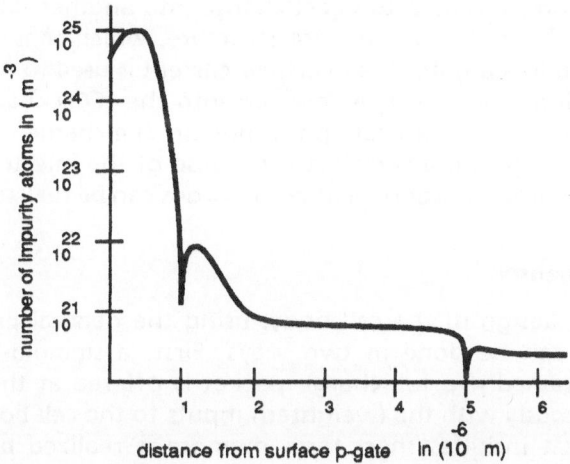

Figure 12   Impurity profile through a p-gate for a JCCD

ARTIFICIAL JCCD NEURON

This section subsequently describes: the concepts of JCCD technology for the design of artificial neural networks, and the design of an artificial neuron, that is, the design of the cell body, the design of the synapse, and the design of the dendrites and axons.

JCCD design concepts for ANNs

JCCD artificial neural circuits combine the properties of CCDs with charge transport vertically through one of the pn junction diodes, induced by surplus charge. The gate under which the surplus charge is created is called a threshold gate. The operation condition for obtaining vertical charge transport is that the gate voltage on the threshold gate is taken above the channel potential of a grounded gate. In this condition, the channel potential under the threshold gate---if the potential well is filled completely---is more positive than the channel potential under the neighboring gates, which have zero gate voltages. In addition, surplus charge will further decrease the local potential in the channel and the pn junction will become forward biased [21]. The surplus charge can, for example, be created by combining different charge packets in a well that can only contain one charge packet. The structure will act as a pnp-transistor; the gate (emitter) will inject holes into the n-layer. This charge flow, called overflow current, will continue until all surplus electrons have been removed.

The overflow current is used to inject charge into another JCCD channel. This can be performed by an injector structure, which consists of a n+-diffusion placed in a p-gate. The overflow current is used to forward-bias this n+p junction; and charge is injected into the JCCD channel, which acts as a collector of this vertical npn transistor. The charge injection can be inhibited by applying a current to the base of the injector structure. Using vertical charge transport, artificial neurons can be realized [12].

JCCD artificial neuron

Cell body. The design of the cell body, using the general properties of JCCD devices, can be done in two ways. First, a sigmoid-like output function is obtained if a full charge packet is offered at the threshold gate simultaneously with the (weighted) inputs to the cell body. Second, a cell body that includes short term memory is realized by adding a charge circulator to the threshold gate. In this charge-circulator the charge from the threshold gate is delayed and added to the following state of the neuron. Figure 13 shows three structures for the cell body. In the first structure the charge Q is shifted towards the cell body, in which the first gate performs a threshold function; the second gate blocks simultaneously the drain. The second structure shows a zero threshold implementation; the charge packet, Q, is combined with a well structure (a one-circulator) always containing one full bias charge packet. In the third structure the previous activation is used to obtain a new activation, thus exhibiting short-term-memory. If STM memory is added then the charge circulator is emptied (resetted) before the STM starts. A part (of the complete) activation is temporarily stored.

Synapse. In the design of the synapse we can distinguish various types of synapses. In principle, all the possible weight values are between zero and one. The synapse can be excitatory by adding charge packets to the charge packet representing the input value to the neuron. This mechanism is called pre-charging: some charge is put in a JCCD channel before a current, representing an input value, is injected into the JCCD channel. In the next gate is tested if the total charge exceeds the amount of charge injected by the input. The result is an overflow current proportional to the product of the weight and the input. Figure 13(d) illustrates the weighting function. The synapse can be inhibitory in the McCulloch-Pitts sense, that is, one inhibitory input cancels all the other inputs. The physical realization of this inhibitory weight is the application of a current, representing the (weighted) input, to the base of the

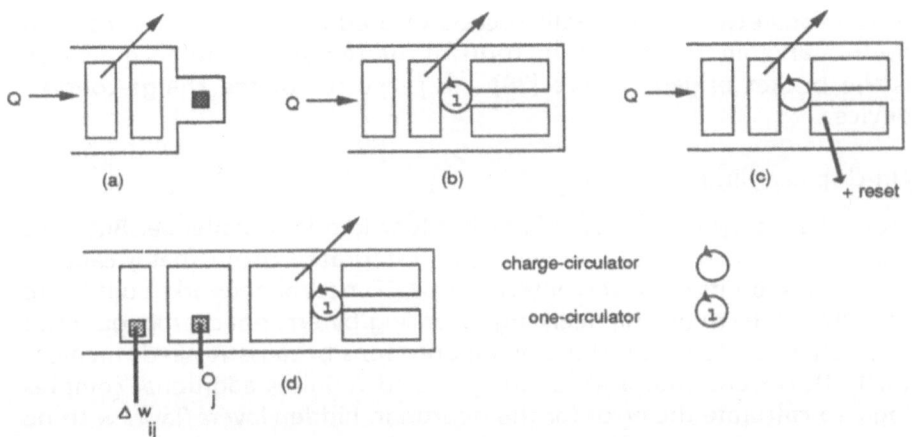

Figure 13   Cell body and weighting mechanism in JCCD

injector gate. This current prevents the charge injection into the channel. And, thirdly, the weight value can be given the value zero (reset).

Dendrites and axons. Dendrites as well as axons can be realized in JCCD technology using the charge transport in JCCD channels or using the overflow current from one channel to another. The combination of charge transport in both ways give rise to the interesting properties of artificial neurons in our technology.

IMPLEMENTATION OF LEARNING RULES

To obtain an artificial neural network, learning rules must be added to the network of artificial neurons. The learning rules generally adapt the weights in the network. The algorithms adapting these weights can be performed off-chip or on-chip.

Off-chip learning

If the adapting procedure is done off-chip, then the time to load or reload the weights on the chip is the most important parameter. When the weights are adapted in a parallel manner then it is possible to load the weights in a CCD weight matrix by parallel CCD input lines. Otherwise, if the weights are adapted serially, the fast serial in-parallel out structure (part of the SPS structure) can be used. Most popular learning algorithms and network paradigms are implemented in this way, for example, the Hopfield network [13] and the back-propagation learning rule [25]. The CCD nature of acting as a delay line makes it also easy to implement the less popular temporal-difference methods [28] and time-delay neural networks [29], which assign credit by means of the difference between temporally successive predictions. It is in this content not remarkable that one of the roots of the temporal difference method is the bucket brigade device [28] ---a precursor of the charge-coupled device.

On-chip Learning

From the viewpoint of speed, on-chip learning is a challenge. But, also from an intellectual point of view is an integrated circuit that is capable of learning on its own, very interesting. CCD neural networks built up to now do not have on-chip learning. Learning by error-backpropagation is not very suitable for on-chip implementations because it needs memory for both current and previous weights and requires additional complex logic to calculate the error for the neuron in hidden layers (layer with no direct connection to off-chip input or off-chip output). Besides this, at this moment only JCCD neural network implementations have a hidden layer.

In the following, various learning schemes for on-chip learning and implementation possibilities using JCCD neural network technology are discussed. They can all be seen as extensions of the Hebb-rule and are in most cases connected with Pavlovian conditioning methods. The Hebb-rule itself is mostly stated as: If two neurons become active at the same time then the connections between these neurons must be strengthened. Pavlovian conditioning is evident in the learning behavior of a range of species. In Pavlovian conditioning an organism learns to link two discrete stimulus events. One event is the 'unconditioned' stimulus, the other the 'conditioned' stimulus. In order to learn the association between conditioned and unconditioned stimuli, an organism must in effect remember the temporal relation of the events [2]. In addition to the temporal relation there is often a relation in the location of the stimuli [2]. These effects can be implemented in JCCD technology.

A Hebbian-like rule for updating the weight can be achieved by pre-charging, with an increasing amount of charge, if both neuron's input and output are active. To make the weight zero a reset is used. Some Pavlovian learning rules utilize lateral inhibition or lateral excitation, which can also be implemented using the inhibit mechanism.

A last remark on learning concerns the use of the pulsed character of the CCD delay line. In principle the CCD neuron can be clocked with a lower clock frequency than the synapses, and thus in effect integrating the incoming pulses over a time period. This effect can be emulated by using one clock frequency but by having a large gate representing a neuron's cell body and a circulator representing short-term-memory. This last system is, however, more powerful because it can also distinguish pulse-trains in which the pulses are differently ordered.

REALIZATION OF A MCCULLOCH-PITTS NEURON

The intrinsic properties of the JCCD devices used for the construction of artificial neurons are illustrated by a McCulloch-Pitts neuron. Figure 14 shows the tested network in terms of neural networks; it consists of 5 neurons with 7 synapses, in which the synapses towards the cell body of the hidden neuron have a fixed value.

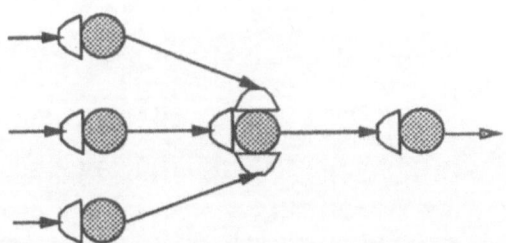

Figure 14   Schematic network for testing the McCulloch-Pitts neuron

The actual layout of the hidden neuron is depicted in Figure 15. The injector gates represent three input dendrites, the threshold gate represents the cell body in which the threshold function is realized.

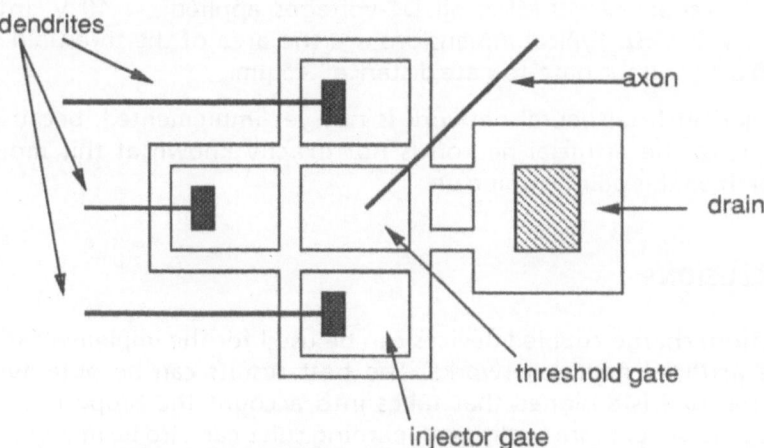

Figure 15   Cell body McCulloch-Pitts neuron

In essence, this cell body is a majority gate to which three inputs are offered. These inputs are weighted charge packets converted to currents. These currents are injected into different injector gates. These gates transfer their charge packets to the central threshold gate, which can contain only one charge packet. The last figure, Figure 16, shows the output of the test device, performing an McCulloch-Pitts neuron. The output of the McCulloch-Pitts neuron is binary, however the analog character of the neuron can also be distinguished.

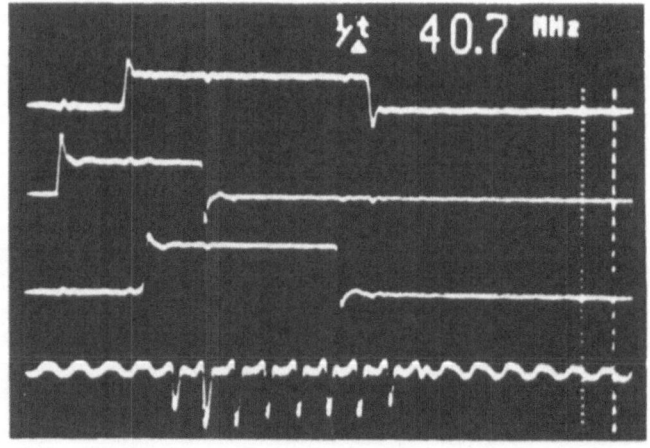

Figure 16   Experimental results on the McCulloch-Pitts neuron

The resulting output function is the analog output of a majority gate. The experimental parameters are: a three phase clock, clock voltage: 7 V,clock frequency: 40 MHz, all DC-voltages applied: < 10 V, internal resistors: 2.2 kΩ. Typical dimensions are the area of the threshold gate: $26 \times 20 \ \mu m^2$ and the gate-to-gate distance: 3.5 μm.

A larger artificial neural network is not yet implemented, because the fan-out of the artificial neuron is not exactly known at this moment; research on this point is going on.

CONCLUSIONS

(Junction) charge-coupled devices can be used for the implementation of some artificial neural networks. The best results can be obtained if a learning rule is designed that takes into account the properties of the device, however more traditional learning rules can also be implemented. In the design of neural networks special properties of CCDs, such as the use of the CCD as a delay line, and the use as a light sensor, can be incorporated. JCCDs offer the possibilities for on-chip learning and the on-chip processing of the output (threshold) function. The bipolar buried-channel JCCD can implement fast processing elements.

REFERENCES

[1] Agranat, A., Yariv, A., 'Semiparallel Microelectronic Implementation of Neural Network Models Using CCD Technology', Electronics Letters, Vol 23, pp 580-581, 1987.

[2] Alkon, D. L., 'Memory Storage and Neural Systems', Scientific American, pp 26-34, 1989.

[3] Allen, R.A., et al., 'Charge-Coupled Devices in Signal Processing Systems, Vol V, Final Report, U.S. Navy Contract no N0014-74-C0068, 1979.

[4] Beynon, J.D.E.; and Lamb, D.R., Charge-Coupled Devices and their Applications, (London: McGraw-Hill), 1980.

[5] Chiang, A., Mountain, R., Reinold, J., LaFranchise, J., Lincoln, G., 'A Programmable CCD Signal Processor', IEEE International Solid-State Circuits Conference, ISSCC 90, pp 146-148, 1990.

[6] Esser, L.J.M., 'Charge Coupled Devices: Physics, Technology and Applications', proceedings of the "International Workshop on the Physics of Semiconductor Devices" ed. SC Jain and S Radhakrishna (New Delhi: Wiley Eastern Limited).

[7] Esser, L.J.M., and Sangster, F.L.J., 'Charge Transfer Devices', in Handbook on Semiconductors, vol4. Device Physics, ed. Hilsum, C. (Amsterdam: North-Holland), 1981.

[8] Goser, K., Hilleringmann, U., Rueckert, U., and Schumacher, K., 'VLSI Technologies for Artificial Neural Networks', IEEE MICRO, December 1989, pp 28-44.

[9] Grossberg, S., ' Nonlinear Neural Networks: Principles, Mechanisms, and Architectures', Neural Networks, Vol 1, pp 16-61, 1988.

[10] Hoekstra, J., 'Simple JCCD Logic at 20 MHz', Electronics Letters, Vol. 23, 246, 1987.

[11] Hoekstra, J., 'Junction Charge-Coupled Devices for Bit-Level Systolic Arrays, IEE Proc. Vol 134, Pt. G., pp 194-198, 1987.

[12] Hoekstra, J., Some Models and Implementations of Digital Logic Functions Using Junction Charge-Coupled Devices, PhD-thesis, Delft University of Technology, 1988.

[13] Hopfield, J. J., 'Neural Networks and Physical Systems with Emergent Collective Computational Abilities', Proc. Natl. Acad. Sci. USA, Vol 79, pp 2554-2558, 1982.

[14] Howes, M.J. and Morgan, D.V., Charge-Coupled Devices and Systems, (Chicester: Wiley), 1979.

44

[15] Kerkhoff, H.G., and Tervoert, M.L., 'Multiple-valued Logic Charge-Coupled Devices', IEEE Trans. on Computers, Vol C-30, pp 644-652, 1981.

[16] Kleefstra, M., 'A Simple Analysis of CCDs Driven by pn Junctions', Solid State Electronics, Vol. 21, pp 1005- 1011, 1978.

[17] Kosonocky, W.F., and Carnes, J.E., 'Charge-Coupled Digital Circuits', IEEE Journal of Solid-State Circuits, Vol SC-6, pp314-322, 1971.

[18] Leonard, J. , Holtz, M., 'System applications Part IV Of the DARPA Neural Network Study', Fairfax: AFCEA International Press, 1988.

[19] Murre, J.M.J., Phaf, R.H., Wolters, G., 'CALM Networks: A modular Approach to supervised and unsupervised Learning', IEEE-INNS Proceedings of the IJCNN, Washington D.C., pp 649-656, June 1989.

[20] Minsky, L. M. , Computation: finite and infinite machines, (London:Prentice-Hall).

[21] May, E.P., van der Klauw, C.L.M., Kleefstra, M., and Wolsheimer, E.A., 'Junction Charge Coupled Logic (JCCL)', IEEE Journal of Solid-State Circuits, 1983, Vol. SC-18, pp 767-772.

[22] Montagne, A.J.M., Kleefstra, M., 'A straightforward Parallel-In, Serial-Out Filter with a Junction Charge-Coupled Device and an Integrated Clock Driver, IEEE Journal of Solid-State Circuits, Vol. 24, 1989, pp 835-839.

[23] Montgomery, J.H., and Gamble, H.S., 'Basic CCD Logic Gates', The Radio and Electronic Engineer, Vol. 50, pp 258-268, 1980.

[24] Psaltis, D. , Sage, J., 'Advanced implementation Technology, Part IV of the DARPA Neural Network Study', Fairfax: AFCEA International Press, 1988.

[25] Rumelhart, D. E. and McClelland, J. L. , Parallel Distributed Processing, Vol. 1: Foundations, MIT-press, Cambridge, Massachusetts, 3rd edition, 1986.

[26] Sage, J.P., Thompson, K., Withers, R.S., 'An Artificial Neural Network Integrated Circuit Based on MNOS/CCD Principles', AIP Conf. Proc. 151, 381, 1986.

[27] Sequin, C.H.S. and Tompsett, M.F., 'Charge Transfer Devices', in Advances in Electronics and Electron Physics, Supplement B, (New York: Academic Press), 1975.

[28] Sutton, R. S.,' Learning by the Methods of Temporal Differences', Machine Learning 3: 9-44, 1988.

[29]   Waibel, A., et. al., 'Phoneme Recognition Using Time-Delay Neural Networks', IEEE Trans. on Acoust. Speech, and Signal Processing, Vol 37, pp328-339, 1989.

[30]   Wolsheimer, E. A., 'Optimization of Potential in Junction Charge-Coupled Devices', IEEE Trans. on Electron Devices, Vol. ED-28, pp 811-818, 1981.

[31]   Zimmerman, T.A., Allen, R.A., Jacobs, R.W., 'Digital Charge-Coupled Logic (DCCL)', IEEE Journal of Solid-State Circuits, Vol SC-21, pp 472-485, 1977.

ANALOG STORAGE OF ADJUSTABLE SYNAPTIC WEIGHTS

E. VITTOZ, H. OGUEY, M.A. MAHER, O. NYS, E. DIJKSTRA
and M. CHEVROULET

INTRODUCTION

The most important specific problem faced in the analog implementation of neural networks is the storage of synaptic weights. Each storage cell should be very dense since one of them is needed for every synapse. The weight must be adjustable to provide an on-chip learning capability. Absolute precision is not usually needed since most learning algorithms are based on successive corrections of the weight values in a closed loop system.

Since the synaptic weighting may be carried out in various manners from the gate voltages of MOS transistors, the most natural method is to store the weight as a voltage across a capacitor. No currents are needed to drive the MOS gates and integrated capacitors have an extremely low value of intrinsic leakage current, thanks to the very high quality of the silicon dioxide used for the dielectric layer. Storage time limitations are essentially due to the need to associate each capacitor with at least one pn junction, if the stored value has to be adjustable by means of a transistor operated as a switch (or as a variable conductance). The leakage current of this junction limits the storage time to short durations and is discussed in section 2.

The storage time can be extended by periodically refreshing the voltage stored on the capacitor. The principle of a new refresh scheme is presented in section 3.

Leakage can be virtually eliminated, to provide long term storage, by placing a floating electrode entirely inside the oxide. The stored voltage can then only be adjusted by injecting carriers across the oxide. Section 4 will present an original method to carry out this analog adjustment by means of the Fowler-Nordheim field effect. Possibilities with UV irradiation will also be presented.

SHORT-TERM STORAGE

Figure 1 shows the most simple way to store the synaptic weight as a voltage V across a capacitor C. The value V_{in} to be stored is sampled by means of a transistor T_S operated as a switch controlled by the gate voltage V_G.

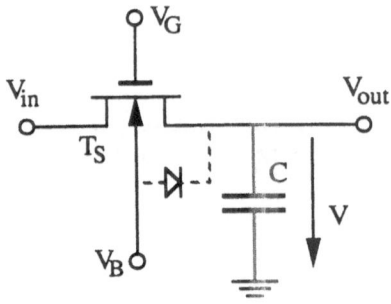

Figure 1   Elementary sample-and-hold

The storage time is limited by the leakage current of the reverse-biased diffusion-to-bulk junction which discharges the storage capacitor C. This current would be cancelled if the local substrate potential V_B would be exactly equal to the stored voltage V, provided the junction does not receive any light.

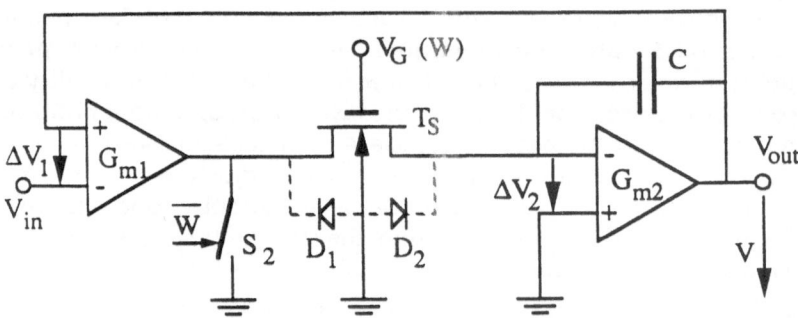

Figure 2   Low-leakage sample-and-hold

Figure 2 shows a possible implementation of a low-leakage sample-and-hold [1]. During sampling (W = 1, V_G positive), any difference between V and V_{in} is corrected by an error current that flows onto C through the conducting transistor T_S. During storage, the voltage across the critical junction D_2 is maintained at a very small value by the virtual ground action of the transconductance amplifier G_{m2}. If the gain of this amplifier is large, the residual voltage is limited to the small offset V_2 and leakage is drastically reduced, provided the channel is perfectly blocked by a sufficiently low value of gate voltage V_G.

During storage, the current delivered by G_{m1} must be diverted by the additional switch S_2 to avoid activating a lateral bipolar action in T_S by forward-biasing the junction D_1.

The accuracy of sampling is limited by the offset ΔV_1 of the first amplifier and by the charge released onto C when the transistor T_S is blocked. This injection of charge does not depend on the stored value V. It can be minimized by limiting the charge stored in the channel in the on-state (by control of the value of V_G) [2] and by using the standard dummy switch compensation technique [3].

Such a scheme is expected to be able to store a synaptic value up to a few seconds at room temperature with conventional CMOS technologies.

MEDIUM TERM STORAGE

To achieve a storage time beyond the limit set by the leakage, the voltage across the capacitor must be regenerated (refreshed) by periodic comparison with predefined quantized levels. This regeneration must be carried out locally to avoid the need for fast traffic on analog busses. The most efficient way to make the quantization levels available to each synapse is to provide them sequentially by means of a staircase signal distributed on a single wire.

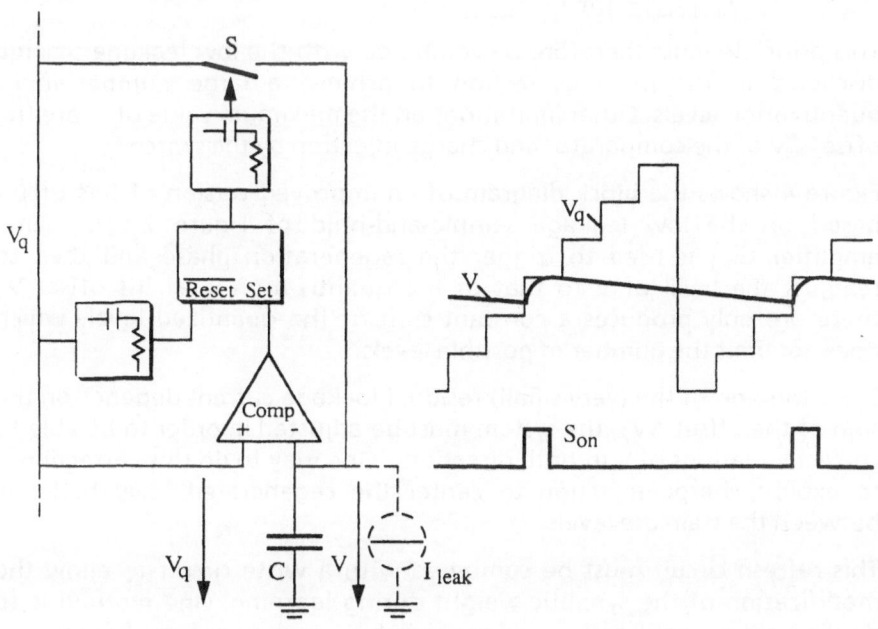

Figure 3   Principle of refreshment

Figure 3 shows the principle of a refresh circuit based on this idea. The stored voltage V is continuously compared with the staircase V_q. When the value V_q crosses that of V, a trigger is set on. A short pulse is then produced, which closes the switch S and thus readjusts the value of V to the staircase level which has just been reached. No further pulses can be produced until the trigger has been reset by the return to zero of the staircase. Experiments on a breadboard implementation have shown that this principle is valid.

This scheme provides the additional advantage of not requiring the distribution of large amplitude pulses to drive the regeneration action, which reduces the problem of induced noise.

The maximum possible value for the number m of quantization levels is strongly dependent on the leakage current I_{leak} to be expected. Indeed, for a given value of the staircase amplitude, the step height is proportional to $1/m$. For a given value of current consumption in the comparator, the time required for comparison is inversely proportional to the step height and thus proportional to m. The total refresh cycle time T_R is thus proportional to m^2. Since refresh must be carried out before leakage has degraded V by more than one staircase step, the maximum acceptable value of I_{leak} is inversely proportional to mT_R and thus

$$I_{leakmax} \sim 1/m^3$$

This principle must therefore be combined with the low leakage scheme discussed in the previous section to achieve a large number m of quantization levels. Other limitations on the maximum value of m are the offset ΔV of the comparator and charge injection by the switch S.

Figure 4 shows the block diagram of an improved version of this circuit based on the low leakage sample-and-hold of Figure 2. The same amplifier G_{m1} is used to trigger the regeneration phase and then to readjust the level of V to that of the quantization level. Its offset V_1 therefore only produces a constant shift of the quantized levels which does not limit the number of possible levels.

Since the sign of the (very small) residual leakage current depends on the sign of the offset ΔV_2, the system must be adjusted in order to be able to correct variations of V in both directions. One way to do this correction is to exploit charge injection to center the regenerated level half-way between the staircase levels.

This refresh circuit must be combined with a write circuit to allow the modification of the synaptic weight during learning. One method is to duplicate the amplifier G_{m1} and the switch T_S, as shown in the figure.

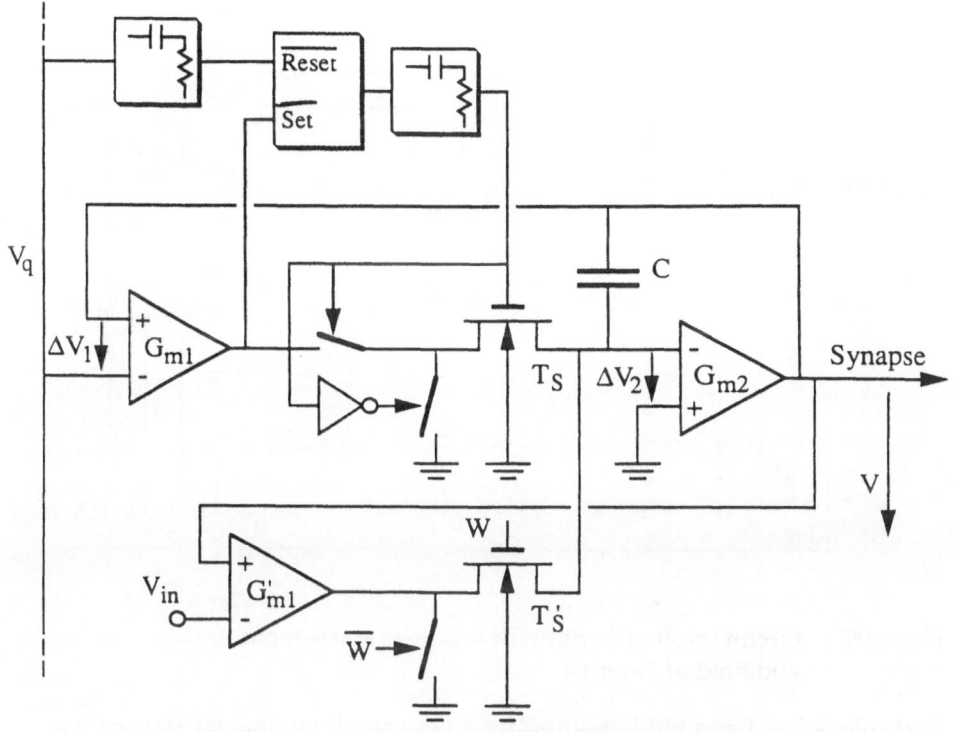

Figure 4   Low-leakage refreshed sample-and-hold

The sample and hold circuit of Figure 4 (which was based on the circuit of Figure 2) can be realized as shown in Figure 5. The first transconductance amplifier G_{m1} consists of transistors M1--M5. Inverter M18--M19 restores the digital level needed at the output of G_{m1} when it is in open loop acting as a comparator.

The drain of transistor Ts is connected to the output of amplifier G_{m1} by means of transistor M15. Instead of applying control signal C_1 to the gate of T_S directly, transistors M16 and M17 allow the gate of T_S to switch between the on and off voltages V_H and V_L, respectively. If these levels are close together, charge injection is reduced.

Transistor M14 connects the drain of Ts to the gate of M10 performing the role of the switch S_2 of Figure 2 which shunts the current delivered by G_{m1}.

The second amplifier G_{m2} consists of a pair of cascoded inverters realized with transistors M6--M13. A copy of the source voltage of T_s is generated at the gate of M10 creating the ground of the amplifier.

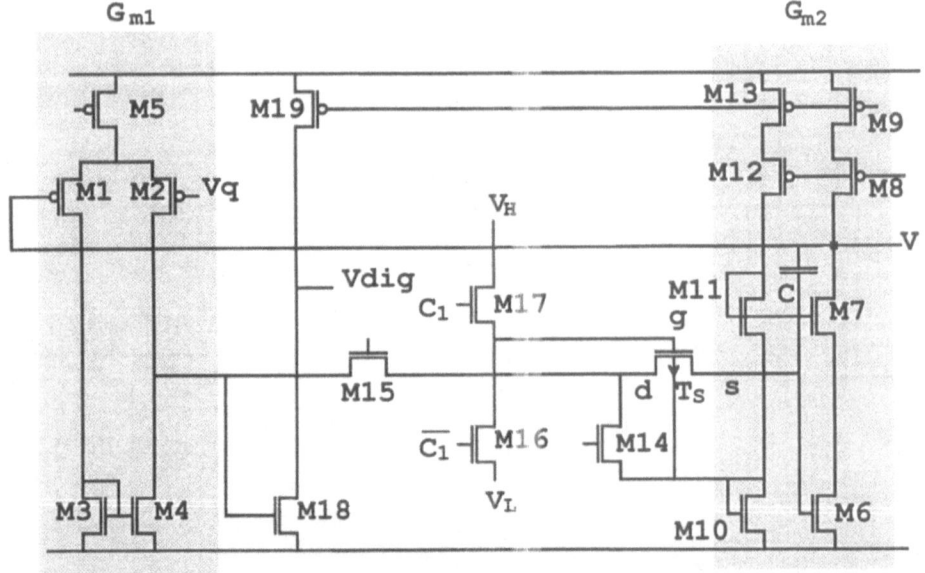

Figure 5   Circuit implementation of low-leakage sample
             and hold of Figure 4

Each block has been implemented by a very small number of transistors. We expect to realize at least 45 synapses/mm^2 in a standard 2 micron CMOS process.

LONG TERM STORAGE

Floating gate (FG) electrically programmable non-volatile memories (EEPROM's) form a mature technology with wide-spread digital applications. Recently, their use for analog applications has been proposed, mainly for trimming [4,5,6]. It has been restricted due to two main reasons: the accurate adjustment of a given voltage has to be done iteratively by a succession of high-voltage pulses and measurements, and a drift is observed shortly after each change.

A single-step programming technique with negligible subsequent voltage drift is introduced here. Figure 6 shows the layout (a) and the cross-section along AA'(b) of a memory cell based upon a polysilicon floating gate MOSFET. A separate injector serves to supply the analog voltage. In a CMOS n-well technology, this injector is a special n+ diffusion in the p- substrate, above which a thin (10 nm) tunnel oxide is grown in a small opening. The floating gate FG is made of a first polysilicon layer. It controls the gate of a separate MOSFET. A control electrode (control gate CG) made of a second polysilicon layer is capacitively coupled to FG. The FG MOSFET will deliver a current as a measure of its gate voltage. In order to recover this gate voltage, a matched comparison MOSFET is provided.

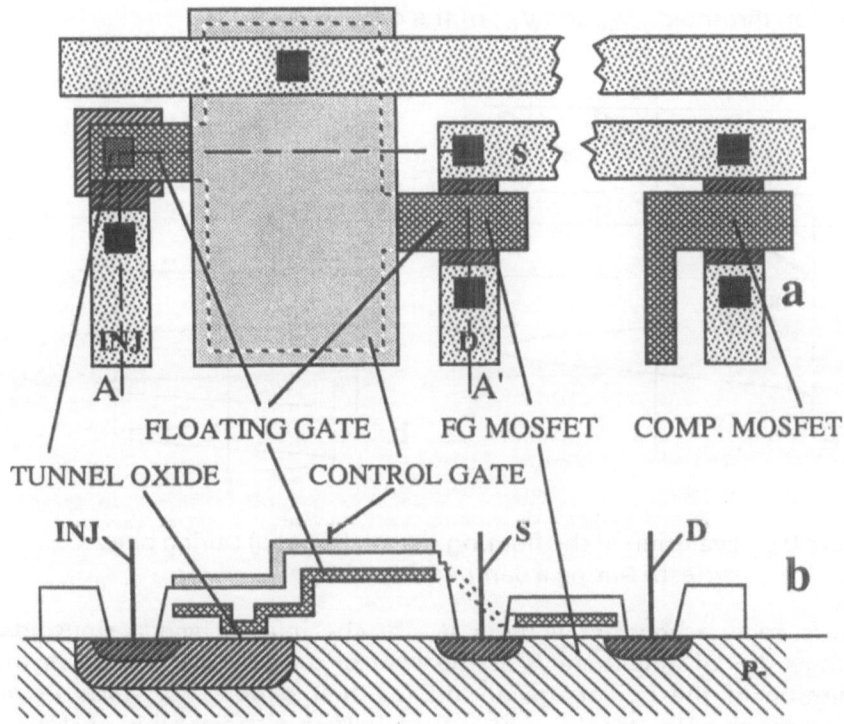

Figure 6   Structure of an analog EEPROM cell: a) Layout, b) Cross-section

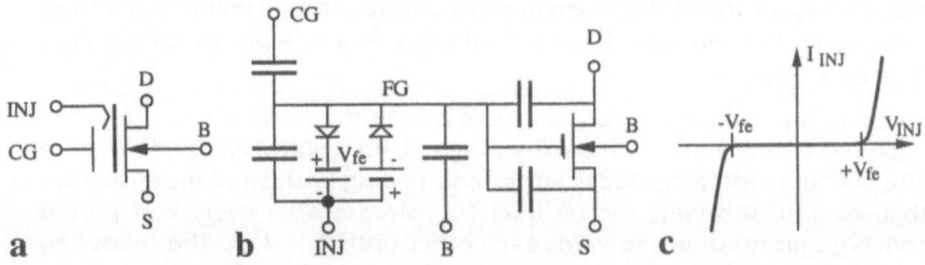

Figure 7   Analog EEPROM cell: a) Symbol, b) Equivalent circuit, c) Tunnel oxide characteristics

In Figure 7, the electrical symbol of this cell is shown (a), together with its equivalent circuit (b). At low voltage, the floating gate potential varies according to the capacitive couplings. If the voltage across the tunnel oxide reaches a certain absolute value, a field-effect current flows by Fowler-Nordheim tunneling. The injector I-V characteristic (c) is almost exponential in both directions, and may be approximated by two field-emission thresholds -V_{fe} and V_{fe}, of the order of 12 V.

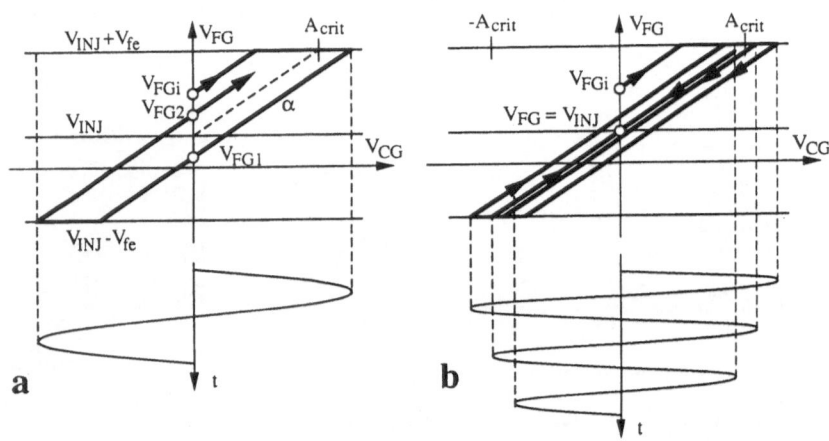

Figure 8   Evalution of the floating gate voltage: a) During one AC cycle, b) During a damped AC voltage

Let us apply a constant voltage V_{inj} at the injector and a sinusoidal voltage V_{CG} of amplitude A at the control gate. Figure 8a shows the evolution of the F_G voltage during the first period. V_{FG} starts at an arbitrary value V_{FGi} for $V_{CG} = 0$, it then follows a straight line of slope a $= C_{CG}/C_{total}$ up to the value $V_{INJ} + V_{fe}$. It is clamped at this value until V_{CG} reaches its maximum A, then it decreases along another line of slope a and crosses the axis $V_{CG} = 0$ at a point $V_{FG1} = V_{INJ} + V_{fe} - aA$. After the negative excursion of V_{CG}, it crosses this axis at a second point $V_{FG2} = V_{INJ} - V_{fe} + aA$. If amplitude A has a critical value $A_{crit} = V_{fe}/a$, we obtain $V_{FG1} = V_{FG2} = V_{INJ}$.

The original technique proposed here stores the injector voltage while applying a slowly decreasing AC voltage at the control electrode (Figure 8b), starting with amplitudes larger and ending with amplitudes smaller than A_{crit}. In this way, the residual FG voltages after every half period converge progressively towards each other until $A = A_{crit}$. The following periods cause reversible changes only, so that the stored F_G voltage should be very close to the injector voltage. During this operation, both the source and the drain of the MOSFET are kept at zero potential.

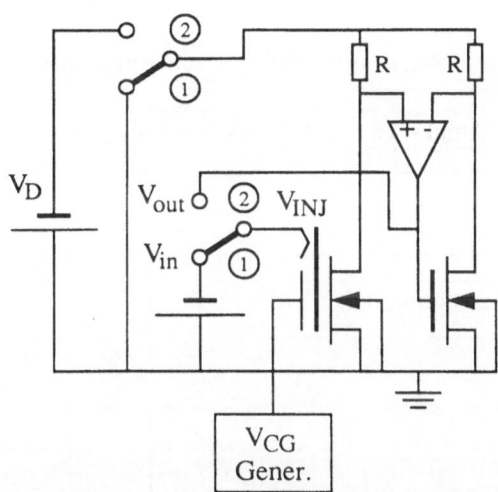

Figure 9   Diagram of the analog EEPROM measuring setup

Experimental verification of this analog EEPROM principle has been proven with the cell of Figure 6 in a CMOS n-well 2 micron technology. The principle of the experimental setup is shown in Figure 9. In the writing phase (1), a constant voltage is applied at V_{INJ}, an AC voltage with a decreasing amplitude is applied at V_{CG}, and the F_G MOSFET has both its source and its drain at ground. During the read phase (2), the current in the read MOSFET is measured. The gate of the comparison transistor is adjusted by a current comparator in order to deliver the same current. Notice that the floating gate voltage during the read operation can vary by capacitive coupling with the voltage of any electrode which is coupled with it. The highest accuracy requires all electrodes coupled to the floating gate to be at the same potential during read as during write. For this reason, the drain voltage is kept at a low value, and the injector is connected to the output, which is close to the potential applied during write.

Figure 10 shows a typical result. The injector voltage has been varied from 0 to 5 V and back to 0 in 0.5 V steps. In Figure 11 the difference between the read and the write voltages is shown. It can be seen that the writing process is very linear, with a slope close to one. The largest error is a systematic offset. This offset could be eliminated by known offset compensation techniques, for example, with two matched memory cells.

56

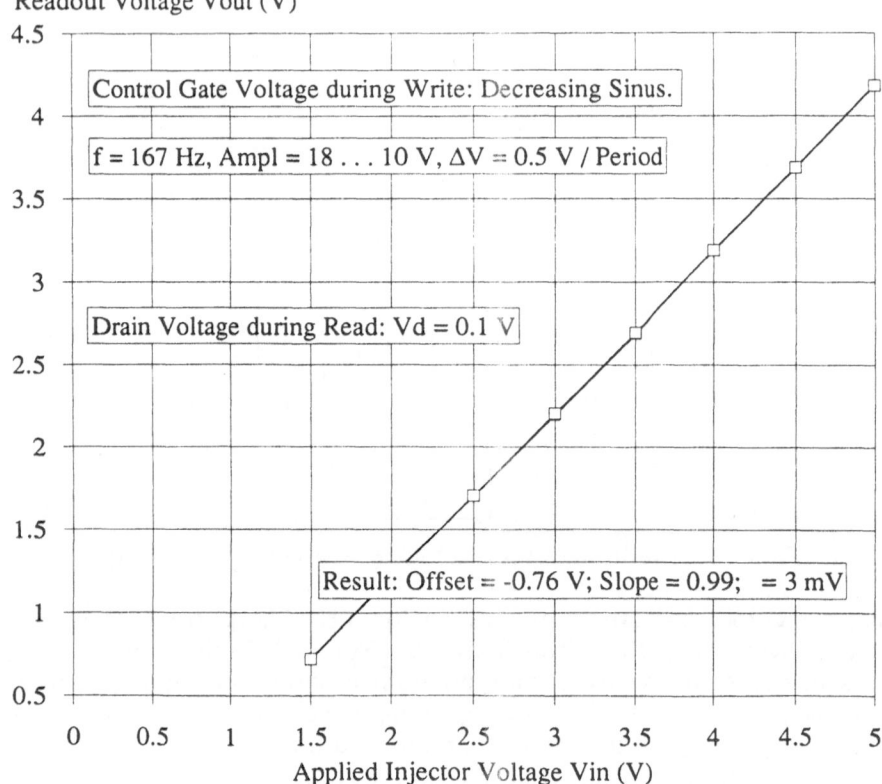

Figure 10   Typical analog EEPROM write-read voltage characteristics

A physical explanation of this offset is shown in Figure 12. In accordance with direct measurements of the tunnel injection characteristics, the offset is attributed to the fact that the n-doped injector surface is in accumulation for positive applied voltages (a), and in inversion for negative voltages (b). In the latter case, the voltage drop across the depletion layer reduces the field across the tunnel oxide and shifts the negative $-V_{fe}$ threshold, which produces the observed offset. It is our hope to reduce this effect by providing a p-doped diffusion near the injector (c).

Further experiments have proven that this analog EEPROM writing procedure is reproducible for a wide range of frequencies, voltages and decrement values of the control voltage. In previous analog memory proposals, a large drift after writing was present. With this new principle, a small relaxation effect, corresponding to a shift of up to 1 mV per time

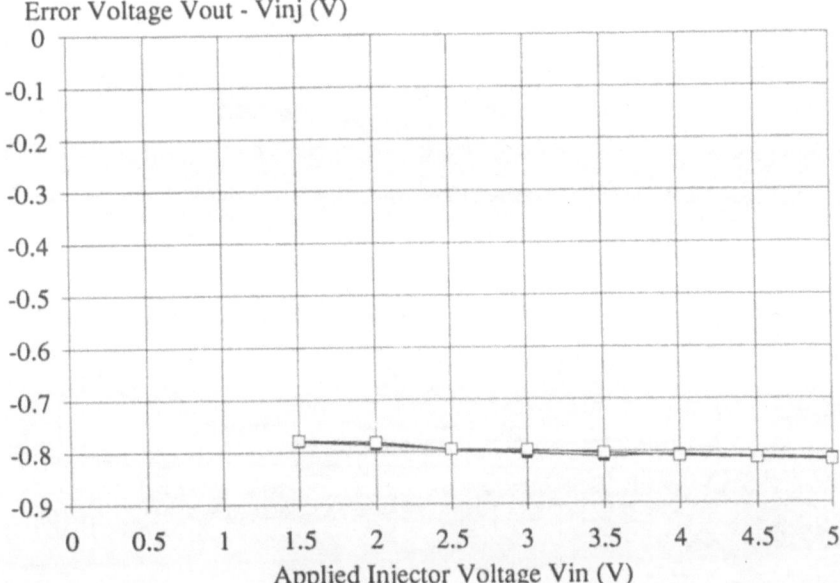

Figure 11   Error in analog EEPROM write-read voltage characteristics

decade, is observed. Endurance of 3000 writing operations has been observed, and is expected to improve in a more mature technology and with optimal writing conditions. Note that the oxide can withstand a larger number of cycles under AC stress than under unipolar stress [7].

This analog memory principle is very promising due to some unique advantages:

--   Single step writing lasts 10 to 100 ms and can be done in parallel for a large number of cells.

--   Complete separation of the signal path (input voltage, output current) and of the control path allows the use of a low-voltage VLSI technology for all signal processing operations. A simple external generator can provide the "high voltage" (15 to 20 V) control signal. Application of analog EEPROMS for neural networks is presently under investigation in many laboratories, and some realizations have already been reported [8]. This analog EEPROM principle promises to be a good solution for accurate and reliable storage of adaptive synaptic weights in analog neural networks.

As an alternative to the use of high voltages, ultraviolet (UV) light can be used to program floating gate non-volatile memories. UV light has been used to remove electrons from a floating gate transistor in the erase

58

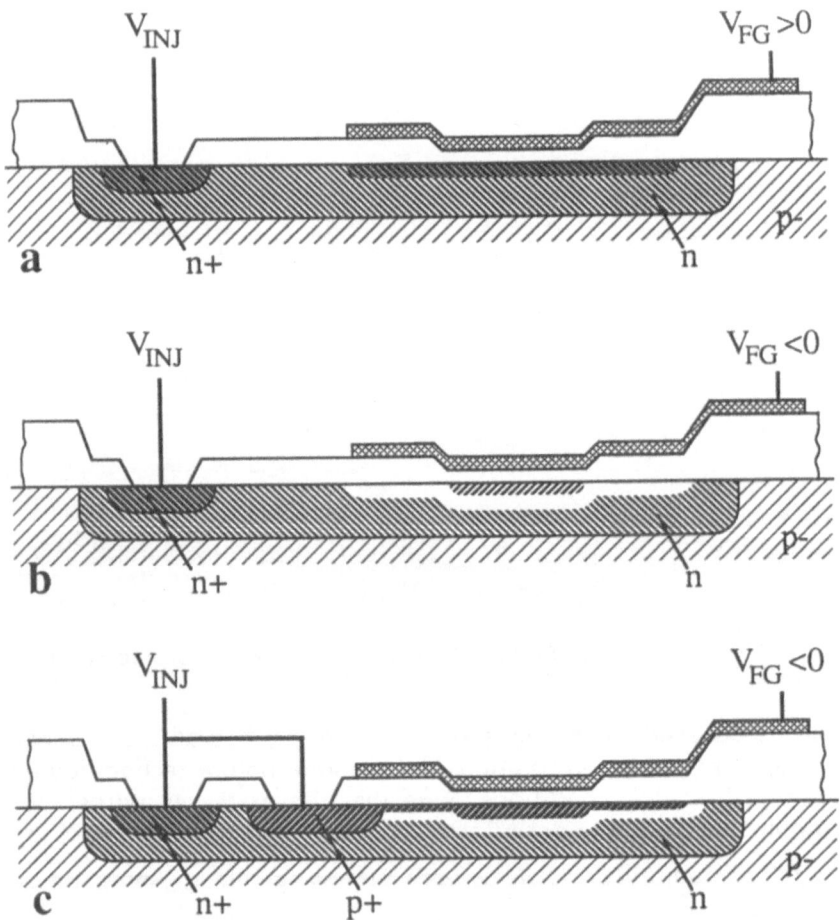

Figure 12   Physical structure of the injector. For a positive FG voltage (a),
              it is in accumulation. For a negative gate voltage (b), a space
              charge layer causes a voltage drop. This can be avoided by
              providing a p-doped diffusion (c)

operations of EEPROMS. UV light can also be used to inject electrons onto
the floating gate from the drain electrode. Glasser [9] described a digital
floating gate memory using this technique. The action of ultraviolet light
on a floating capacitor is to shunt it with a small conductance. The
conductance is zero when the light is turned off. This conductance can be
used to adapt the charge on a floating gate in analog steps. Mead [10]
employs this principle for adapting offsets in his retina chips and other
structures currently being developed at Caltech [11]. The principle can be
used to design an analog memory that can be adapted by small steps
needed by many neural network algorithms. Since silicon dioxide is an

excellent dielectric, the charge will remain on the floating gate for years and the memory is then suitable for long term storage.

We have designed a series of test structures in a double polysilicon CMOS technology to characterize the UV adaptation process and determine the geometrical design rules for building circuits utilizing it. One structure is shown in Figure 13 along with a cross-section taken along the line AA'. It

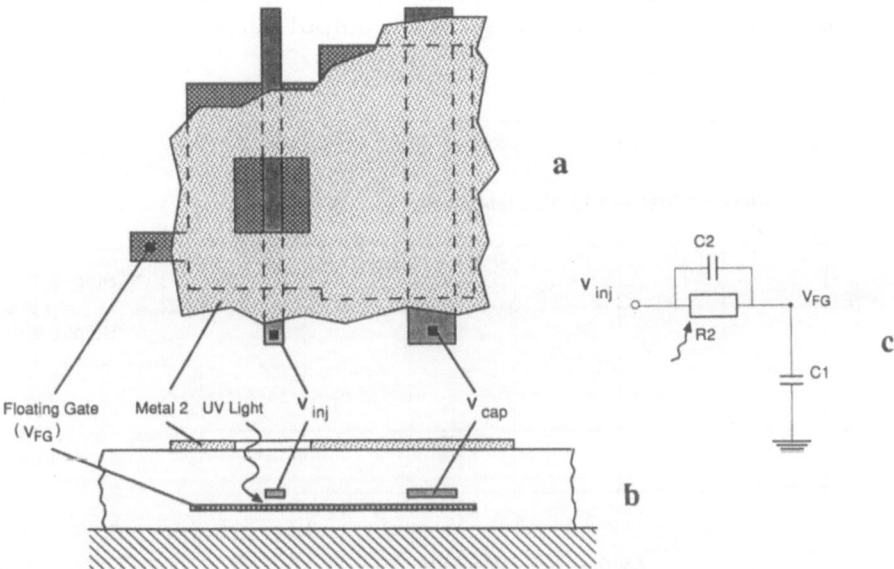

Figure 13   Structure of an analog UV memory test cell: a) Layout, b) Cross-section, C) Equivalent circuit

consists of a floating capacitor made from first poly which is covered with two electrodes in second poly (labeled V_{inj} and V_{cap} in the figure). The entire structure is shielded from light with second metal except for a small window where adaptation to the electrode V_{inj} is desired. The voltage on the floating node is monitored through a voltage follower constructed from a five transistor differential transconductance amplifier.

The electrical model also shown in the figure consists of a coupling capacitor and resistor between the first poly and second poly layers and a capacitor from first poly to the substrate. Capacitors to other layers are not shown. The value of the resistor is set by the device geometry and strength of the UV irradiation. A voltage step applied to the injector electrode results in a first order charging of the floating node.

Test chips have been fabricated in a CMOS p-well two micron technology by MOSIS, the ARPA silicon foundry [12]. The chip was tested inside an

EEPROM eraser for a source of UV light and as a consequence operated at high temperature. A typical test result is shown in Figure 14. The voltage on the injector has been varied from 1 to 4.5 volts and back to 1 volt in 0.5 volt steps and was tied to the V_{cap} line. Voltages below 1 volt and above 4.5 volts were not used due to the output voltage limitations of the follower. The error voltage (shown in Figure 15) is a function of absolute injector voltage and suggests that there is a parasitic resistor to ground formed by adaptation to other layers. The offset is believed to be from the high temperature voltage offset of the output follower.

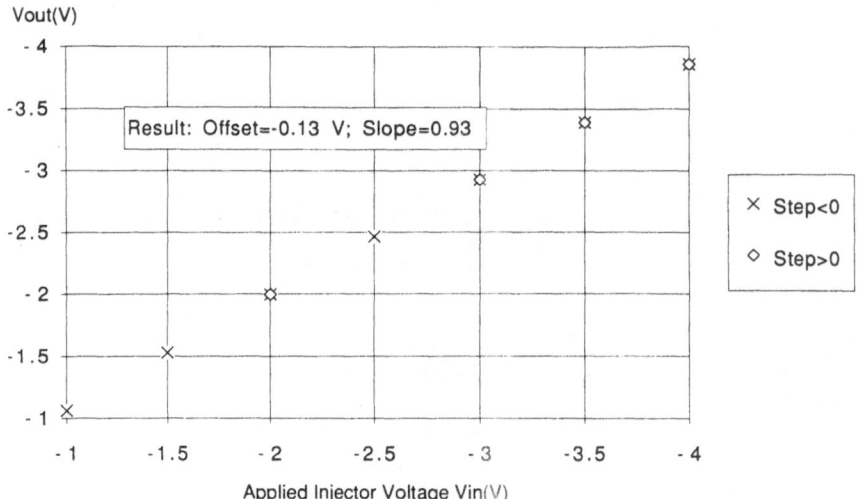

Figure 14   Typical analog UV memory test cell write-read voltage characteristics

Other experiments on the test chips show that positive and negative steps applied to the injector voltage result in the same equilibrium floating gate voltage. In addition, this voltage was independent of step size and was repeatable. However, for our test chips the time constant of the adaptation depended on the direction of the step. Further experiments showed that the model of Figure 13 is valid for small steps and the deviation from this resistive model increases with step size due to saturation of the UV process. The value of the resistance extracted from our test data was 3000 TΩ due to the use of a minimum geometry test structure. Using larger cells will decrease the resistance and increase speed performance.

In other structures the distance to the V_{cap} line and to the substrate is varied to determine the parasitic adaptation to these nodes and hence the minimum design rules. Structures using the drain of a floating gate transistor as an injector have also been fabricated in both single poly and double poly processes and are currently being tested. Annular structures have also been investigated to minimize parasitic adaptation. In addition, we have fabricated structures to estimate the dependence of the time constant of the adaptation on the device geometry.

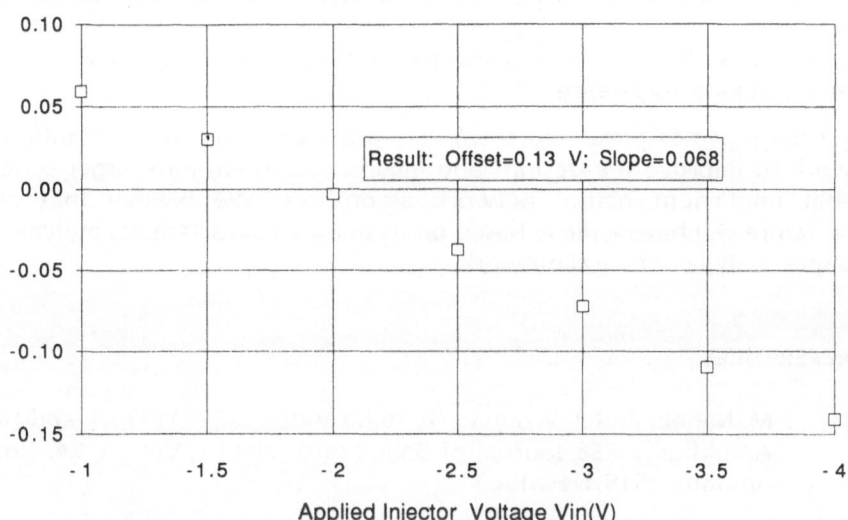

Figure 15   Error in analog UV memory test cell write-read voltage
                characteristics

A major advantage of UV analog memories are their compatiblity with standard CMOS processes. Large voltages and ultra thin tunnel oxides used with EEPROM type floating gate memories are avoided. However, the speed of adaptation is dependent on the geometry of the structure and is slow for small sized devices. In addition, a source of UV light must be provided.

CONCLUSIONS

In this paper, we have explored four methods for analog storage of adjustable synaptic weights. The first method extends the storage time of a capacitor with a low-leakage sample and hold. The periodic refresh scheme compares the stored voltage with levels sequentially provided by a staircase. The novel implementation of the EEPROM floating gate memory provides single step writing of the weights in parallel in 10-100 ms and allows separation of the signal processing paths and the high voltage control signal. The UV analog memory uses UV irradiation to inject electrons through the oxide to adapt the charge on a floating gate.

The low leakage capactor provides short term storage, the periodic refresh scheme provides medium term storage and the UV and EEPROM analog memories provide long term non-volatile storage. The first three of this schemes can be implemented in a standard CMOS process while the EEPROM requires a special process with a tunnel oxide and a high voltage line. The UV memory requires the use of a source of UV light and is very slow for minimum sized geometries. Both the UV and EEPROM memories are very dense.

All the schemes presented seem very promising and we are continuing work to improve the designs and incorporate them into larger systems that implement neural network algorithms. We believe that each structure will have a niche based on its unique characteristics matched to a specific type of neural network.

REFERENCES

[1]   M. Nahebi and B. Wooley, "A 10-bit Video BiCMOS Track-and-Hold Amplifier", IEEE Journal of Solid-State Circuits, Vol. SC-24, No. 6, pp. 1507-1516, Dec. 1989.

[2]   E.Vittoz and G.Wegmann, "Dynamic Current Mirrors", in Advances in Analog Integrated Circuit Design, C. Toumazou, F. Lidgey and D. Haigh eds., Peter Perebrinus, London, 1990.

[3]   G. Wegmann, E. Vittoz and F. Rahali, "Charge Injection in Analog MOS Switches", IEEE Journal of Solid-State Circuits, Vol. SC-22, No. 6, pp. 1091-1097, Dec. 1987.

[4]   E. Säckinger and W. Guggenbühl, "An Analog Trimming Circuit Based on a Floating-Gate Device", IEEE Journal of Solid State Circuits, Vol. SC-23, No. 6, p. 1437-1440, Dec. 1988.

[5]   Chr. Bleiker et al, "EEPROM-Speicher-Elemente in der Analogen Schaltungstechnik", Bulletin SEV/VSE 80, 15, pp. 941-946, Aug. 1989.

[6]   L. Richard Carley, "Trimming Analog Circuits Using Floating-Gate Analog MOS Memory", ISSCC Digest of Technical Papers, p. 202-203, Febr. 1989.

[7]   P. Fazan, M. Dutoit, J. Manthey, M. Ilegems and J. Moret, "Dielectric Breakdown in Thin Films of SiO_2 Used in EEPROM", ECS Fall Meeting, October 1986.

[8]   Marc Holler et al., "An Electrically Trainable Artificial Neural Network (ETANN) with 10240 'Floating Gate' Synapses", International Joint Conference on Neural Networks, Proc. pp. II-191-196, Washington, June 1989.

[9]   L. A. Glasser, "A UV Write-Enabled PROM.", in 1985 Chapel Hill Conference on VLSI, Computer Science Press, Rockvill, MD, pp. 61-65, 1985.

[10]   C. A. Mead, "Adaptive Retina", in Analog VLSI Implementation of Neural Systems, Carver Mead and Mohammed Ismail eds., Kluwer Academic Publishers, Boston,MA, pp. 239-246, 1989.

[11]   C.A. Mead, Private Communication.

[12]   D. Cohen and G. Lewicki, "MOSIS - the ARPA Silicon Broker", in Proceedings from the Second Caltech Conference on VLSI, California Institute of Technology, Pasadena,CA, pp. 239-246, 1981.

PRECISION OF COMPUTATIONS
IN ANALOG NEURAL NETWORKS

M. VERLEYSEN, P. JESPERS

INTRODUCTION

VLSI implementations of analog neural networks have been strongly investigated during the last five years. Except some specific realizations where the precision and the adaptation rule are more important than the size of the network [1] [2], most applications of neural networks require large arrays of neurons and synapses. The fan-out of the neuron is not the crucial point: digital or analog neuron can be easily designed so that they can drive a large number of synapse inputs (in the next layer in the case of multi-layered networks, in the same layer in the case of feedback networks). Fan-in is more important: whatever is the transmission mode of information between synapses and neurons (voltage, current, pulses,...) the neuron input must have a large dynamics if it is connected to hundreds of synapses. Digital neurons are of course the solution: if the dynamics of the neuron inputs has to be increased, more bits will be used and the required precision will be obtained. However, digital cells are in general much larger than their analog counterpart: for example, a neuron connected to 100 synapses must contain a digital adder with 100 inputs, each of them coded in several bits. The silicium area occupied by the cells and the connections between cells will be incompatible with the integration of a large number of synapses and neurons on a single chip.

ANALOG NEURONS AND SYNAPSES

In order to compensate for such lack of efficiency, analog cells are used in VLSI neural networks. Several techniques have been proposed to transmit information between synapses and neurons. A.Murray [3] proposed a method inspired by biological mechanisms, where the partial sum-of-products are coded by the frequency of a pulse stream. This solution presents several advantages about the precision of computations, but requires in general a complex circuitry in comparison with other methods.

One of the first VLSI analog neural networks was realized by AT&T Bell Labs [4]. This circuit was based on the sum of positive and negative currents for excitatory and inhibitory synapses (figure 1). A simple logic realizes the product of the synapse input by its internal weight; depending on the sign of this product, a current is sourced to or sunk

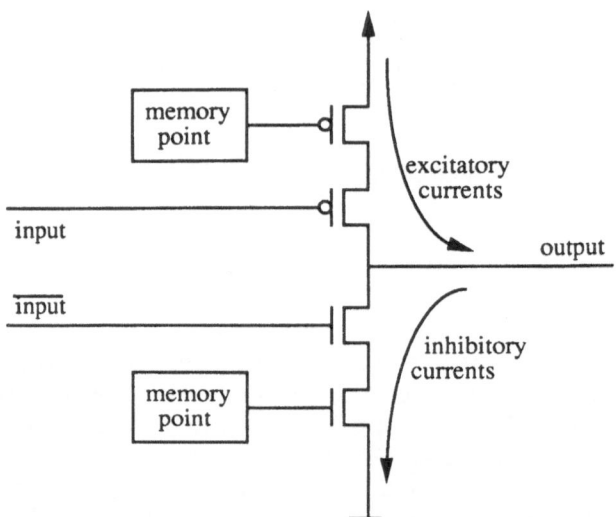

Figure 1   Positive and negative currents

from the input line of the connected neuron. The drawback of this architecture is that one can never assume the excitatory and inhibitory currents to be exactly the same; even with adjustments of the size of the p- and n-type transistors, there is always a risk of mismatching between the sourced and sunk currents because of the technological mobility differences between the two types of charges.

On the other hand, the neuron realizes a non-linear function of its input; in the case of a simple threshold function, the neuron determines the sign of the sum of all synaptic currents. The mismatching between excitatory and inhibitory currents are also summed, and can rapidly become greater than a typical synaptic current. For example, if the error between currents from n-type and p-type transistors is about 20 %, a computation error may occur if only 5 synapses are connected to the same neuron!

In order to compensate for such lack of efficiency, a two-line system can be used: all the excitatory currents are summed on one line, all the inhibitory ones on another line (figure 2). All currents are thus generated by the same type of transistors, and the mismatchings between current sources decrease. In the following, a two-line system is described where the neuron values are binary, and where the synaptic weights can only take three values: + 1, 0 and -1. Such restrictions of course limit the use of the neural network; however, for some architectures, like the Hopfield network, learning algorithms can be found where these restrictions have almost no decreasing effect on the performances of the network [5]. Moreover, the complexity of the synapses is strongly reduced due to the

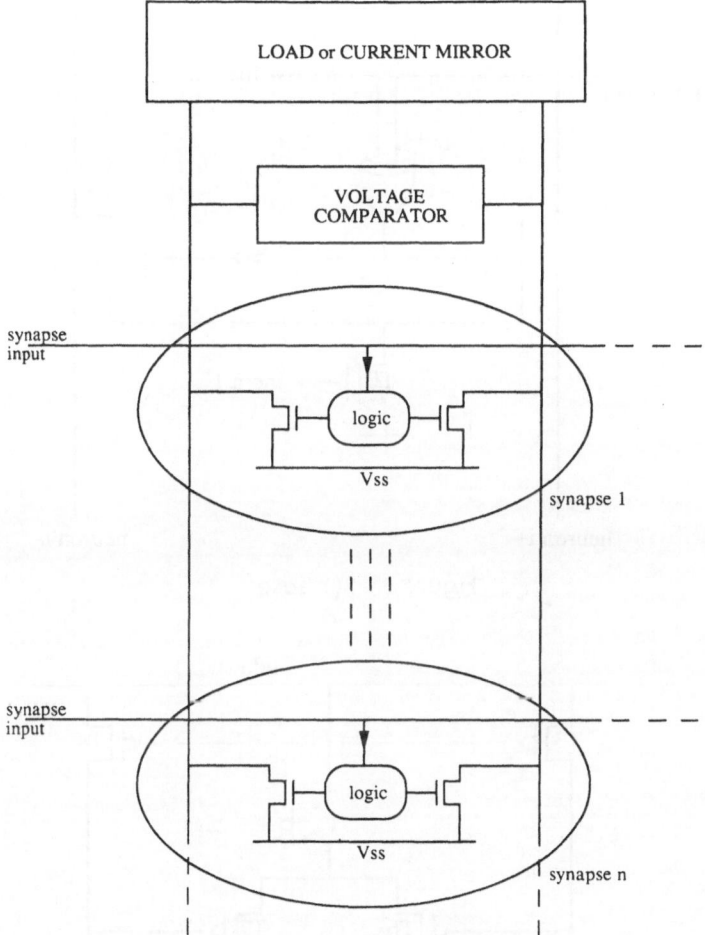

Figure 2   Two-line system

restricted dynamics, and for the same area of silicon a greater number of synapses can thus be integrated.

Each synapse is a programmable current source controlling a differential pair (figure 3). Three connection values are allowed in each synapse. If "mem1" = 1, current is delivered to one of the two lines with the sign of the connection determined by the product of "mem2" and the output of the neuron to which the synapse is connected. If "mem1" = 0, no connection exists between neurons i and j, and no current flows neither to the excitatory and inhibitory lines.

Depending on the state of the XOR function, the current may be sourced either on the line i + or on the line i-. In the neuron, the comparison of the two total currents on the lines i + and i- must be achieved. This is

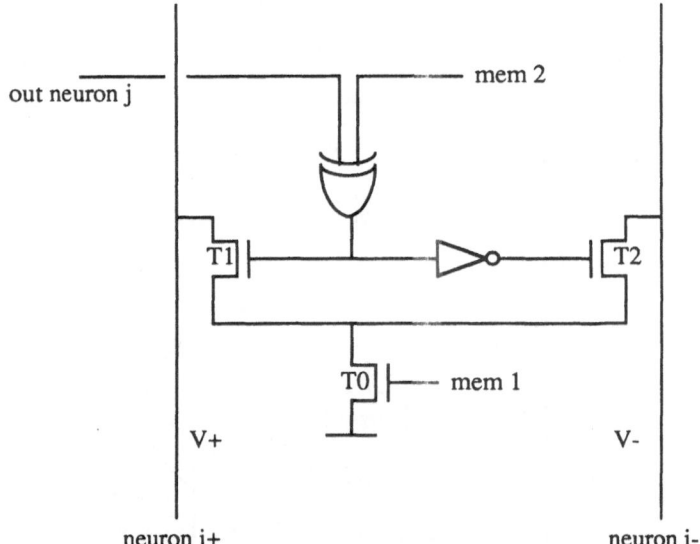

Figure 3   Synapse

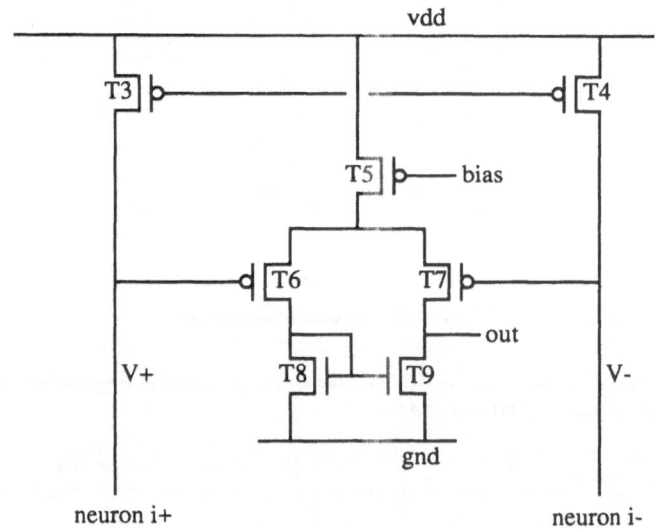

Figure 4   Neuron

done by means of the current reflector shown in figure 4. The currents on the lines are converted into voltages across transistors T3 and T4; these voltages themselves are compared in the differential input reflector formed by transistors T5 to T9. Because of the two-stage architecture of the neuron, the gain may be very large and the output (out) is either 5V if the current in neuron i- is greater than the one in neuron i + , or OV in the opposite case.

Two-transistor load

In order to convert the two currents into voltages in a two-transistor load, two simple solutions can be considered: two loads (figure 5.a) or a current mirror (figure 5.b).

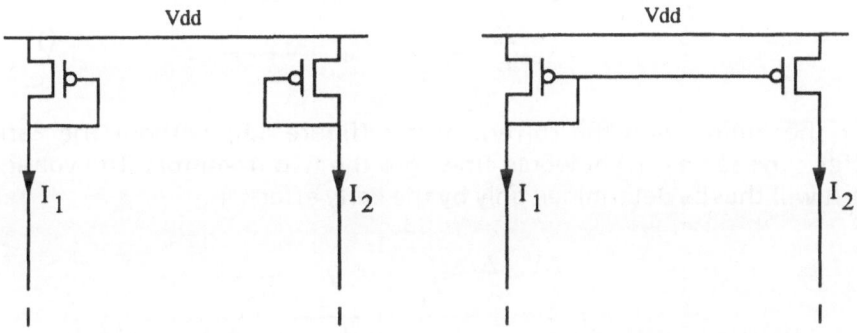

Figure 5a Two loads   Figure 5b Current mirror

When many current sources are connected to the neuron, the load must be able to discriminate small currents (i.e. one synaptic current) between the two lines, whatever is the common-mode current in these lines (for example, the neuron must have the same behaviour if one excitatory and two inhibitory currents are connected, or with 500 excitatory and 501 inhibitory ones). The ability to discriminate currents in the neuron will of course be enhanced with the differential gain of the load (differential gain is here defined as the voltage difference between the two lines for a given current difference at the input of the load). This gain can easily be computed, assuming the transistors are in saturated mode, and neglecting second-order effects; current can then be expressed by:

$$I = \mu C_{ox} \frac{W}{L} \frac{(V_{gs} - V_t)^2}{2}$$

For the option with the two loads (figure 5.a) we have:

$$\Delta I = \mu C_{ox} \frac{W}{L} \left[\frac{(V_{gs1} - V_t)^2}{2} - \frac{(V_{gs2} - V_t)^2}{2} \right]$$

$$= \mu C_{ox} \frac{W}{L} \left| \frac{(V_{gs1}^2 - V_{gs2}^2)}{2} + V_t (V_{gs2} - V_{gs1}) \right|$$

$$= \mu C_{ox} \frac{W}{L} (V_{gs1} - V_{gs2}) \left[\frac{(V_{gs1} + V_{gs2})}{2} - V_t \right]$$

$$G = \frac{\Delta V}{\Delta I} = \frac{1}{\mu C_{ox} \frac{W}{L}} \frac{1}{\frac{V_{gs1} - V_t}{2} + \frac{V_{gs2} - V_t}{2}} \tag{1}$$

For the option with the current mirror (figure 5.b), without the Early effect, the same current would flow into the two transistors. The voltage shift will thus be determined only by the Early effect:

$$\Delta V = \Delta I \frac{1}{\frac{I_1}{V_{EAp}} + \frac{I_2}{V_{EAp}}}$$

where V_{EAp} is the Early voltage of p-type transistors, and V_{EAn} the Early voltage of the current sources driving I_2. Gain is thus given by:

$$G = \frac{\Delta V}{\Delta I} = \frac{1}{\frac{I_1}{V_{EAp}} + \frac{I_2}{V_{EAn}}} \frac{1}{\mu C_{ox} \frac{W}{L}} \frac{1}{\frac{(V_{gs1} - V_t)^2}{2 V_{EAp}} + \frac{(V_{gs2} - V_t)^2}{2 V_{EAn}}} \tag{2}$$

Comparing (1) and (2), and assuming

$$\frac{V_{gs} - V_t}{V_{EA}} << 1 \ ,$$

the gain in figure 5.b is clearly larger than the gain in figure 5.a.

Once the gain has been computed, the precision of the mirror must be examined. Due to the oxide gradient or other technological imperfections, the threshold voltages of two transistors are never exactly the same; β factors (β = m C_{ox} W/L) can also differ. The impact of the differences in the threshold voltages can be expressed by:

$$I_2 = \frac{\beta}{2}(V_{gs1} - V_t + \Delta V_{tm})^2$$

where V_{tm} is the threshold voltage of the transistors in the current mirror and ΔV_{tm} the possible difference between the V_t of the two transistors. But

$$V_{gs1} = V_t + \left(\frac{2I_1}{\mu C_{ox}\frac{W}{L}}\right)^{\frac{1}{2}}$$

thus

$$I_2 = \mu C_{ox}\frac{W}{L}\frac{1}{2}\left|\left(\frac{2I_1}{\mu C_{ox}\frac{W}{L}}\right)^{\frac{1}{2}} + \Delta V_{tm}\right|^2$$

$$\simeq I_1 + \mu C_{ox}\frac{W}{L}\left(\frac{2I_1}{\mu C_{ox}\frac{W}{L}}\right)^{\frac{1}{2}}\Delta V_{tm}$$

(neglecting second-order effects). The error in the current I_{tm} is thus given by:

$$\Delta I_{tm} \simeq \left(2\mu C_{ox}\frac{W}{L}I_1\right)^{\frac{1}{2}}\Delta V_{tm}$$

$$\simeq \frac{2}{V_{gs}-V_t}\Delta V_{tm}I$$

The effect of β variations is expressed by:

$$\Delta I_\beta = \frac{\Delta\beta}{\beta}I_1$$

A third error to consider is the mismatching between the two transistors at the input stage of the differential amplifier which will measure the voltage difference between the V_{gs} of the two transistors in the mirror (figure 6). This error is given by:

72

$$\Delta I_{td} = \frac{\Delta V_{td}}{V_{EA}} I_1$$

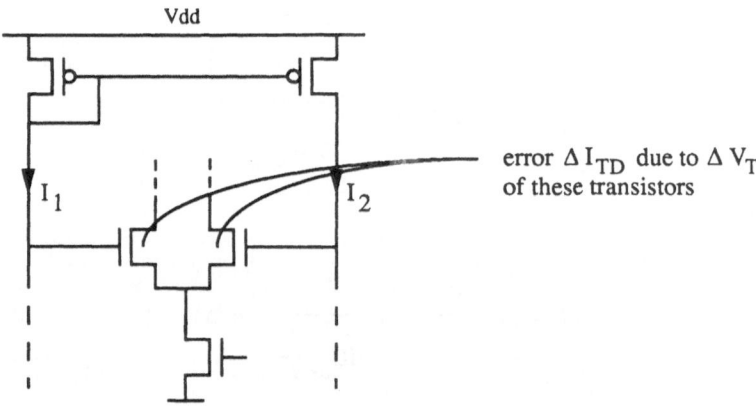

error ΔI_{TD} due to ΔV_T of these transistors

Figure 6   Mismatching of the voltage comparator

In order to compare these three errors, realistic values are chosen:
I: from 0 to 500 µA
$\mu_p C_{ox}$: 1.5 10⁻⁵ A/V² (standard CMOS process)
W/L: must be chosen to cope with the maximum current (500 µA).
If $V_{gs}-V_t$ can be up to 1.5V, then

$$\frac{W}{L} = \frac{I_{max}}{\mu_p C_{ox} \dfrac{(V_{gs} - V_t)^2}{2}} \approx 30$$

V_{EAn}: 20 V
ΔV_{tm}: 10 mV
$\Delta\beta/\beta$: 0.01
ΔV_{td}: 10 mV
(these three last values can be reached with careful design of the mirrors and comparators).

The three currents ΔI_{tm}, ΔI_β and ΔI_{td} are given in figure 7.

The error due to the threshold voltage difference in the current mirror (ΔI_{tm}) is obviously the most important one, especially for small currents. This is due to the fact that this error is proportional to the square root of the size of the transistors in the mirror. Even for small currents, this error

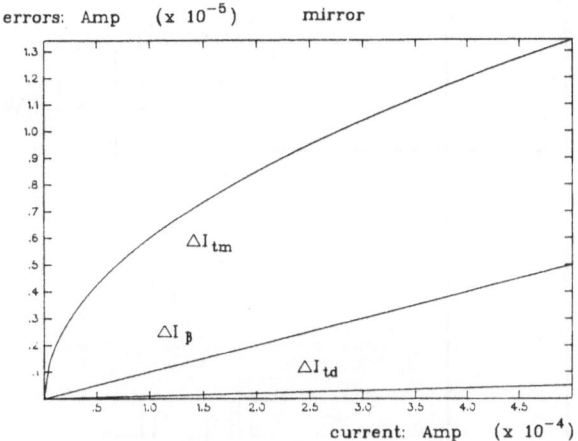

Figure 7   Errors for current mirror

is thus important because these two transistors must be large enough to drive the maximum current (here 500 µA).

One solution to this problem would be to connect several mirrors in parallel, each of them being active only when necessary to drive the total current. This solution is considered in section "multi-transistor load".

Multi-transistor load

A load where several mirrors are connected in parallel through switches is now considered (figure 8). The switches are supposed to be active sequentially, depending on the value of the greatest current among I_1 and I_2; in other words, if this current is I_M, we have:

$$0 < I_M \leq I_{ref} \qquad \rightarrow 1 \text{ load active}$$
$$I_{ref} < I_M \leq 2 I_{ref} \qquad \rightarrow 2 \text{ loads active}$$
...

where I_{ref} is a given current which will be estimated at the end of this paper.

In order to focus on the advantages of such solution, the same three errors computed in section "two transistor-load" will be estimated, but this time for a load as described in figure 8, with two current mirrors. We suppose first that the switches have no influence on these errors, and that $I_{ref} = I_{max}/2$, where I_{max} is the maximum current in the load (here 500 µA). Since the maximum current I_{max} is supposed to be the same, the size of

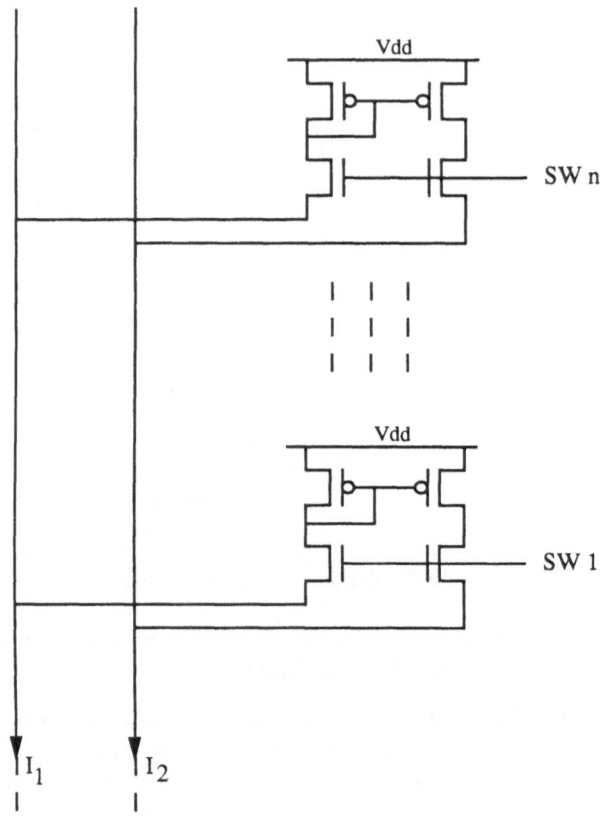

Figure 8   Load with several stages

each current mirror can be reduced to W/L = 15. The values of all other parameters are identical as in section "two transistor-load". Errors ΔI_β and ΔI_{td} do not change; error ΔI_{tm}, however, depends on the W/L of the transistors in the mirrors. As far as only one load is active, ΔI_{tm} is given by:

$$\Delta I_{tm} = \left(2\mu C_{ox} \left(\frac{W}{L} \right)_2 I_1 \right)^{\frac{1}{2}} \Delta V_{tm} \text{ , where } \left(\frac{W}{L} \right)_2 = 15$$

When the two loads are active, error ΔI_{tm} is given by:

$$\Delta I_{tm} = 2 \left(2\mu C_{ox} \left(\frac{W}{L} \right)_2 \frac{I_1}{2} \right)^{\frac{1}{2}} \Delta V_{tm}$$

The three errors ΔI_{tm}, ΔI_β and ΔI_{td} are illustrated in figure 9.

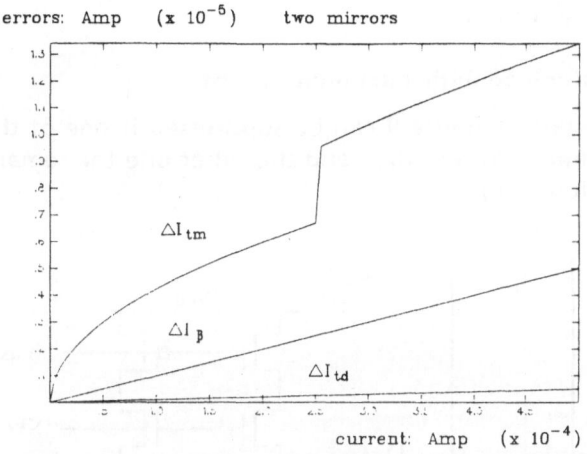

errors: Amp (x 10^{-5})   two mirrors

ΔI_{tm}

ΔI_β

ΔI_{td}

current: Amp (x 10^{-4})

Figure 9   Errors for a load with two mirrors

Two remarks have to be made. First, the error ΔI_{tm} when $I_1 = I_{max}$ does not change between the devices from figure 5.b and figure 8. If the W/L of the transistors in the mirrors are indeed respectively $(W/L)_1$ and $(W/L)_2$, we have for the first case

$$\Delta I_{tm} = \left| 2\mu C_{ox}\left(\frac{W}{L}\right)_1 I_{max} \right|^{\frac{1}{2}} \Delta V_{tm}$$

and for the second one

$$\Delta I_{tm} = 2\left| 2\mu C_{ox}\left(\frac{W}{L}\right)_2 \frac{I_{max}}{2} \right|^{\frac{1}{2}} \Delta V_{tm}$$

Since $(W/L)_1 = 2 \cdot (W/L)_1$, these errors are identical. However, the error ΔI_{tm} for $I_1 < I_{ref}$ is smaller in the second case, due to the fact that the transistors are more efficiently used (a greater current flows in the transistors with respect to their size).

Secondly, the error ΔI_{tm} for $I_1 \geq I_{ref}$ is identical in the two situations, the two devices being equivalent if all the switches are on (the switches are still considered to have no influence on the errors). Furthermore, figure 9 shows a discontinuity in the curve ΔI_{tm} when $I_1 = I_{ref}$. The diminution of the error ΔI_{tm} can thus be improved if this discontinuity is suppressed; a solution to this problem is presented in section "multi-transistor load with maximum current".

Multi-transistor load with maximum current

The discontinuity in figure 9 can be suppressed if one of the two loads gets his maximum current (I_{ref}) and the other one the remaining current ($I_1 - I_{ref}$) (see figure 10).

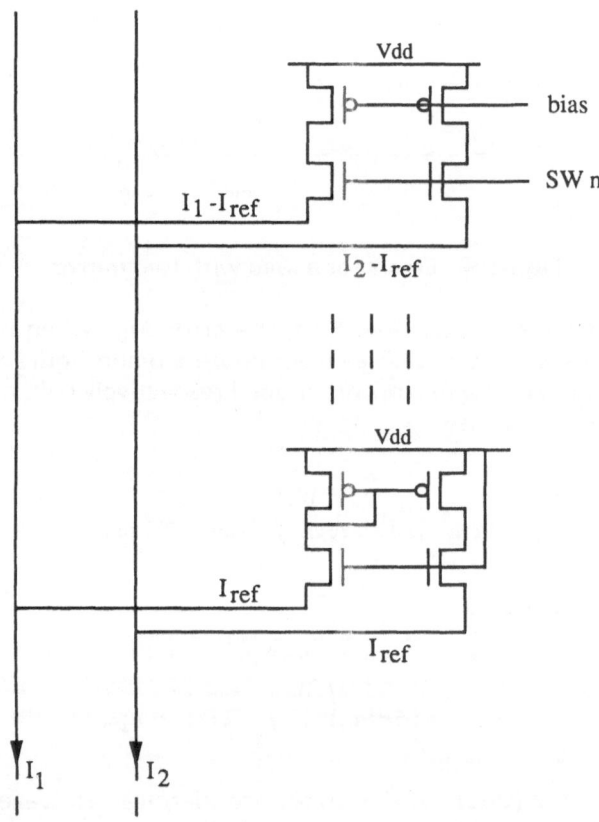

Figure 10   Loads with maximum currents

Error ΔI_{tm} is then given by:

$$\Delta I_{tm} = \left| 2\mu C_{ox} \left(\frac{W}{L} \right)_2 I_1 \right|^{\frac{1}{2}} \Delta V_{tm} \qquad\qquad if\, I_1 \leq I_{ref}$$

$$\Delta I_{tm} = \left| 2\mu C_{ox}\left(\frac{W}{L}\right)_2 I_{ref}\right|^{\frac{1}{2}}\Delta V_{tm} + \left| 2\mu C_{ox}\left(\frac{W}{L}\right)_2 (I_1 - I_{ref})\right|^{\frac{1}{2}}\Delta V_{tm}$$

$$if I_2 \geq I_{ref}$$

The three errors ΔI_{tm}, $\Delta I\beta$ and ΔI_{td} are illustrated in figure 11.

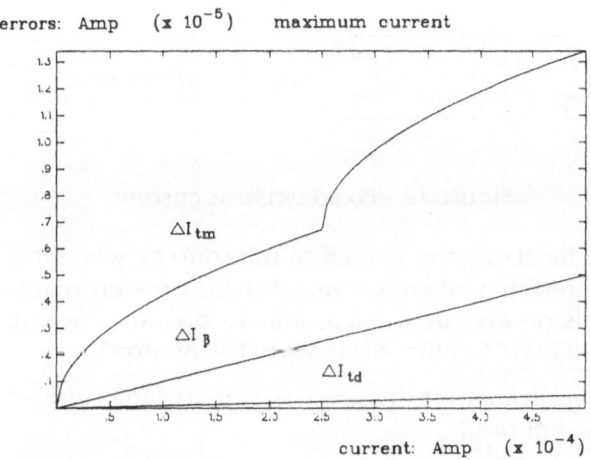

Figure 11   Errors for a load with maximum currents

VLSI NEURON

The solution of section "multi-transistor load with maximum current" can be used to implement a VLSI neuron for an artificial neural network where the number of synapses connected to a single neuron is important. In order to avoid changes in the current flowing through the synaptic current sources, an operational amplifier in a feedback loop is introduced as shown in figure 12. By this way, the drain voltage of the current sources is kept fixed, and the synaptic currents remain identical whatever is the total current in the load. Furthermore, with this feedback loop, the current in the two lines is directly determined by the synapses, and parameter variations in the switches have thus only second-order effect.

The principle explained in section "multi-transistor load with maximum current" can be expanded to more than two stages. If this is the case, the

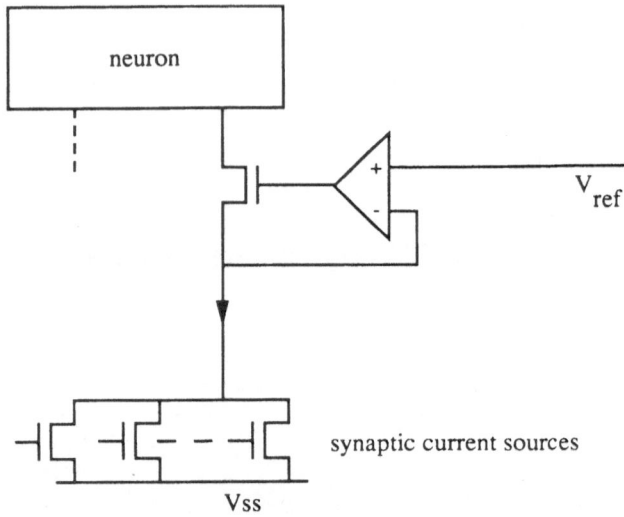

Figure 12   Fixed synaptic current

first stage will be always connected to the sources, while the second one will only be connected when $I_M = max(I_1,I_2)$ is greater than I_{ref}, the third one when I_M is greater than $2.I_{ref}$, and so on. An efficient I_{ref} will be computed in section "Number of stages in the neuron".

The switches which connect the successive loads can be driven by a device as illustrated in figure 13.

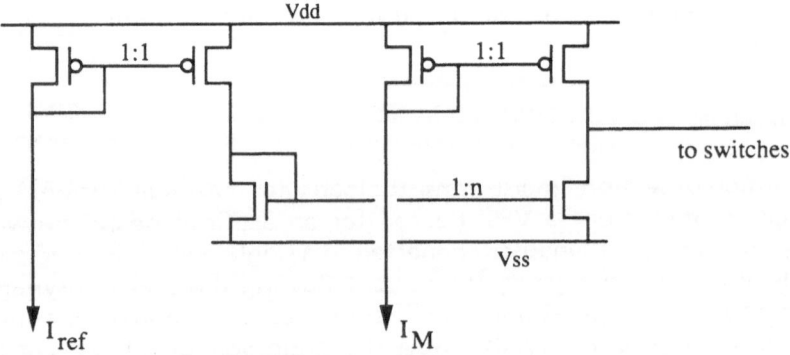

Figure 13   Command for switches

The ratio n:1 used in the N-type current mirror depends on the current $n.I_{ref}$ to which I_M must be compared. This device needs the current I_M as input; this could be done by inserting another cell which generates the maximum of the currents I_1 and I_2. However, I_M can be replaced by I_1 or I_2 without loss of performance. It is not very important, indeed, if the loads are not activated exactly at I_{ref}, $2I_{ref}$, $3I_{ref}$,... but well at an approximation of these values. If the currents I_1 and I_2 are thus nearly identical, one of these two currents can be used for I_M. If I_1 and I_2 are quite different, the circuit will work properly if, by chance, the current I_1 and I_2 which is chosen to drive the cell of fig.10 is greater than the other. If this is not the case, say $I_M = I_1$ and $I_1 < < I_2$, only the number of loads necessary to drive properly I_1 will be active. The transistors driving I_2 will not be sufficient for such a current, and their drain voltage will thus spectacularly decrease. In this case, it will not be difficult to discriminate between I_1 and I_2 with a simple comparator; the current I_M can thus be replaced for example by I_1. An important point is to avoid to duplicate the current I_1 at the output of the synapses; the advantage of the circuit would indeed be lost because of the imperfections in the mirrors used to duplicate the current. A second, less precise, current I_1 has thus to be generated directly in the synapses. The complete circuit is shown in figure 14; all the mirrors have unity ratios, except those indicated in the figure.

NUMBER OF STAGES IN THE NEURON

The last question to solve is the choice of the current I_{ref}, and so to decide the optimum number of stages in the neuron. First, the number n of stages in the neuron and the current I_{ref} are related by:

$$n I_{ref} \leq I_{max} < (n + 1) I_{ref}$$

where I_{max} is the maximum current the neuron has to drive. The link between this architecture to compute analog sum-of-products and neural networks can now be restored. The function to realize is the sum of fixed synaptic currents, and the logic comparison between the total excitatory and inhibitory currents. If I_{syn} is one single synaptic current, the error $\Delta I_{tot} = \Delta I_{tm} + \Delta I b + \Delta I_{td}$ has no influence on the logic comparison as long as $\Delta I_{tot} < I_{syn}$. This relation can be developed:

$$\left| 2\mu C_{ox} \frac{W}{L} I \right|^{\frac{1}{2}} \Delta V_{tm} + \frac{\Delta \beta}{\beta} I + \frac{\Delta V_{td}}{V_{EA}} I < I_{syn}$$

In section "multi-transistor load", the fact was proven that the introduction of several mirrors does not change the error when $I = I_{max}$. It can also be shown that the errors computed at values of the current which switch on a new stage are proportional to this current. The optimum number of stages will thus be:

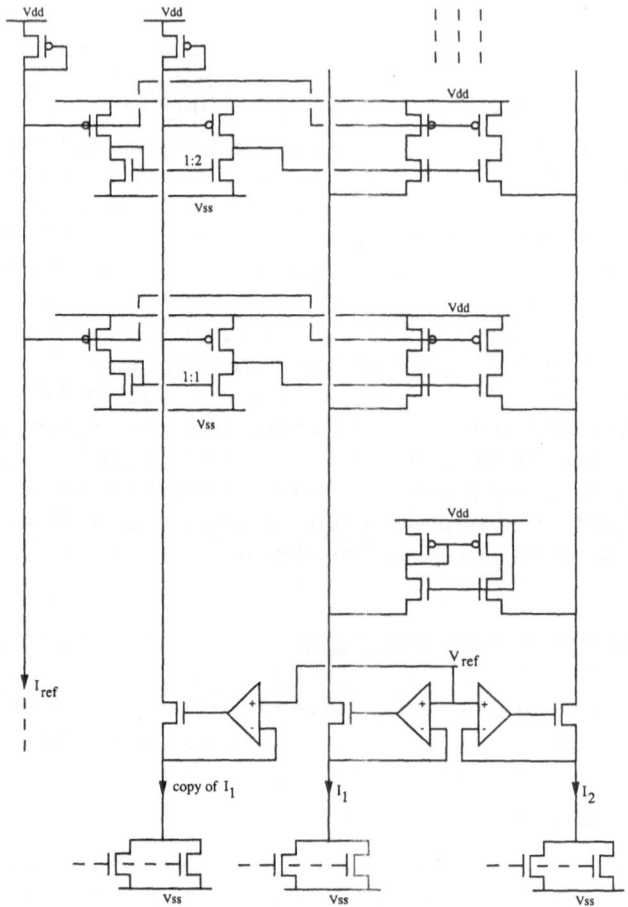

Figure 14   Complete neuron

$$n = int \left\{ \frac{\left[2\mu C_{ox} \left(\frac{W}{L} \right)_{tot} I_{max} \right]^{\frac{1}{2}} \Delta V_{tm} + \frac{\Delta\beta}{\beta} I_{max} + \frac{\Delta V_{td}}{V_{EA}} I_{max}}{I_{syn}} + 1 \right\}$$

where $(W/L)_{tot}$ is the size of a transistor which could drive the total current I_{max}. The value of I_{ref} is then given by:

$$I_{ref} = \frac{I_{max}}{n}$$

This value of I_{ref} corresponds to a maximum error of I_{syn}. It would probably be useful to have a security factor on the allowed error, i.e. to replace I_{syn} by 0.9 I_{syn}. It can easily be verified that if the error with a current I is less than I_{syn}, then the error with a current 2I will be less than 2I_{syn}, and so on. This value of I_{ref} is optimum, because the current for which the error is less than I_{syn} is maximum (and also for 2I_{syn},...). It would be unprofitable to enhance the number of stages, and thus to decrease I_{ref}; the error current indeed would decrease in absolute value, but not in terms of integer multiples of I_{syn}; this would thus have no effect on the logic comparison between the total excitatory and inhibitory currents.

TEST CHIP

A test chip has been realized according to the design of figure 14. The cells implement a 4-stage neuron, with automatic switching of the different stages. A microphotograph of the cell is given in figure 15. The large dimensions of the cell (600 x 886 µm) are due to the fact that it has not been optimized (for example, very large transistors were used in the current comparators).

Figure 15   Microphotograph of the cell

CONCLUSION

A method is presented to reduce the errors due to mismatching of components in a VLSI neuron used in a neural network where the information is transmitted by currents. Since the number of neurons in a chip is much less than the number of synapses, the loss of area due to this neuron is not very significant. However, the errors in the decisions taken by the neurons are reduced, especially when relatively few synapses are connected (for example is sparsely-coded memories).

Furthermore, another improvement of this neuron, but which cannot be precisely predicted, is the fact that if the mirror is splitted into several parts, the probability that one mismatching between components will be compensated by another mismatching is enhanced. The neuron must of course be carefully designed, for example by physically inverting half of the mirrors, in order to compensate for oxide gradients.

ACKNOWLEDGEMENTS

All our acknowledgements go to Brigitte Wénin-Dupont, who developed and helped us to use Bananas, a graphical software used to plot the simulations of this paper. This work has been partially financed by the ESPRIT-BRA project NERVES.

REFERENCES

[1]   E. Vittoz and X. Arreguit, "CMOS integration of Herault-Jutten cells for separation of sources", Analog implementation of neural systems, C. Mead and M. Ismail eds., Kluwer Academic Publishers, Norwell, MA, 1989.

[2]   M. Sivilotti, M. Mahowald and C. Mead, "Real-time visual computation using analog CMOS processing arrays", Proceedings of the 1987 Stanford conference on advanced research in VLSI, P.Losleben ed., MIT Press 1987.

[3]   A. F. Murray, "Pulse arithmetic in VLSI neural networks", IEEE Micro, vol.9, n°6, December 1989.

[4]   H. P. Graf and P. de Vegvar, "A CMOS implementation of a neural network model", roceedings of the 1987 Stanford conference on advanced research in VLSI, P.Losleben ed., MIT Press 1987.

[5]   M. Verleysen, B. Sirletti, A. Vandemeulebroecke and P. Jespers, "Neural networks for high-storage content-addressable memory: VLSI circuit and learning algorithm", IEEE Journal of Solid-State Circuits, vol.24, n°3, 1989.

ARCHITECTURES FOR A
BIOLOGY-ORIENTED NEUROEMULATOR

S. J. PRANGE, H. KLAR

ABSTRACT

Today's electronic neural networks consist of very simple processing elements, such as weighting multipliers and a sigmoid transfer function. The electrical processes at the membrane of a biological neuron are much more complicated. This paper describes the steps towards the development of an emulator for a more biology-oriented neural network.

Based on the electrical phenomena at a nerve cell's membrane a biology-oriented neuron model was established and a simulation program was implemented. Evolution strategy instead of a dedicated learning algorithm was used to optimise the network behavior. Since the simulation was extremely time-consuming even for simple tasks, an electronic neuron model was integrated on chip and a hardware emulator consisting of these chips was put into practice.

Based on the experience with this emulator two new chip architectures for an integrated emulator with a totally parallel analog and with a semi-serial digital biology-oriented neural network are discussed. Such an emulator will contribute to dynamic neural signal processing and can especially be used by neurophysiologists to emulate biological processes.

INTRODUCTION

In neural networks used today the synapses are modelled by multipliers and the neurons by adders followed by a static transfer function. With these kinds of networks considerable results for the recognition and associative storage of static patterns have already been achieved [LIPP87]. However, with these networks dynamic signal processing like speech recognition, e.g., can only be achieved by an additional transformation of time-dynamic speech signals into static spectra [KOHO88].

The neurons in biological nervous systems are working with pulse-pause-modulated signals and dynamic transfer functions at the synapses. By investigating biological prototypes, sound statements about simplified model conceptions can be deduced. A better understanding of the association mechanisms in biology will also contribute to new applications of artificial neural networks in information technology.

Performing simulations is one way to do research on biological neural systems. However, the simulation of massively parallel, asynchronous, dynamic processes, as to be found in biological neural systems, on current digital computers based on the principles of the von-Neumann-machine needs enormous computing time. There is a need for specific hardware for a realistic simulation and emulation of neural networks [KLAR89].

The biology-oriented neuron model established at our institute respects a dynamic synaptic transmission function with two parameters per synapse and an output function that creates pulses. An integrated emulator for biology-oriented neural networks on chip is under development.

This paper starts with the description of the neuron as found in biology and the characteristics of the model used at our institute. Then a specific simulation program for this model is described. It is shown how evolution strategy was used to adjust the network parameters. Since simulation times were too long even for simple tasks, a hardware emulator was developed. A neuron model as an analog integrated circuit and an emulator consisting of 16 such circuits are presented. The next chapter shows architectures for an integrated neural network emulator and explains which requirements have to be met for the circuit design. One fully parallel analog and one neuron-parallel synapse-serial digital emulator architecture are shown as examples. The last chapter describes possible applications for such an emulator.

THE NEURON IN BIOLOGY

The smallest data processing unit in the nervous system is the neuron, the nerve cell. The human brain contains about 15 billion neurons. Each neuron possesses 1000 to 10000 inputs and one output, the axon, that further branches to 1000 to 10000 inputs of other neurons. The data transmission between neurons is done with electrical pulses on the nerve fibers, the axons. The connections between neurons are called synapses. A synapse consists of the sending cell's presynaptic terminal, the synaptic cleft and the receiving cell's postsynaptic membrane [STEV87].

Inside a nerve cell a resting-potential is maintained and restored after a disturbance. A nerve pulse arriving at a synapse makes it pour out transmitting substances into the synaptic cleft. By this, selective ion channels are opened at the postsynaptic membrane, changing the postsynaptic membrane's potential. The time course of this potential change depends on the size of the synapse, the amount and kind of the transmitting substances, and characteristics of the postsynaptic membrane. The postsynaptic potential changes are electrotonically conducted along the cell membrane to the axon hillock. With increasing spatial distance between a synapse and the axon hillock the influence on the sum potential at the axon hillock decreases [KATZ87] (Fig. 1).

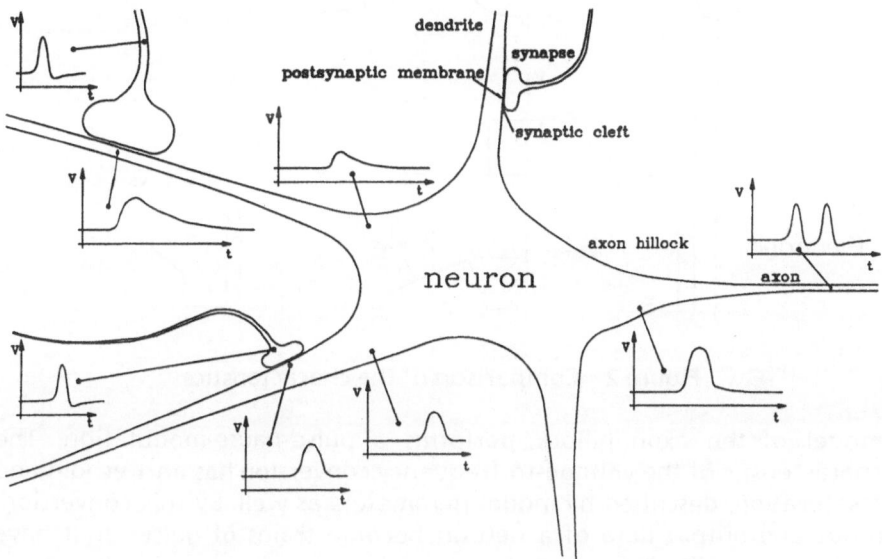

Figure 1   Schematic diagram of a nerve cell with all potentials during the arrival of two nerve pulses via excitatory synapses

The sum of these preprocessed postsynaptic potentials is converted into a train of nerve pulses at the axon hillock and conducted along the axon to other neurons. Amplitude and duration of these pulses are constant. Information is coded in the arrangement of pulses in time. The conversion characteristic from potential to frequency shows a threshold and a saturation as well as a transient response and fatigue symptoms [FOHL80, NELS80, STEI74].

Nowadays, the transmitting characteristics of a synapse are generally accepted to be the responsible quantities for learning [BLAC87]. A short but strong excitation of a synapse leads to an exhaustion of the stock of transmitting substance and in this way to a fatigue or a desensitisation. Feedbacks to the cell body can stimulate the production of transmitting substances and therefore lead to an increase of a synapse's influence on a receiving cell. The search for the biological learning algorithm is one of the research activities in biology [ALKI89]. One difficulty is to measure postsynaptic potentials. A suitable simulation tool would be helpful to test and prove theories about learning.

The model used at our institute respects a time-dynamic pulse response of the synapse with two parameters, namely height and decay (Fig. 2). The

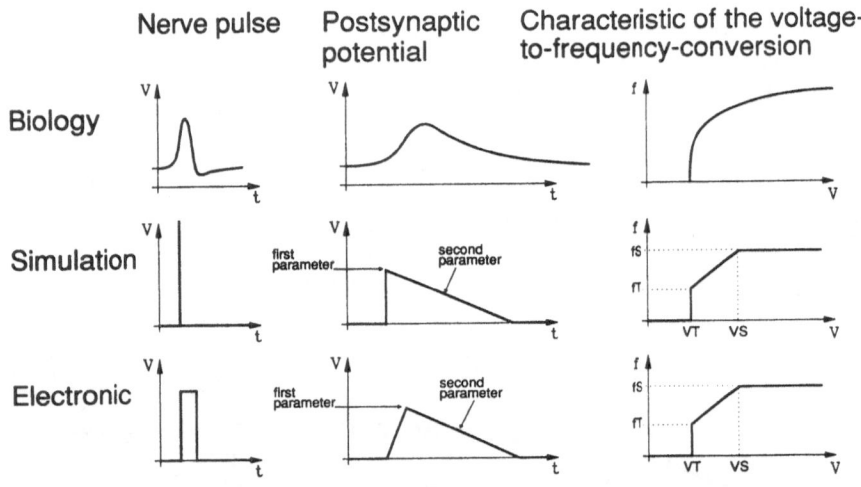

Figure 2   Comparison of the characteristics

model of the axon hillock performs a pulse-pause-modulation. The characteristic of the voltage-to-frequency conversion has a threshold and a saturation, described by model parameters as well. By this conversion, input and output data of a neuron become trains of pulses that have constant amplitude and duration. Adaptive phenomena due to a transient response or fatigue symptoms are not considered. However, this missing feature is no restriction for the emulation of short term behavior. The transmission characteristics of the axon are not imitated, but idealised. Learning behavior can be imitated by adjusting the parameters [PRAN90a].

SIMULATION OF BIOLOGY-ORIENTED NEURAL NETWORKS

The first step towards the examination of biology-oriented neural networks was the implementation of a simulation program. In the software simulation, nerve pulses are assumed to be infinitesimally short, that means that they are only represented by a discrete point in time like a Dirac pulse. The postsynaptic potential is assumed to jump instantaneously and then to descend linearly to zero (Fig. 2). Therefore, the transmitting characteristics of a synapse are exactly defined by the (positive or negative) height of the jump and the slope of the trailing edge. The duration of a postsynaptic potential can be predicted, if the input pulse sequence is known.

If the input of a neuron remains unchanged, the sum of one neuron's postsynaptic potentials can completely be described by a starting value and an event list, containing the durations of the synapses' postsynaptic potentials and the according slopes (Fig. 3). These slopes are the sum of the particular active synapses' slopes.

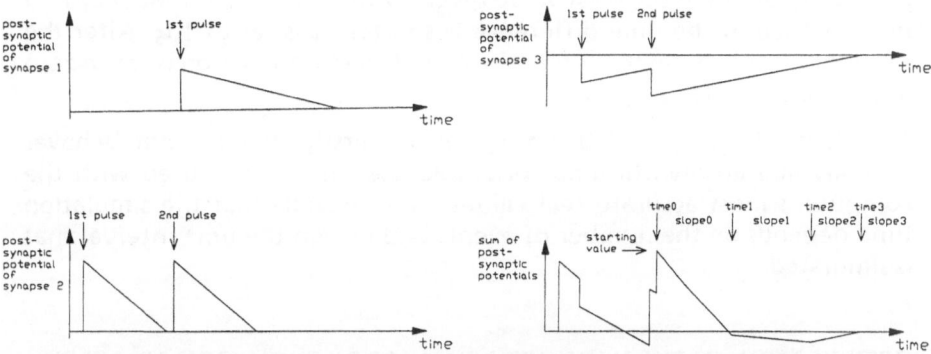

Figure 3   Sum of postsynaptic potentials in the simulation

With respect to a stable input state of the whole network, a probable point of time for the next nerve pulse to be fired is calculated for each neuron. The firing times of all neurons are compared. The nearest point of time is the actual next pulse. If a pulse is fired the calculation for the whole net is revised.

The voltage-to-frequency conversion of the simulation program uses a linearised characteristic (Fig. 2) and the potential's mean value over time to calculate the next pulse. Assume that V(t) is the sum of the postsynaptic potential and T the time between two pulses. The characteristic describes the firing frequency $f = 1/T$ over the sum of postsynaptic potentials V. For the interval between threshold (V_T and f_T) and saturation (V_S and f_S) there shall be $f = 1/T = k \cdot V'$ with the mean potential value over time V' and a fitting parameter k. It follows that:

$$\overline{V} = \frac{1}{T} \int_{T_x}^{T_x+T} V(t)\,dt \quad and \quad \overline{V} = \frac{1}{k \cdot T} \,,$$

$$\rightarrow \int_{T_x}^{T_x+T} V(t)\,dt = \frac{1}{k}\,.$$

T_x is the starting point of the integration and is given by the point of time when the potential sum is greater than the threshold value V_T, or by the point of time of the last pulse fired by the neuron under consideration.

For one neuron with undisturbed inputs the procedure to calculate the firing time of a nerve pulse is as follows: First the potential sum $V(t)$ has to exceed the threshold value V_S. Then the integration of the potential sum starts until a second threshold value $1/k$ is exceeded. If the potential is still greater than the threshold V_T and the time difference to the last pulse is longer than the least value given by the saturation frequency f_S, a pulse is fired. If the time difference is shorter, it is set to $1/f_S$. After the firing of a nerve pulse the integral value of the firing neuron is set to zero and calculation starts again.

The clear advantages of the program are firstly that the net behaves virtually like an asynchronous one, because time is quantised with the computer's most accurate real values, and secondly that the simulation time depends on the number of events and not on the time interval that is simulated.

OPTIMISATION OF THE NETWORK BEHAVIOR BY EVOLUTION STRATEGY

To achieve a desired transmitting characteristic of the network according to a specific task, two parameters per synapse (height and decay) and three additional global parameters for all neurons (first and second threshold value, saturation value) have to be adjusted. Existing learning algorithms like the error-back-propagation cannot be applied to the biology-oriented model due to the two parameters per synapse and the non-linear and non-differentiable transfer function of the neuron. A dedicated learning algorithm for this model is not known up to now. Since the mathematical description of the network is impractical because of the nonlinear behavior, random optimisation strategies are used to adapt the network.

The principles of the used evolution strategy [RECH73] are those of mutation and selection, i. e., the survival of the fittest. All parameters are changed randomly with a medium step factor. The step factor itself is changed randomly with a log-normal distribution and is hereditary, too. Evolution strategy cannot guarantee to avoid local quality maxima, but there always is the probability to leave such a local maximum.

The optimisation task is to make the time course of some output neurons equal to an output pattern exactly defined by the user. To be able to use evolution strategy or any other optimisation strategy, a quality measure has to be found. The quality measure used respects firstly the number of output pulses of a neuron compared to the desired number, and secondly the squared temporal deviation of the pulses from the desired time course. Problems occur due to the weighting of the number of pulses, which causes jumps within the quality landscape. Those jumps influence the optimisation behavior negatively.

Evolution strategy was also used to show whether some of the model parameters are useless. It is possible that, in terms of evolution, biology had to come up with the possibilities given by organic chemistry so that, e. g., there had to be pulses instead of static signals. If there were useless parameters of the model, they might converge to zero or infinity, e. g., the saturation described by a minimum pause between two pulses. However, such convergences could not be observed.

The evolution strategy used permits a rough approximation to the optimum if the output pulse train is exactly defined. Experiments with a large number of neurons considering, e.g., only synchronous activities or statistical coherences are utopic up to now, since they are too time-consuming on current computers. Here is a need for hardware improvement to speed up simulations.

NEURON MODEL AS AN ANALOG INTEGRATED CIRCUIT

The first step towards a hardware emulator was the development of an electronic neuron model [PRAN89]. In the electronic model the nerve pulses are assumed to be square pulses with a constant amplitude and duration. A nerve pulse arriving at a synapse makes the postsynaptic potential jump with an analogously adjustable postitive or negative height and decay subsequently. The time-constant of the decay is analogously adjustable, too. An adaptive or learning behavior can be imitated by adjusting the parameters. The time-dependent postsynaptic potentials are summed up and applied to a voltage-to-frequency converter with a threshold and saturation characteristic. Threshold, saturation and gradient of the characteristic are model parameters as well.

The electronic realisation of a synapse (Fig. 4) consists of a current source that is switched on by an input pulse charging an off-chip capacitor. The current of a discharging current source is adjustable by an external voltage. The capacitor voltage is applied to a two-quadrant transconductance multiplier that multiplies it with a second external parameter voltage. The summation of these postsynaptic signals is done by an operational amplifier (Fig. 5).

The voltage-to-frequency converter is an astable multivibrator with a preprocessing stage (Fig. 6). This preprocessing stage consists of two competitive current sources. The first current is proportional to the input voltage. The second is a constant current that is substracted from the first one. The resulting current is connected to the base of one of the multivibrator transistors. If the current becomes positive, the transistor opens and the base-emitter voltage switches off the constant current source. This results in the threshold characteristic of the voltage-to-

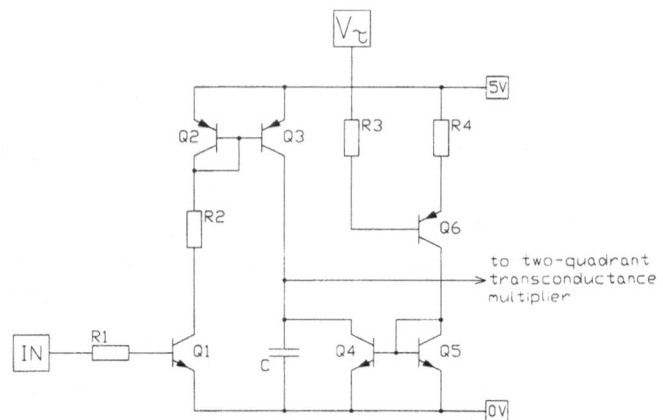

Figure 4 Schematic of the synapse building block

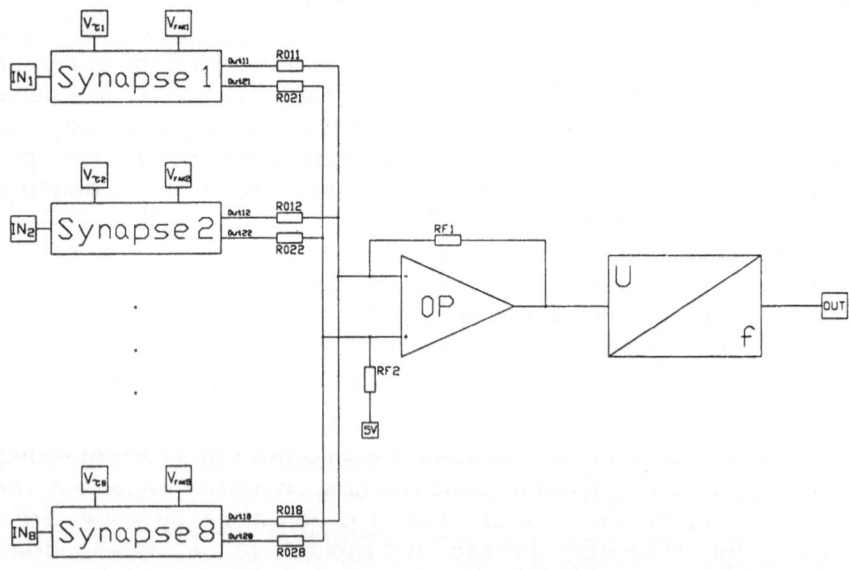

Figure 5 Block diagram of the electronic neuron model

frequency conversion. The saturation is given by the saturation of the operational amplifier's summation. A change of the conversion characteristics can be achieved by changing the off-chip capacitors, adding an off-chip resistor, or adjusting the offset voltage of the operational amplifier off-chip.

This neuron model was implemented as an integrated analog bipolar circuit, designed on the B1000-Transistor array of AEG, fabricated with

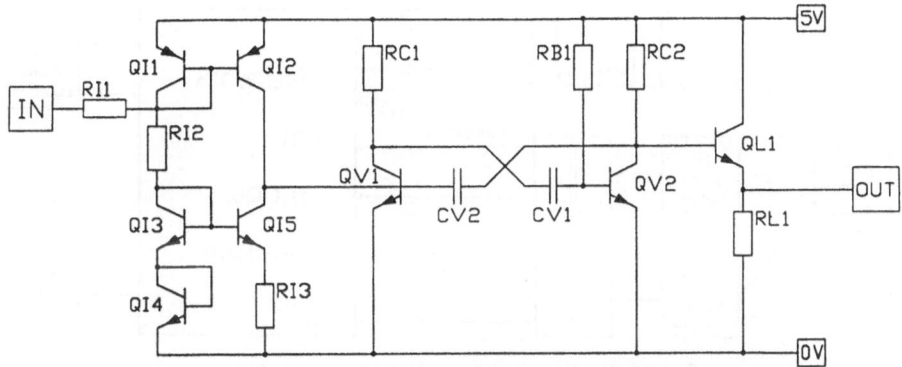

Figure 6   Schematic of the voltage-to-frequency-converter

one metal layer with a line width of 12 microns, and tested on a GenRad GR16. The chip-area of this array is 4.5·5mm² with about 200 npn- and 100 pnp-transistors. One chip contains one neuron with eight synapses, for which, altogether, about 200 transistors are used. A synapse building block needs 10 npn- and 5 pnp-transistors, 12 resistors, one external capacitor, and four pads for input, two parameter voltages and the capacitor. The chip is packaged in a 48 pin dual-in-line package. The saturation frequency with external capacitors of 50nF is about 1kHz. The chip works purely analog and asynchronous.

NEUROEMULATOR ON PRINTED CIRCUIT BOARDS

Consisting of the integrated circuits described above, a hardware emulator with 16 neurons was built up on printed circuit boards [PRAN90b]. In an emulation the computer is only used to adjust parameters, to provide input data and to read out output data. The emulation is running on an own model hardware, and not algorithmically on a computer as in a simulation. Additional external input sources and output devices besides the computer can easily be applied to the model hardware. The block diagram of our emulator, consisting of an analog neural network together with a controlling personal computer for adjusting the synaptic parameters, specifying input pulses and reading out output pulses, is shown in figure 7.

The distribution of the synaptic transmission parameters to the synapse building blocks is done with one digital-to-analog converter by an analog multiplex method and analog sample-and-hold storage units. The resolution of the parameter values is 8 bit. The variation during a discharge phase lies below 1%. The emulation process itself is done analogously and asynchronously in realtime on the specific model

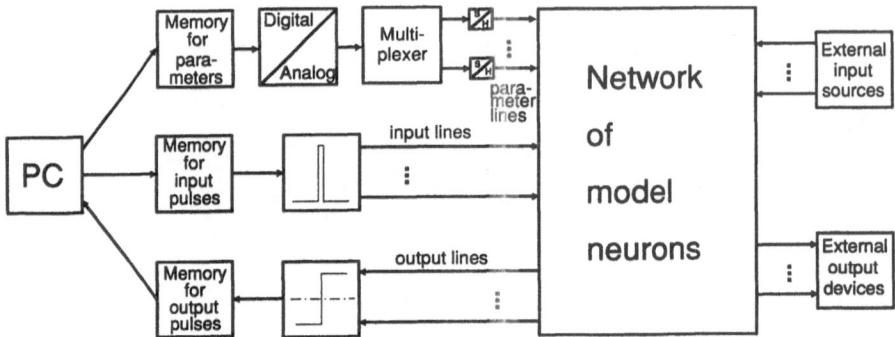

Figure 7 Block diagram of the emulator

hardware. The frequency range of the net is defined by external capacitors. A change of the frequency range can easily be achieved by a change of these capacitors.

One printed circuit board with a $32 \cdot 10 cm^2$ area contains four model neurons with sample-and-hold units and the multiplex path. The emulator consists of four such boards, a connection board and another board inserted into the PC that contains the input/output port, the parameter-, input-, and output-RAMs, the clock and counter circuitry, and the digital-to-analog converter. Specific software to control the emulator was developed, too.

CHIP ARCHITECTURES FOR A BIOLOGY-ORIENTED NEUROEMULATOR

Based on the experiences with the hardware emulator described above, chip architectures for an integrated emulator are developed. The development of the neural network architecture has to respect three functional building blocks: the synaptic transmission function (called synapse in the following), the summation block, and the block with the neural output function (called neuron) (Fig. 8).

The aim of the research activities is to emulate a number of neurons as large as possible in a time as short as possible with a hardware as cheap as possible. VLSI technologies have to be used to achieve the first goal [KLAR89]. The limitations of VLSI to be faced in connection with neural networks are limited area and limited number of inputs and outputs. The number of pins per chip may not exceed a number given by today's bonding techniques. Multiplexing techniques or serial processing have to be taken into consideration to overcome this limitation.

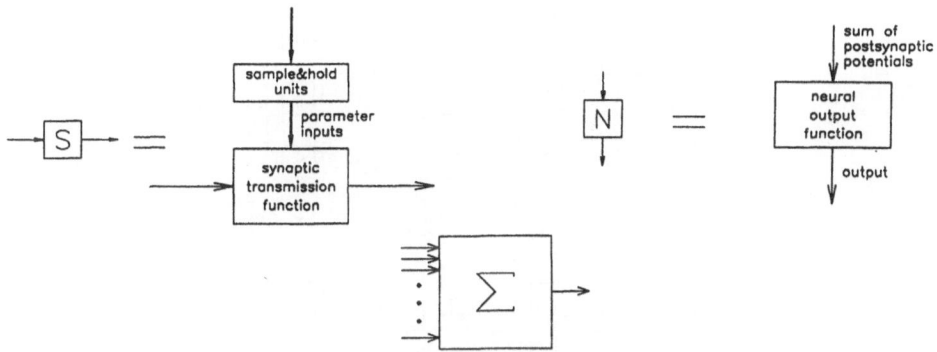

Figure 8   Functional building blocks for neural networks

Due to the limited area and, by this, the limited number of elements on chip, the chip architecture has to be cascadable. It should be possible to build up a fully interconnected network with cascaded chips. By this, every possible network architecture can be built. However, it is not guaranteed that every architecture can be built optimally by full interconnection, i. e., the number of unused synapses could be very large. An optimal scheme to build up multilayer network architectures - with feedback if possible [ECKH89] - should at least be taken into consideration, too.

A third problem is that of memory. In an emulator synaptic and other parameters will be adjusted by a personal computer or a workstation. The parameter memory necessary for a network consisting of 1000 fully interconnected neurons, each parameter digitised with 8 bit resolution, needs 16Mbit. At least 2Mbit are needed for the storage of the input and output patterns. It is important to find out the minimum number of necessary resolution bits by simulations.

Two architectures facing the problems described above are proposed. The first one is an emulator with a fully parallel analog neural network. The other one uses a digital neural network. The digital network works serially with only one physically present synapse per neuron.

Emulator with an analog biology-oriented neural network

One of the important advantages of the analog network design is that the summation of postsynaptic signals can simply be done by current summing in a node if the output signals of the synapses are currents. A network with N nodes and connections between the nodes can always be described by an N*N matrix containing the characteristics of the connections from one node to another. By this, the obvious architecture for an absolutely flexible, fully parallel neural network is a row of N neurons with a square field of N*N synapses (Fig. 9) [RÜCK88]. Such an

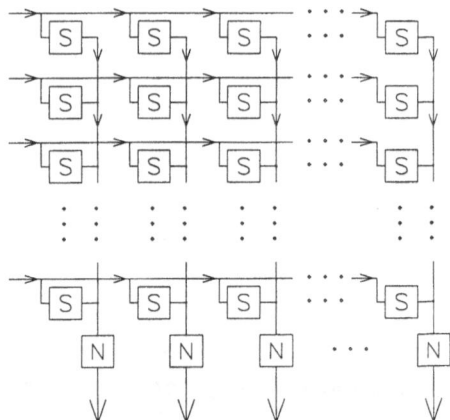

Figure 9   Obvious architecture for an analog neural network

architecture becomes fully interconnected if the output of each neuron is connected to one synaptic input line. It becomes multilayered, if the input lines are connected to the outputs of another matrix representing the former layer.

The proposed adjustment of the network parameters has already been described above (Fig. 7). That means that the architecture has to include digital memories, a digital-to-analog converter, an analog multiplexing path switched by a row and column decoder, and analog sample-and-hold units (Fig. 10). A chip solution should at least contain the neural network with the row and column decoder. The digital-to-analog converter could be and the parameter memories surely have to be off-chip.

A first solution to reach cascadability is to have two kinds of chips, one with N neurons and one with N ·N synapses including summation. Since the neuron circuit is not as demanding in area as N synapses, it makes more sense to have one standard chip which has the possibility to override the neurons so that the synapses of a chip can also be used for neurons of another chip. Such a solution would be useful for fully interconnected networks but also for multilayer architectures with an equal number of neurons per layer.

The cascadability puts high demands on the circuit design of the analog synapse and neural output building blocks. The neuron building block has to be adapted to the number of synapses that determine its minimum and maximum input current. In a cascadable chip this adaption has to be adjustable off-chip.

If the number of neurons on a chip is greater than about one third of the number of possible pins, multiplexing techniques have to be taken into

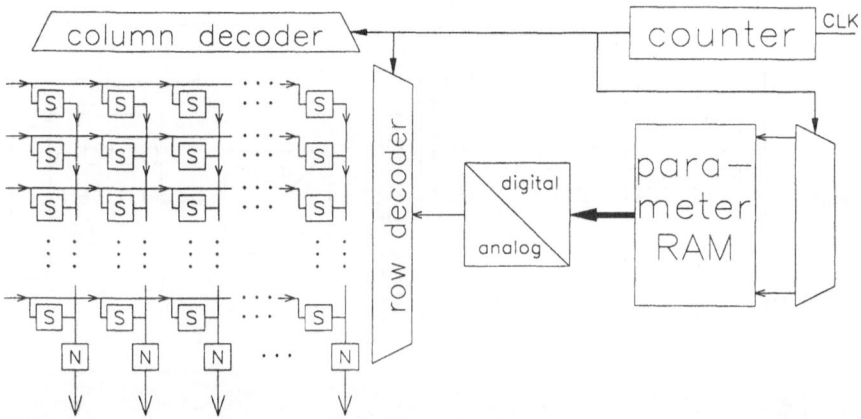

Figure 10   Architecture for an analog neuroemulator (without decoder
lines for reasons of clarity)

account. The multiplexing of the outputs and the demultiplexing for the
inputs can be done with standard methods. As described above, the
summation of postsynaptic signals is simply done by a node. For
cascadability it must be possible to connect these nodes with the same
nodes from other chips to give neurons additional synapses. That means
that - if necessary - multiplexing circuits for currents in both directions
have to be developed, if synapses and neurons are on one chip (Fig. 11).

The multiplexing of a time-varying signal has to be done with respect to
the quantisation theorem. That means that the multiplexing frequency
has to be high or the signal frequency has to be low. Low frequency for
the synapse building block means low currents or high capacitances for
the realisation of the synaptic low pass function. High capacitances on
chip occupy large areas. Since the power dissipation on chip is a problem,
too, subthreshold circuits for low currents have to be taken into
consideration. However, lower currents are more susceptible to noise.
The noise tolerance of these networks has to be shown by simulations.

The advantage of such an analog emulator architecture is the absolutely
parallel, asynchronous, and analog processing of all neurons and
synapses in realtime.

Emulator with a digital biology-oriented neural network

An important advantage of digital circuits compared to analog ones is
their easier circuit design. Secondly, a digital solution offers the
possibility to interrupt and continue the emulation at any point of time.
By this, an additional cascadability can be achieved by means of software,
e. g., a multilayer network can be emulated serially layer for layer. And

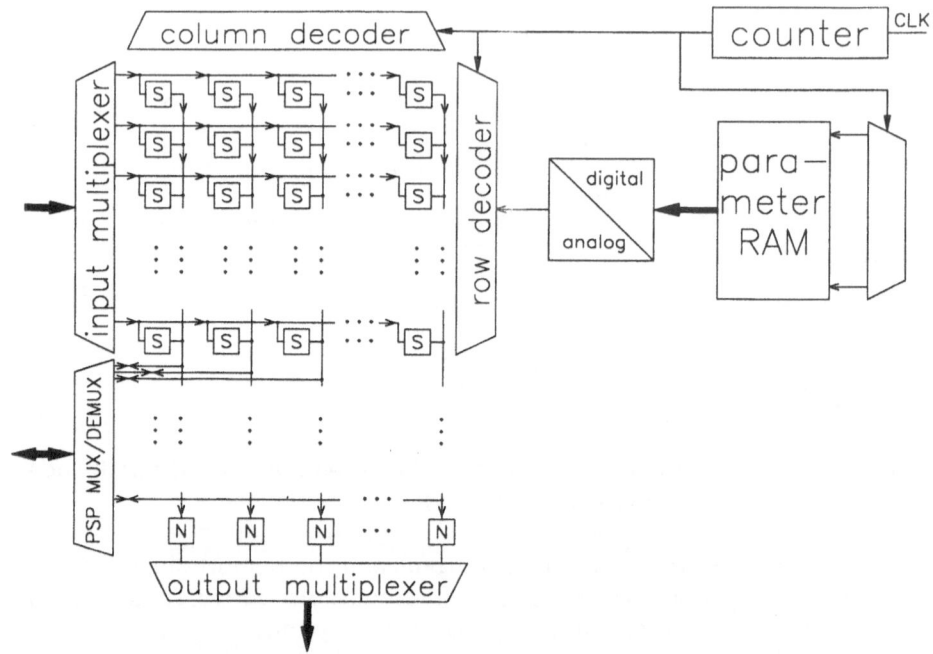

Figure 11   Fully flexable and cascadable analog neuroemulator
architecture

thirdly, in the digital solution the noise problem is reduced to a problem
of necessary quantisation bits.

Since firstly a serial communication between neurochips does not need
massive parallel bus systems, secondly the input data are fed and the
output data read serially by the controlling digital system anyway, and
thirdly a parallel adder for many input values requires a large area, a
neuron-parallel synapse-serial digital architecture is proposed (Fig. 12).

The leaky integration of the synaptic transmission function can be
achieved using a first-order digital filter (Fig. 13). In a parallel concept the
multiplier would be the delay bottleneck. The synapse-serial concept
offers the possibility to pipeline the multiplier, to let every part of it work
the whole time and to place it only once per neuron. The emulation time
of the serial concept is slower than in a fully parallel concept by a factor
equal to the number of synapses per neuron.

Many of the proposed neurons including synapses can be pipelined to a
neural network chain. Then only one input and one output pin per
neuron chain have to be considered (Fig. 14). The rest of the pins is used
for the adressing of the parameter memories. Cascadability of chips can
be achieved by adding neuron circuits and expanding the synaptic

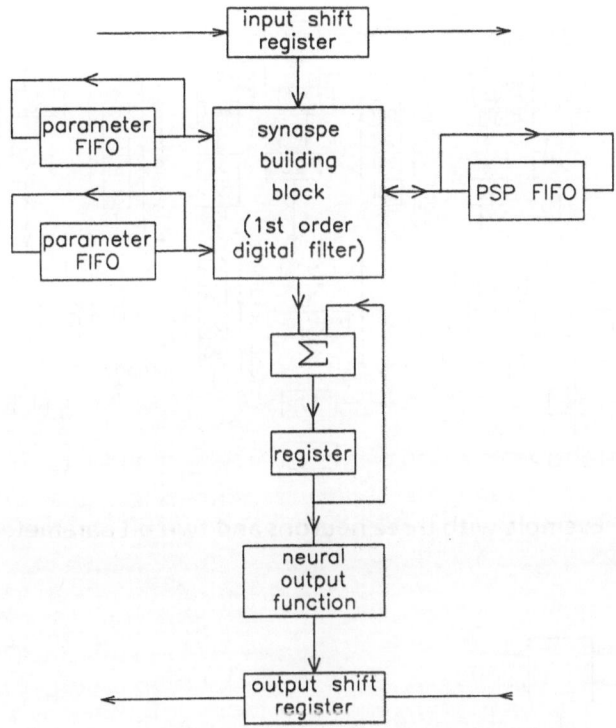

Figure 12   Architecture for a digital neuron model

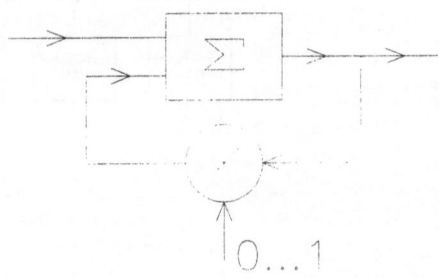

Figure 13   First-order digital filter

memories. That means that a clever concept to handle off-chip memories has to be implemented. An optimal clocking and pipelining scheme is guaranteed if the number of neurons is equal to the number of synapses per neuron (Fig. 14), as needed in a fully interconnected network or a multilayer feedforward architecture with the same number of neurons per layer (Fig. 15).

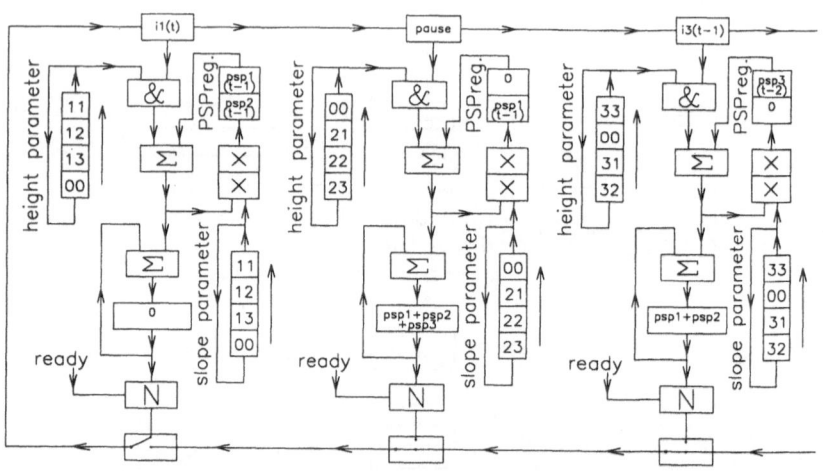

Figure 14 Example with three neurons and two bit parameters

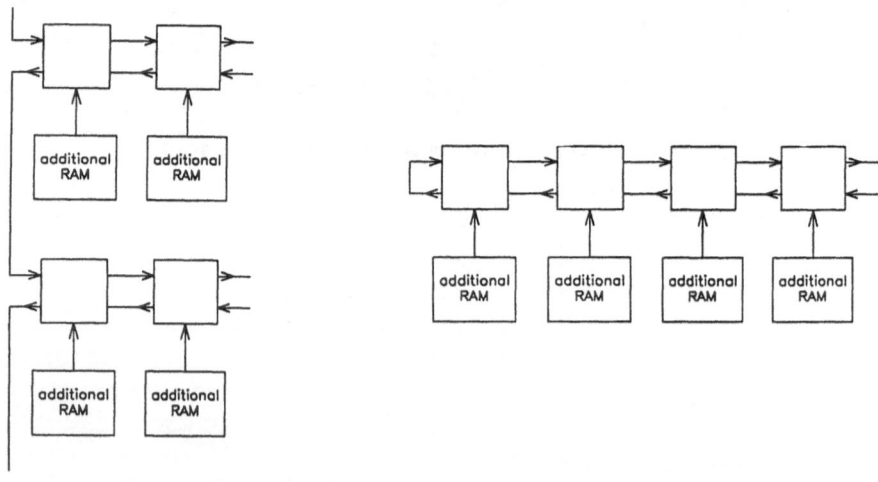

Figure 15 Fully interconnected and multilayer feedforward architecture with the proposed digital concept

APPLICATIONS

The great international interest in neural networks results from the astonishing efficiency of even simple biological nervous systems in processing sensory data and controlling complex movements or other processes. Research activities on neural networks for information processing should therefore be accompanied by the examination of neural systems in biology, since only in that way sound statements about possible simplifications can be made.

The presented neuron model is much more oriented towards biology than existing artificial neural networks. The advantageous applications of such a network model are in one respect research on biological processes inside the nervous system. However, the model promises to be more suitable for applications in dynamical signal processing than existing models. This biology-oriented neural network with dynamic synaptic transmission characteristics and two parameters per synapse is a starting point for fault-tolerant dynamic signal processing, e.g., auditory signal processing, instead of using static neural network models with extensive preprocessing. Applying biology-oriented neural networks, a massive preprocessing can probably be dispensed with, without loosing general advantages of neural networks like fault tolerance or learning abilities.

A neuron model that relies on pulses instead of steady signals is used in simulations at the University of Marburg [REIT89], too. Synapses are modeled as leaky integrators. A special feature is the use of two kinds of inputs, namely feeding and linking inputs (Fig. 16). The linking inputs have additional modulatory influence on the phase of the output spike

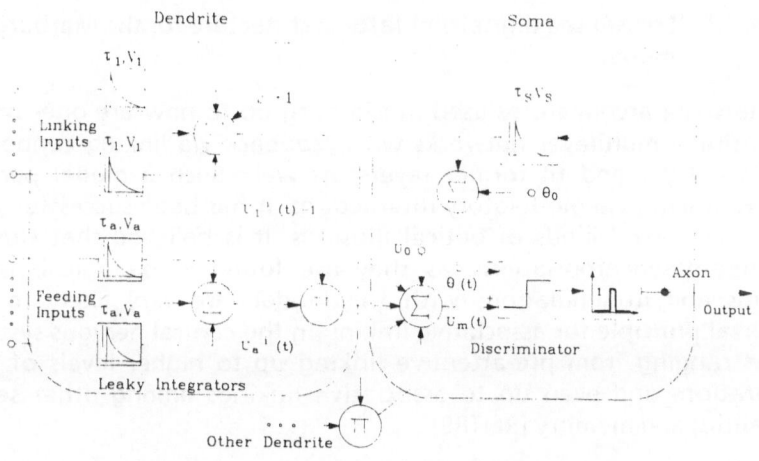

Figure 16   The Marburg model [ECKH89]

train. The biological background is that synapses at the beginning of a dendrite (a branch of a nerve cell, Fig. 1) have a stronger influence on the sum potential than others. There are hints that they have, e. g., the ability to cut off the whole dendrite. The output function used is the one described by French and Stein [FREN70].

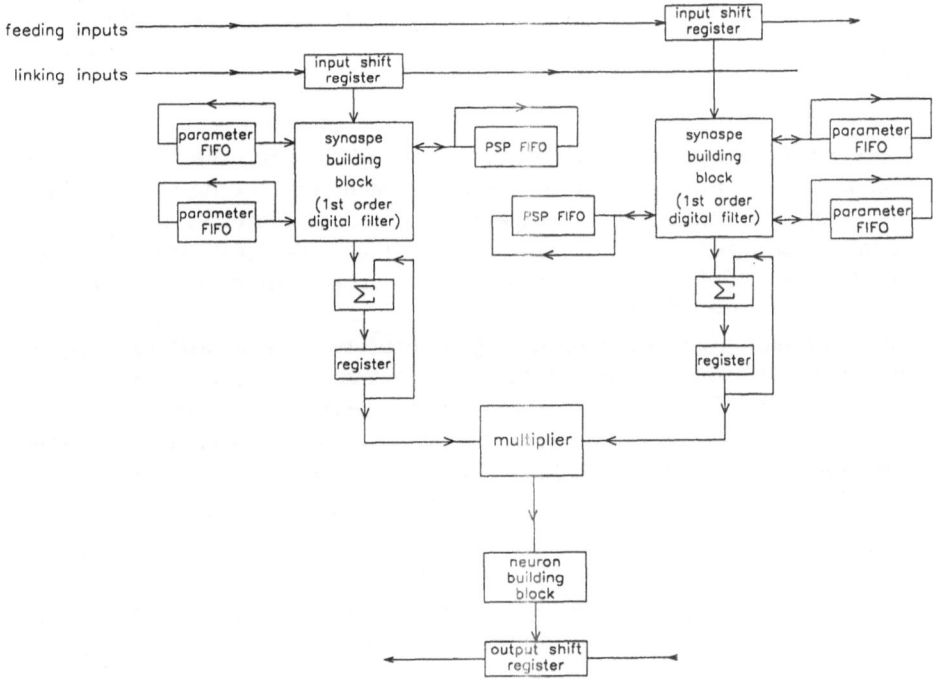

Figure 17   Specialised digital emulator architecture for the Marburg model

The network architectures used in Marburg up to now are one- or two-dimensional multilayer networks with feedback via linking connections within a layer and to former layers, as well. Such a model performs feature linking via modulatory interactions. It has been successfully used to model several kinds of optical illusions. It is believed that stimulus-dependent synchronisations, as they are found in biological nervous systems and in simulations with the model, are very likely to be a universal principle for associative linking in the central nervous system at stages ranging from pre-attentive linking up to higher levels of visual integrations and even up to associative linkings among other sensory modalities and memory [REIT89].

Specialised architectures for the Marburg model should contain two blocks of N*N synapses, one for feeding and one for linking inputs. The neural output function has to take into account an additional multiplier.

These specialised architectures can easily be deduced from the concepts proposed above (Fig. 17).

SUMMARY

This paper describes preparatory works and new concepts for a biology-oriented neuroemulator. Existing neural networks are working with much simpler neuron models than the one described here. The use of dynamic asynchronous neural networks seems to be much more suited to dynamic signal processing. To cover and understand the power of biological nervous systems, biology has to be the prototype for the examinations. The emulator for biology-oriented neural networks presented will be a first step towards this direction.

ACKNOWLEDGEMENTS

The authors would like to thank Prof. Rechenberg for giving the idea of this work, all the members of the Department of Biophysics at the University of Marburg for the friendly and close collaboration, Dipl.-Ing. Dirk Reuver for writing the simulation program, and the students Stefan Quandt, Christian Mischeff, and Kemal Bati for their part of this work.

REFERENCES

[ALKI89]   Alkin, "Memory Storage and Neural System", Scientific American, July 1989

[BLAC87]   Black, Adler, Dreyfus, Friedman, LaGamma, Roach, "Biochemistry of Information Storage in the Nervous System", Science, June 87, Vol. 236, p. 1263ff

[ECKH89]   Eckhorn, Reitboeck, Arndt, Dicke, "Feature Linking via Stimulus-Evoked Oscillations: Experimental Results from Cat Visual Cortex and Functional Implications from a Network Model", IEEE/INNS Int. Joint Conf. on Neural Networks, Washington 1989, Vol. I, p. 723-730

[FOHL80]   Fohlmeister, "Electrical Processes Involved in the Encoding of Nerve Impulses", Biol. Cybernetics 36, p. 103-108 (1980)

[FREN70]   French, Stein, "A Flexible Neural Analog Using Integrated Circuits", IEEE Transactions on Bio-Medical Engineering, Vol. BME-17, No. 3, July 1970, p. 248-253

[KATZ87]   Katz, 'Nerv, Muskel und Synapse', Stuttgart, New York, Thieme 1987

[KLAR89]   Klar, Ramacher, 'Microelectronics For Artificial Neural Networks', Düsseldorf, VDI-Verlag 1989

[KOHO88]   Kohonen, "The 'Neural' Phonetic Typewriter", IEEE Transactions on Computers, March 88, p. 11-22

[LIPP87]   Lippmann, "An Introduction to Computing with Neural Nets", IEEE ASSP Magazine, 04/87, p. 4-22

[NELS80]   Nelson, "General Input-Output Relations for the Electrical Behaviour of Neurons", Kybernetes 1980, Vol. 9, p. 123-131

[PRAN89]   Prange, "Neuronenmodell als analoge integrierte Schaltung", '4. Treffpunkt Medizintechnik', 17.2.1989, p. 42-43

[PRAN90a] Prange, "Modeling the Neuron", in 'Pharmacological interventions on central cholinergic mechanisms in senile dementia', Berlin, Springer 1990

[PRAN90b] Prange, "Emulation of Biology-Oriented Neural Networks", in Eckmiller e.a., 'Parallel Processing in Neural Systems and Computers', North-Holland 1990, p. 79-82

[RECH73]   Rechenberg, 'Evolutionsstrategie', Stuttgart, Frommann-Holzboog 1973

[REIT89]   Reitboeck, Eckhorn, Arndt, Dicke, "A Model of Feature Linking via Correlated Neural Activity", in: Haken, 'Synergetics of Cognition', Springer, 1989

[RÜCK88]   Rückert, Goser, "VLSI Architectures for Associative Networks", 1988 IEEE Int. Symp. on Circuits and Systems, Helsinki

[STEI74]   Stein, Leung, Mangeron, Oguztöreli, "Improved Neuronal Models for Studying Neural Networks", Kybernetik 15, 1-9 (1974)

[STEV87]   Stevens, "Die Nervenzelle", in 'Gehirn und Nervensystem', Spektrum der Wissenschaft: Verständliche Forschung, Heidelberg 1987

PULSED SILICON NEURAL NETWORKS
- FOLLOWING THE BIOLOGICAL LEADER -

A. F. MURRAY, L. TARASSENKO, H. M. REEKIE, A. HAMILTON,
M. BROWNLOW, S. CHURCHER, D. J. BAXTER

BACKGROUND

Much has been written regarding the merits and demerits of analog and digital techniques for the integration of neural networks. Proponents of both approaches are often scathing about alternatives, and blind to the shortcomings of their own work. In this chapter, we will attempt to give an honest account of a body of work that builds VLSI neural circuitry using digital voltage and current pulses to carry information and perform computation, in a manner similar to the biological nervous system. Although this parallel did not, and does not, form the motivation for the use of this technique, it is an interesting aspect of the work, and was in part its inspiration.

In 1986, we developed the first pulse stream neural chip [1,2]. The motivation for the pulse stream method was initially a desire to build essentially analog, asynchronous networks using a standard digital CMOS process. We observed that biological neural systems use pulses, and that controllable oscillator circuits can be implemented easily in silicon. From these two simple facts, we evolved pulse stream neural circuits that were initially crude, and served only to prove the viability of the technique [3]. We have since uncovered many other useful features of the pulse technique, and developed a set of alternative approaches to design using pulse rate encoding. Other workers have also adopted this style of computation for some or all of their network implementation work [4-9]. In this chapter we will present the different alternatives in pulse stream implementations, and thus provide enough information to allow an interested reader to begin to develop optimised pulse stream circuits. Towards the end of the chapter, we will also draw out the biological parallels, more to stir the reader's imagination than to attempt to use them as a justification for the technique!

Terminology

The state S_i of a neuron i is related to its activity x_i stimulated by the neuron's interaction with its environment through an activation function F(), such that $S_i = F(x_i)$. The activation function generally ensures that $x_i \rightarrow \infty$ corresponds to $S_i \rightarrow +1$ (ON) and that $x_i \rightarrow -\infty$ corresponds to $S_i \rightarrow 0$ (OFF), although other forms are sometimes used. Almost all involve a "threshold" activity that defines the level of activity around which the neural state changes from OFF to ON. The activation function is sometimes referred to as a "squashing function", representing as it does the "squashing" of an unbounded activity into a state whose range is constrained. The dynamical behaviour of any network of n neurons during "computation" may then be described by a generic "equation of motion" [10]:-

$$\frac{\delta x_i}{\delta t} = -A_i x_i + \sum_{j=0}^{j=n-1} T_{ij} S_j + I_i(t) \qquad (1)$$

Changes of state result from input stimuli $\{I_i(t)\}$ and interneural interactions $\{T_{ij}\}$ and $\{A_i\}$. The Synaptic Weight T_{ij} quantifies the weighting from a signalling neuron j to a receiving neuron i imposed by the synapse. During the course of learning, this term will change to alter the stored "information" in the network. The Self Term A_i represents passive decay of neural activity in the absence of both synaptic input and direct external input. The solution to (1) under those conditions is then $x_i(t) = x_i(0) \times \exp(-A_i t)$. I_i is the Input to neuron i, and the details of I_i are dependent on the network's function and environment. However, in principle, I_i can be made to force a state on the network, or may be switched off completely to allow the network to settle.

Neurons may be viewed as signalling their states $\{S_j\}$ into a regular array of synapses, which gate these presynaptic states to increment or decrement the total activity of the receiving neuron, which accumulates down the column. In a pulse stream implementation a neuron is a switched oscillator. The level of accumulated neural activity is used to control the oscillator's firing rate. As a result, the neuron's output state S_i is represented by a pulse frequency v_i, such that for $S_i = 0$, $v_i = 0$, and for $S_i = 1$, $v_i = v_{max}$. Each synapse stores a weight T_{ij}, which is used to determine the proportion of the input pulse stream v_j that passes to the post-synaptic summation, modulating the effect of the presynaptic sending neuron's state S_j on the activity x_i, and thus on the receiving neuron's state S_i.

The passing of pulses may take the form of passing voltage pulses to the postsynaptic column, which are subsequently OR'ed together, or passing current pulses, which may be summed as charge on a capacitor.

PULSE STREAM NEURONS : ALTERNATIVES

Although the neuron is, in a sense, the fundamental component in a network, it has received less attention than the synapse from all neural VLSI researchers, including ourselves. This has occurred largely because networks are seen as having N^2 synapses for every N neurons. Furthermore, the neural function has been seen as a simple one, that is easily abstracted - that of performing the "squashing" or activation function to produce a bounded neural state S_i from an essentially unbounded level of activity x_i. In fact, the details of the neural function are important in such diverse neural systems as the navigational system of the bat [11], which is clearly a dynamical system, and in the correct functionality of the back-propagation learning algorithm in multilayer perceptron networks [12], which are not.

In this section we address three issues in the context of pulse stream neurons. These are smoothness of activation, the details of short-term dynamics, and the uncomfortable issue of scaling in large networks, where the "neuron" is required to drive a large number of synapses.

Smoothness of Activation

As stated above, it is easy to build a silicon neural oscillator, as shown in Fig. 1. In a pulse stream implementation a neuron is such a switched

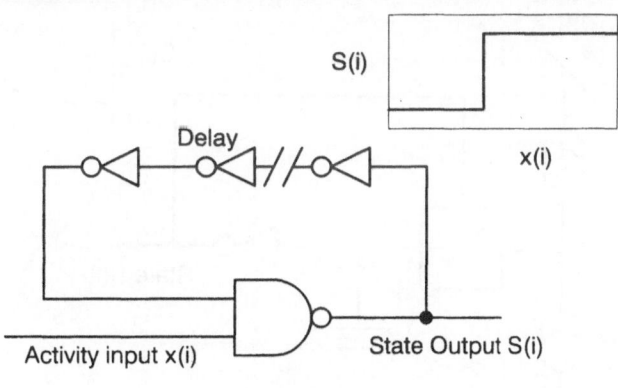

Figure 1

oscillator. The level of accumulated neural activity x_i is used to control the oscillator's firing rate. As a result, the neuron's output state S_i is represented by a pulse frequency v_i , such that for $S_i = 0$, $v_i = 0$, and for $S_i = 1$, $v_i = .v_{max}$ The circuit in Fig. 1 is simple, and therefore can be made to occupy a small silicon area. However, the "switch on" characteristic is, as Fig. 1 shows, inelegant. As the activity input voltage rises through the switching threshold of the NAND gate the oscillator switches on suddenly, and the resultant activation function is equivalent to a "hard limiter" form. For some primitive neural architectures, such as the original Hopfield network [13], this does not matter. However, for more useful networks forms such as the multilayer perceptron, an activation function of this nature is highly unsuitable, as it is not

106

differentiable, and violates the preconditions for reliable back-propagation learning [14].

Techniques for the design of Voltage Controlled Oscillators (VCOs) are well established, and may be found elsewhere [15]. The operating principle is the same as the simple ring oscillator in that feedback is used to create oscillation. However, extra devices are included to remove the discontinuities in the oscillator's input-output characteristic. Fig. 2 shows schematically how the neural activity x_i is used to control a current sink, used to discharge a capacitor at a rate controlled by the difference between the activity and a reference voltage. The oscillator output switches the capacitor between charge and discharge, and the output frequency is thus controlled by the input activity via the discharge rate. The additional complexity of this oscillator is offset by the truly differential sigmoidal activation function. Additional devices may also be added to the feedback path to further "round off" the edges of the activation curve, at the cost of extra area and simulation time.

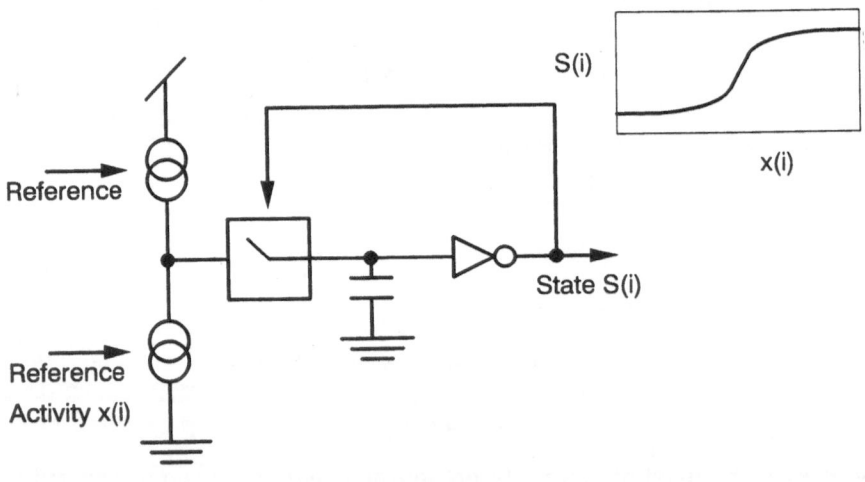

Figure 2

Self-Depletion?

The above discussion assumes that the presynaptic activity input is an analog voltage generated by the synapses to represent $x_i = \Sigma T_{ij} S_j$, and that a neuron will fire at a frequency determined by this value activity, without affecting the activity voltage. Biological neurons are known to be essentially "one-shot" devices, however, only firing a sequence of pulses if the synapses provide a replenishing stream of stimulation. The details of this process are complex and irrelevant in this context, but give rise to a different view of pulse neural activation, removed from the

Voltage Controlled Oscillator form. In fact, the associated oscillator form is a Current Controlled Oscillator, where the ability to fire a repeating sequence of pulses depends on the net ability of the synapses to supply current to replenish a reservoir of charge at the oscillator input.

Fig. 3 shows a neural oscillator based on this principle. The feedback loop from the neural output to the positive input of the comparator is, as before, responsible for the delay, and thus the pulse length. However, a further layer of feedback serves to block the input current from the synapse during an output pulse and simultaneously bleed charge from the activity capacitor, thus depleting the neuron's reservoir. A further pulse cannot occur until the synapses have replenished this charge, and the neuron's pulse repetition rate is controlled by the input current - x(i) on Fig. 3. The different dynamical behaviour of such a network may be visualised easily by considering the effect of an enforced total cessation in all presynaptic activity in a feedback network. In a network using VCO's, the stored activity voltages will immediately cause all the neural oscillators to commence firing again, and normal activity will resume. In a network using the self-depleting structures in Fig. 3, however, "brain death" will occur, whereby, in the absence of any spatio-temporal integration, a single pulse from each neuron will deplete its own reservoir of charge, and no further pulse activity will occur.

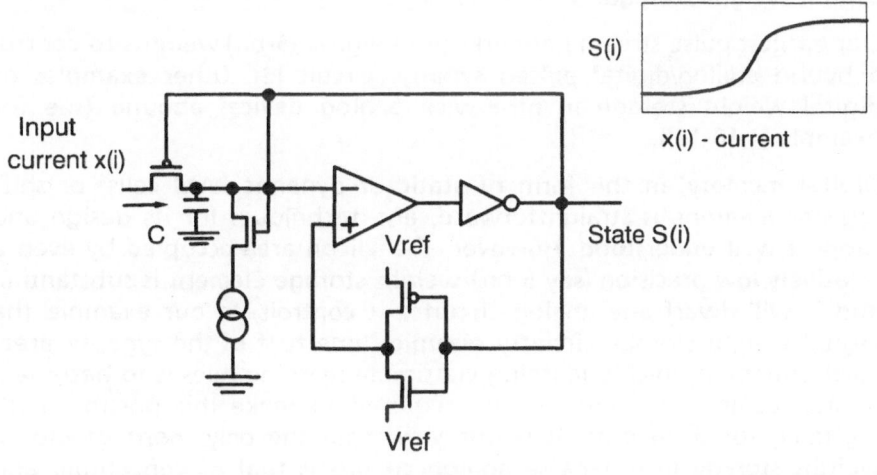

Figure 3

The detailed dynamical behaviour of such neurons outside these extreme conditions is complex, and as yet no detailed study has been made of the effect (if any) on computation. At this stage it is impossible to do more than highlight the difference - that this neural circuit moves away from the abstracted neural function in the majority of synthetic networks, and toward a more biologically related form. The importance of this issue may not appear until attempts are made to use time delays between individual pulses either as parts of a computation or to introduce temporal changes.

SYNAPTIC WEIGHT STORAGE : ALTERNATIVES

This is perhaps the biggest problem in analog neural VLSI, and four generic approaches have emerged. These are:-

(i) Fixed weights → non-programmable networks

(ii) Digital weight storage

(iii) Dynamic, analog weight storage

(iv) Extra technology - floating gate/EEPROM/a-S$_i$

We have tested (ii) and (iii) in silicon, in the context of pulse stream networks. Our interests do not include fixed-function networks, and we are currently in the earliest stages of a program involving a- (amorphous) silicon devices. Accordingly, we offer here only our views and experiences on the success of (ii) and (iii).

Digital Weight Storage

Our earliest pulse stream networks used digital (5-bit) weights to control a hybrid analog/digital pulsed synapse circuit [3]. Other examples of digital weight storage in otherwise analog devices abound (see for example [4,16-18]).

Digital memory, in the form of static or dynamic RAM cells, or shift register memory, is straightforward, and techniques for its design and support well understood. However, the silicon area occupied by even a relatively low precision (say 5 bit) weight storage element is substantial, and it will dwarf any analog circuitry it controls. In our example the digital weight storage circuitry occupied one half of the synapse area. Furthermore, if on-chip learning via simple learning rules is to become a reality, complex support will be required to make this possible with digitally stored weights. It is our view that the only merit of digital weight storage in otherwise analog circuits is that of conceptual and circuit simplicity.

Analog, Dynamic Weight Storage

Capacitors are used widely to store charge, representing binary values in a conventional DRAM. There is no fundamental reason why they may not alternatively store charge representing an analog voltage. This technique has found favour with several groups, including our own. The problem is that, as well as noise from the conventional (environmental) sources, dynamically stored weights must be robust against leakage through the transistor switch network used to program the analog voltage in the first place. Fig. 4 shows the main leakage mechanisms. They are I(1), the subthreshold current through the address MOSFET [19], and I(2), the reverse bias saturation current through the diode formed by the drain of the address transistor. In most cases, I(1) can be made much smaller than I(2). Both of these can be reduced by cooling the device, and I(1) may be minimised by turning the address MOSFET firmly off (by setting $V_{GS} \ll V_T$). In the end, capacitor area must be traded off against noise- and leakage-immunity, as the charge representing a given weight voltage is directly proportional to the capacitor value.

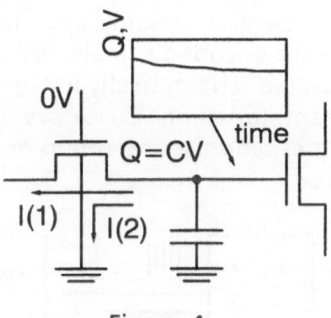

Figure 4

Alternatively, methods can be sought which partly cancel the leakage. Both subthreshold leakage in a MOSFET and reverse saturation current in a diode are approximately independent of the voltage across the element. Methods have been proposed [20] that store synapse weight as the difference between two capacitively stored voltages - in other words, as

$$\Delta V = V_1 - V_2 = \frac{Q_1}{C_1} - \frac{Q_2}{C_2} .$$

Since $\delta Q / \delta t = I_{discharge}$ is constant, and C_1 and C_2 are chosen to be equal, ΔV is also constant. In fact, such a technique can only hope to extend the viable hold time, rather than render it infinite, and we hope to achieve hold times of minutes, to 1% accuracy, using this method. To achieve truly non-volatile storage, novel technology must be employed.

Analog Weight Storage : Summary

In applications where long-term information or some other form of "knowledge" is to be stored as synapse weights, technological

110

innovation must be resorted to, unless fixed functionality can be tolerated. Dynamic weight storage can only be used in such an application with complex support circuitry for refreshing capacitively-stored values. However, where learning is to be truly dynamic, the invasive programming techniques (such as the application of 35V to transistor gates in MNOS [21]) are not suitable, and dynamic weight storage is a more obvious solution.

This dichotomy will persist until technological innovation produces a non-volatile storage technique that is easily programmable, "on the fly".

SYNAPTIC MULTIPLICATION : ALTERNATIVES

We have now tried several techniques for multiplication and accumulation of pulse-rate encoded signals. We do not believe that our early attempts [3], illustrated schematically in Fig. 5 represent good use of the power of pulse stream signalling. Excitatory and inhibitory pulses are signalled on separate lines, and used to dump or remove charge packets from an activity capacitor.

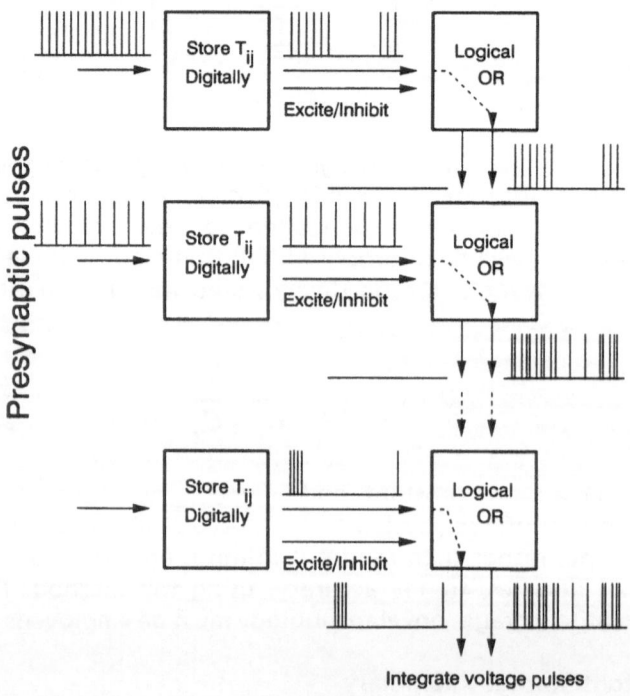

Figure 5

Synaptic gating is achieved by dividing time artificially into periods representing 1/2, 1/4..... of the time, by means of "chopping clocks", synchronous to one another, but asynchronous to neural activities. In other words, clocks are introduced with mark-space ratios of 1:1, 1:2, 1:4, etc., to define time intervals during which pulses may be passed or blocked. These chopping clocks therefore represent binary weighted bursts of pulses. They are then "enabled " by the appropriate bits of the synapse weights stored in digital RAM local to the synapse, to gate the appropriate proportion (i.e. 1/2, 1/4, 3/4,.....) of pulses S_j to either the excitatory or inhibitory accumulator column, as shown in Fig. 5. Multiplication takes place when the presynaptic pulse stream S_j is logically ANDed with each of the chopping clocks enabled by the bits of T_{ij}, and the resultant pulse bursts (which will not overlap one another for a single synapse) ORed together. The result is an irregular succession of aggregated pulses at the foot of each column, in precisely the form required to control the neuron circuit of Fig. 1.

A small $\simeq$ 10 - neuron network was been built around the 3 μm CMOS synapse chip developed using this technique [1-3,22]. The small network, while not of intrinsic value, owing to its size, served to prove that the pulse stream technique was viable, and could be used to implement networks that behaved similarly to their simulated counterparts.

The introduction of pseudo-clock signals to subdivide time in segments is inelegant. Furthermore, ORing voltage pulses together to implement the accumulation function incurs a penalty in lost pulses due to overlap [23]. We have therefore abandoned this technique, in favour of techniques using pulse width modulation and pulse amplitude modulation via transconductance multiplier-based circuits.

Pulse Width Multiplication

The principle here is simple. Pulses of a constant width arrive at a silicon synapse at a rate v_j determined by the signalling neuron state S_j. Multiplication by an analog synapse weight T_{ij} is implemented by reduce the pulse's length in time by a fraction that depends on T_{ij}. These postsynaptic voltage pulses are subsequently used to dump or remove charge from an activity capacitor C_i, whose voltage represents the receiving neuron's activity, and controls the firing rate of S_i.

This is illustrated in Fig. 6. By altering the characteristics of the dump/remove circuit, the dynamic range of synapse weights $\{T_{ij}\}$ may be split , such that the upper half represents excitation, and thus an increase in postsynaptic activity (here, charge on the distributed activity capacitor), while the lower half represents removal of charge, and thus inhibition [24]. This technique avoids placing requirements on the neuron other than the ability to drive a capacitive load, and is fully cascadable in

112

that the activity capacitance is distributed, and thus automatically reduces the significance of each synaptic contribution as more synapses are added. While transconductance multiplier techniques reduce the raw transistor count for pulse stream synapses, they place heavier requirements on the fanout capability of the neuron, as will be discussed below.

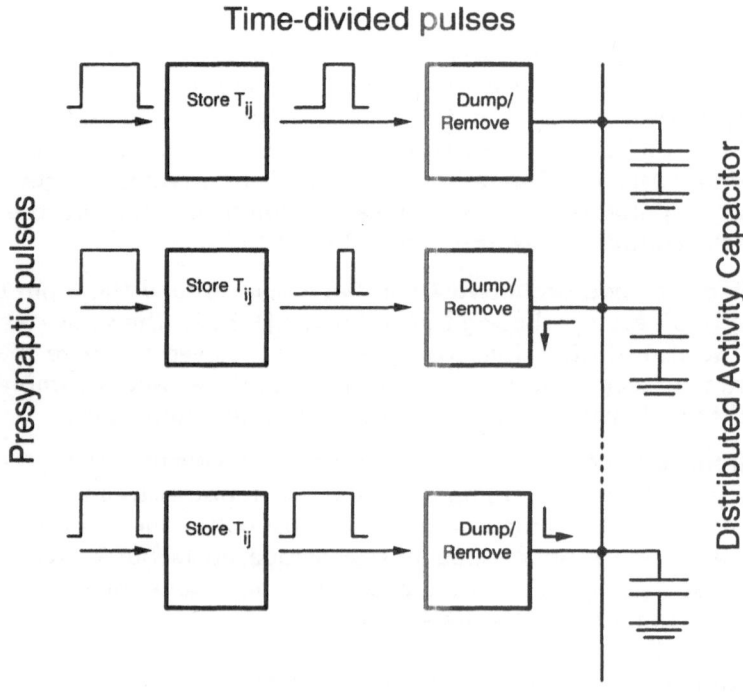

Figure 6

Transconductance Multiplier Synapses

The underlying principle here is rather different, as is shown in Fig. 7. A circuit is used that produces a current proportional to the synapse weight, and this output is pulsed at a rate controlled by the signalling neuron.

The equation for the drain-source current, I$_{DS}$, for a MOSFET in the linear or triode region is

$$I_{DS} = \frac{\mu C_{ox} W}{L} \left| (V_{GS} - V_T) V_{DS} - \frac{V_{DS}^2}{2} \right| \qquad (2)$$

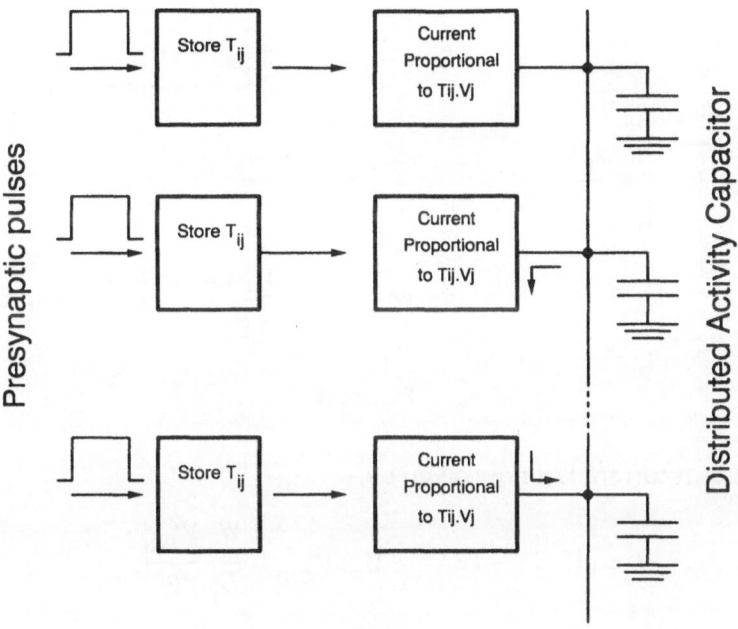

Figure 7

Here, C_{ox} is the oxide capacitance/area, μ the carrier mobility, W the transistor gate width, L the transistor gate length, and V_{GS}, V_T, V_{DS} the transistor gate-source, threshold and drain-source voltages respectively. This expression for I_{DS} contains a useful product term:

$$\frac{\mu C_{ox} W}{L} \times V_{GS} \times V_{DS}$$

However, it also contains two other terms in V_{DS} x V_T and V_{DS}^2.

One approach might be to ignore this imperfection in the multiplication, in the hope that the neural parallelism renders it irrelevant. We have chosen, rather, to remove the unwanted terms via a second MOSFET, as shown in Fig. 8.

114

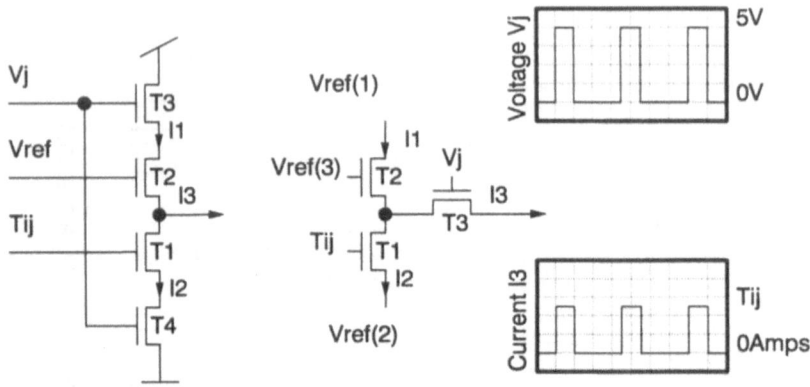

Figure 8

The output current I_3 is now given by:-

$$I_3 = \mu C_{ox} \left| \frac{W_1}{L_1} (V_{GS1} - V_T) V_{DS1} - \frac{W_1}{L_1} \frac{V_{DS1}^2}{2} \right.$$

$$\left. - \frac{W_2}{L_2} (V_{GS2} - V_T) V_{DS2} - \frac{W_2}{L_2} \frac{V_{DS1}^2}{2} \right| \qquad (3)$$

The secret now is to select W_1, L_1, W_2, L_2, V_{GS_1}, V_{GS_2}, V_{DS_1} and V_{DS_2} to cancel all terms except

$$\mu C_{ox} \frac{W_1}{L_1} V_{GS1} \times V_{DS1} \qquad (4)$$

This is a fairly well-known circuit, and constitutes a Transconductance Multiplier. It was reported initially for use in signal processing chips such as filters [25,26]. It would be feasible to use it directly in a continuous time network, with analog voltages representing the { S_i }. We choose to use it within a pulse-stream environment, to minimise the uncertainty in determining the operating regime, and terminal voltages, of the MOSFETs.

Fig. 8 shows two related pulse stream synapses based on this technique. The presynaptic neural state S_j is represented by a stream of 0-5V digital, asynchronous voltage pulses V_j. These are used to switch a current sink and source in and out of the synapse, either pouring current to a fixed voltage node (excitation of the postsynaptic neuron), or removing it (inhibition). The magnitude and direction of the resultant current pulses are determined by the synapse weight, currently stored as a dynamic, analog voltage T_{ij}.

The fixed voltage V_{fixed} and the summation of the current pulses to give an activity $x_i = \Sigma T_{ij} S_j$ are both provided by an operational amplifier integrator circuit. The feedback capacitor in the operational amplifier circuit is distributed across the synapse array in the interests of cascadability. The transistors T1 and T4 act as power supply "on/off" switches in Fig. 8a, and in Fig 8b are replaced by a single transistor, in the output "leg" of the synapse, Transistors T2 and T3 form the transconductance multiplier. One of the transistors has the synapse voltage T_{ij} on its gate, the other a reference voltage, whose value determines the crossover point between excitation and inhibition. The gate-source voltages on T2 and T3 need to be substantially greater than the drain-source voltages, to maintain linear operation. This is not a difficult constraint to satisfy.

The attractions of these cells are that all the transistors are n-type, removing the need for area-hungry isolation well structures, and in Fig. 8a, the vertical line of drain-source connections is topologically attractive, producing very compact layout, while Fig. 8b has fewer devices. It is not yet clear which will prove optimal.

There are three major issues to be addressed in all analog neural VLSI. These are interconnect (particularly between devices), cascadability and process variations between sections of synapse array in a multi-chip system. Interconnection is dealt with in the next section, where pulse stream signalling displays one of its more impressive, if obvious advantages. In this section we illustrate how the more detailed issues of cascadability and process variations may be dealt with in a pulse stream system.

Cascadability

Here the problem is one of amplifier drive capability, and is particularly worrying in the case of the transconductance multiplier synapse, where a rigid voltage has to be maintained at the synapse output. This means that the current drive capability of the operation amplifier maintaining such a rigid voltage is directly proportional to the number of synapses driven. In the time-division synapse, the neuron oscillator only drives gate capacitance, and nowhere is there a requirement to maintain a virtual reference voltage. This is an easier requirement to satisfy, via a high digital drive capability at the VCO output. Steps must be taken to render cascadable a network whose synapses place requirements on the neuron's analog drive capability. Fig. 9 shows conceptually how this may be achieved. In effect, the output transistors, which give rise to an operational amplifier's current drive capability, are distributed amongst

116

the synapses [1]. This increases the number of transistors at each synapse via a concomitant decrease in the size of the neuron circuit, and renders the system far more cascadable. The operational amplifier no longer has to sink and source large currents. Rather it has to provide a signal to the gates of the distributed output transistors that is capable of reacting quickly to changes in the network. This brings the transconductance multiplier synapse into line with the time modulation version, where the neuron's ability to drive capacitive loads is important, while its raw current drive capability is not.

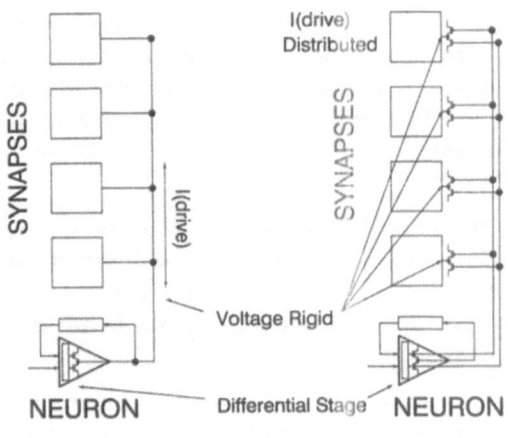

Figure 9

Process Dependence

Chips from the same source can differ in such characteristics as the transistor threshold, the body effect parameter, doping densities etc. All of these variations cause disparities in the analog behaviour of identical circuitry on different devices. Some of these differences can be taken account of in a neural network via an appropriate adjustment of the synaptic weights. However, variations should be minimised if intolerably complex pre-processing of the weight set is to be avoided.

In a true digital device, restricted process variation is tolerable, as it does not affect the correctness of the digital signal - only, perhaps, its speed. In an analog device, however, correct voltage/current levels are of paramount importance, and must be maintained. This is nowhere more important in neural VLSI, where we may be expecting several cascaded neural devices to agree on the meaning of voltages and currents passed across chip boundaries. In a purely analog device, this implies either that individually adjustable current/voltage references must be set for each

1. "This is a simplified explanation, and the implementation details are more complicated"

device, or that reference voltage/current signals must accompany actual voltage/current signals passed between devices. This is clearly a non-trivial overhead, restricting both bandwidth and pinout.

When pulse frequencies/lengths are used to signal and perform calculations, however, the matter is more easily dealt with. Consider as an example the issue of pulse length. If pulse frequency is used to signal neural states, the length of individual pulses fired by on-chip neurons must be constant across a multi-chip network. However, pulse length in a VCO will vary with several process-dependent factors, such as threshold voltage. Fig. 10 shows one way in which this may be dealt with. A reference pulsed signal, whose frequency is irrelevant, is provided to all chips. On-chip circuitry compares the pulse width generated by the on-chip neurons with the reference pulses, and uses the error signal to trim an on-chip voltage controlling a current that sets the VCO pulse length. This "phase-locking" technique removes all dependence on process (at least for the pulse width parameter). Similar techniques are employed in our networks to preserve the integrity of neural states passed through a multiplexed (hybrid analog/digital) serial bus, as the next section describes.

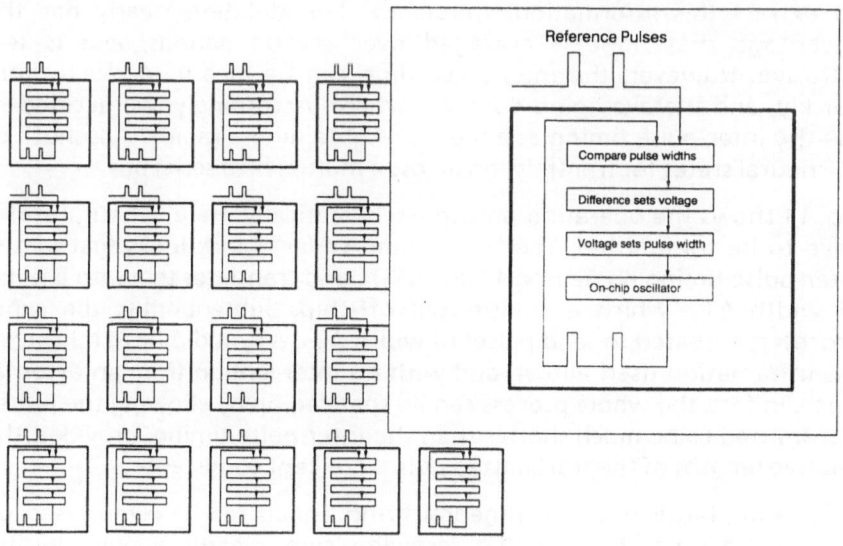

Figure 10

INTER-CHIP COMMUNICATION

We are, with the techniques described in summary above, and in detail elsewhere [20] in a position to integrate around 10,000-15,000 fully

118

programmable analog synapses on a single chip. Even within the architecture with the highest neuron:synapse ratio (i.e. totally interconnected, $N : N^2$), this represents at least 100 analog (or in our case, pulsed) neural states. To include these, along with power supplies, weight inputs, weight address lines etc. as physical pads stretches the pinout capability of even the most exotic bonding technologies. In, say, a multilayer perceptron (MLP), where the neuron:synapse ratio is much less that $N : N^2$, the problem is exacerbated, and chips become severely pad-limited. For instance, in one of our current demonstrator networks, currently under construction, the MLP architecture is $100 \rightarrow 80 \rightarrow 100$ neurons, giving 16,000 synapses (split between two devices). The total number of neural input/output signals involved is 200, and when the extra signals required to "split" the network between the two devices is accounted for, the chip pin count is significantly greater. We have therefore addressed this problem now, rather than allowing it to take over from raw synapse count as a limitation, via a novel, self-timed, asynchronous intercommunication scheme, which we describe in this section.

The neural state information contained in a pulse-rate encoded system may be seen as being either the pulse frequency or the inter-pulse time. To extract this information, frequency demodulation clearly has the advantage that noise is averaged over several periods, and is less intrusive. However, the inter-pulse time can be measured much more quickly, and the averaging done elsewhere. Accordingly, we decided to use the inter-pulse timing as a means of obtaining a rapid "snapshot" of the neural states for transmission across a multiplexed serial bus.

Fig. 11 shows the operating principle schematically. Here, 4 neural states have to be transmitted. The "pulse time coder" block looks first at the inter-pulse timing on neuron 1 (say, ΔT_1), and translates that into a pulse of width ΔT_1, which is transmitted off-chip. Subsequently, the same process is repeated to send pulses of width ΔT_2, ΔT_3 and ΔT_4, as quickly as the information itself allows, and with no intervention from an external clock. In fact, the whole process can be speeded up by allowing the pulses transmitted to be much shorter than the inter-pulse timing, provided the relative lengths of the transmitted pulses are kept consistent[2].

As a result, large numbers of neural firing signals can be compressed on to a single serial bus. On the receiving chip, clearly, a pulse width-frequency demodulator must be included, and this can be achieved independent of process variations using techniques related to those described in the previous section.

2. "It is interesting to note, although not strictly relevant, that inter-pulse timing is used in many biological systems, rather than pulse frequency - see for example [11]."

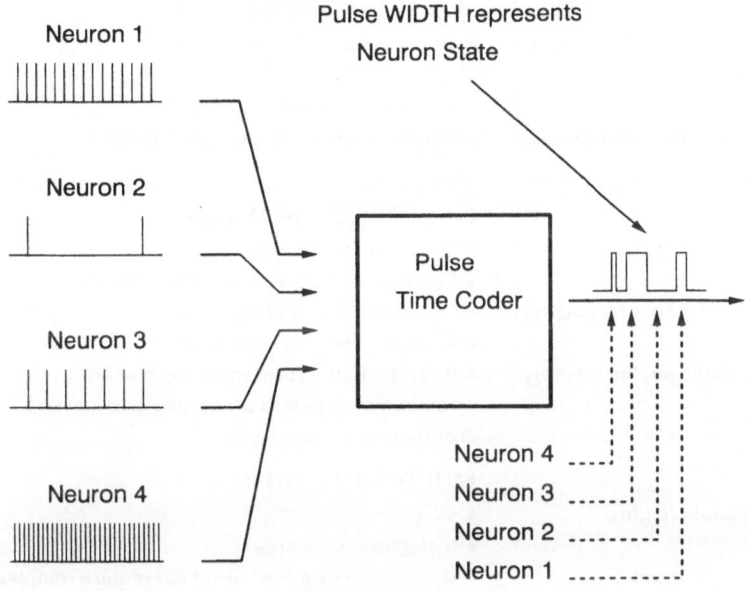

Figure 11

CONCLUSIONS AND POINTERS

It was the explicit aim of this chapter to provide the interested reader with the necessary conceptual toolkit to allow him to explore the pulse stream technique independently, without reinventing too many wheels. Details of the individual components of the pulse stream systems developed by the Edinburgh/Oxford team, can be either found in the references cited above, or will be published when appropriate. The choices as we perceive them have been presented above, and summarised in Table 1 below.

Finally, what of the biological parallel? Certainly, it is an intriguing and stimulating thought that silicon devices based on pulse stream techniques use the same signalling and essentially computational structure as biological neurons and synapses. However, one must not get carried away, and the biological analogy is certainly no justification in itself for the use of the same technique in 2-D silicon (possibly the most hostile medium imaginable for building neural structures).

- NEURAL STATE REPRESENTATION -

Frequency of constant width pulses	• Power requirement averaged
	• Requires good VCO
Width of constant frequency pulses	• Power requirement periodic ("spikey")

- WEIGHT STORAGE METHOD -

Digital weights	• Simple, accurate, non-volatile
	• Clumsy, large area, on-chip adaptation complex
Analog, dynamic (capacitive)	• Small area, easily programmed
	• Volatile - needs refresh or on-chip learning
Nonvolatile analog memory	• Robust, holds information "permanently"
	• Needs non-standard processing, and invasive programming techniques

- MULTIPLICATION -

Via digital Weights	• Simple
	• Inelegant, large area
Via Time Division	• Operates "per pulse", does not require complex operational amplifier, relatively compact, cascades elegantly
	• > 10 transistors/synapse
Via transistor characteristic for multiplication (transconductance multiplier)	• Small number of transistors, operates "per pulse".
	• Cascading needs care - voltage references must be maintained.

- SUMMATION -

Via Voltage ANDing	• Simple
	• Pulses (and thus activation) lost via overlap
Via capacitive integrator	• Simple, cascadable, low area
Op-Amp integrator	• Accurate - permits small synapse
	• Current drive requirement on op-amp→cascadability?

- NEURON OSCILLATOR -

Ring oscillator	• Simple, small
	• Implements hard limiter only
Op-amp based VCO	• Smooth activation, analog output
	• Large area
Comparator-based CCO (self-depleting)	• Small, biologically supported
	• Altered network dynamics

Many of the advantages we perceive in pulsed operation - of power averaging, ease of pulse regeneration, robustness of signals - could well have driven the evolution of this system for much biological signalling. The more detailed implementational advantages, such as the ability to compress pulsed signals on to a serial bus, are totally irrelevant in the context of biological structures.

The question most often asked regarding pulsed operation is "how fast is it - in comparison to a digital DSP chip implementation?". The answer we give is generally evasive, and for good reason. In any analog, asynchronous network (voltage- or pulse- based), the answer is data dependent, and any direct comparison subject to the worst kind of statistical trickery. For a small number of neurons, clearly digital techniques will always be superior in speed. As network sizes grow, however, the actual parallelism of an analog system begins to tell, and we could produce fastest supercomputers. Such a comparison would, however, be both spurious and invidious, and we will leave it out!

ACKNOWLEDGEMENTS

The authors are grateful to the UK Science and Engineering Research Council, and the European Community (ESPRIT Basic Research Action NERVES) for its support of this work.

REFERENCES

[1]   A. F. Murray and A. V. W. Smith, "Asynchronous Arithmetic for VLSI Neural Systems," Electronics Letters, vol. 23, no. 12, pp. 642-643, June 1987.

[2]   A. F. Murray and A. V. W. Smith, "A Novel Computational and Signalling Method for VLSI Neural Networks," European Solid State Circuits Conference, pp. 19-22, VDE- Verlag, Berlin, 1987.

[3]   A. F. Murray and A. V. W. Smith, "Asynchronous VLSI Neural Networks using Pulse Stream Arithmetic," IEEE Journal of Solid-State Circuits and Systems, vol. 23, no. 3, pp. 688-697, 1988.

[4]   D. Del Corso, F. Gregoretti, C. Pellegrini, and L.M. Reyneri, "An Artificial Neural Network Based on Multiplexed Pulse Streams," Proc. ITG/IEEE Workshop on Microelectronics for Neural Networks, Dortmund (Ger many)., pp. 28-39, June 1990.

[5]   A. Siggelkow, A. J. Beltman, J. A. G. Nijhuis, and L. Spaanenburg, "Pulse-Density Modulated Neural Networks on a Semi-Custom Gate Forest," Proc. ITG/IEEE Workshop on Microelectronics for Neural Networks, Dortmund (Germany)., pp. 16-27, June 1990.

[6]   J. Meador, A. Wu, and C. Cole, "Programmable Impulse Neural Circuits," IEEE Transactions on Neural Networks, 1990. to be published

[7]   N. El-Leithy, M. Zaghloul, and R. W. Newcomb, "Implementation of Pulse-Coded Neural Networks," Proc. 27th Conf. on Decision and Control, pp. 334-336, 1988.

[8]   S. Ryckebusch, C. Mead, and J. Bower, "Modelling Small Oscillating Biological Networks in Analog VLSI," Neural Information Processing Systems Conference, pp. 384-393, Morgan Kaufmann, December 1988.

[9]   P. M. Daniell, W. A. J. Waller, and D. A. Bisset, "An Implementation of Fully Analogue Sum-of-Product Neural Models," 1st IEE Conf. on Artificial Neural Networks, pp. 52-56, 1989.

[10]   S. Grossberg, "Some Physiological and Biochemical Consequences of Psychological Postulates," Proc. Natl. Acad. Sci. USA, vol. 60, pp. 758 - 765, 1968. Reprinted 1980 - American Psychological Association

[11]   J. A. Simmons, "Acoustic-Imaging Computations by Echolocating Bats : Unification of Diversely-Represented Stimulus Features into Whole Images," Neural Information Processing Systems Conference, pp. 2-9, Morgan Kaufmann, 1989.

[12]   D. E. Rumelhart, G. E. Hinton, and R. J. Williams, "Learning Representations by Back-Propagating Errors," Nature, vol. 323, pp. 533-536, 1986.

[13]   J. J. Hopfield, "Neural Networks and Physical Systems with Emergent Collective Computational Abilities," Proc. Natl. Acad. Sci. USA, vol. 79, pp. 2554 - 2558, April 1982.

[14]   R. S. Sutton and A. G. Barto, "Toward a Modern Theory of Adaptive Networks : Expectation and Prediction," Psychological Review, vol. 88, pp. 135 - 170, 1981.

[15]   P. E. Allen and D. R. Holberg, in CMOS Analog Circuit Design, Holt, Rinehart and Winston, New York, 1987.

[16]   A. Moopenn, T. Duong, and A. P. Thakoor, "Digital- Analog Hybrid Synapse Chips for Electronic Neural Networks," Neural Information Processing Systems Conference, pp. 769-776, Morgan Kaufmann, 1989.

[17]   D. Del Corso and L.M. Reyneri, D. Del Corso, F. Gregoretti, C. Pellegrini, and L.M. Reyneri, "A Pulse Stream Synapse Based on a Closed Loop Register," Proc. Third Int. Conf. on Parallel Architectures and Neural Networks, Vietri sul Mare (Italy)., May 1990.

[18]   Y. P. Tsividis and D. Anastassiou, "Switched - Capacitor Neural Networks," Electronics Letters, vol. 23, no. 18, pp. 958 - 959, August 1987.

[19]   E. Vittoz, H. Oguey, M. A. Maher, O. Nys, E. Dijkstra, and M. Chevroulet, "Analog Storage of Adjustable Synaptic Weights," Proc. ITG/IEEE Workshop on Microelectronics for Neural Networks, Dortmund (Germany)., pp. 69- 79, June 1990.

[20]   A. F. Murray, M. Brownlow, A. Hamilton, Il Song Han, H. M. Reekie, and L. Tarassenko, "Pulse-Firing Neural Chips for Hundreds of Neurons," Neural Information Processing Systems (NIPS) Conference, pp. 785-792, Morgan Kaufmann, 1989.

[21]   J. P. Sage, K. Thompson, and R. S. Withers, "An Artificial Neural Network Integrated Circuit Based on MNOS/CCD Principles," Proc. AIP Conference on Neural Networks for Computing, Snowbird, pp. 381 - 385, 1986.

[22]   A. V. W. Smith, "The Implementation of Neural Networks as CMOS Integrated Circuits," PhD. Thesis (University of Edinburgh), 1988.

[23]   D. Del Corso, A. F. Murray, and L. Tarassenko, "Pulse- Stream VLSI Neural Networks - Mixing Analog and Digital Techniques," IEEE Trans. Neural Networks, 1990. to be published

[24]   A. F. Murray, L. Tarassenko, and A. Hamilton, "Programmable Analogue Pulse-Firing Neural Networks," Neural Information Processing Systems (NIPS) Conference, pp. 671-677, Morgan Kaufmann, 1988.

[25]   P. B. Denyer and J. Mavor, "MOST Transconductance Multipliers for Array Applications," IEE Proc. Pt. 1, vol. 128, no. 3, pp. 81-86, June 1981.

[26]   S. Han and Song B. Park, "Voltage-Controlled Linear Resistors by MOS Transistors and their Application to Active RC Filter MOS Integration," Proc. IEEE, pp. 1655-1657, Nov., 1984.

ASICS FOR PROTOTYPING OF
PULSE-DENSITY MODULATED NEURAL NETWORKS

P. RICHERT, L. SPAANENBURG,
M. KESPERT, J. NIJHUIS, M. SCHWARZ, A. SIGGELKOW

ABSTRACT

Two alternative ways are presented for creating dedicated neural hardware realizations based on pulse-density modulation. The first approach emphasizes fast prototyping of neural systems in a conventional digital microprocessor environment. It uses an ASIC cell library in combination with a Sea-Of-Gates template to produce testable integrated neural circuits with off-chip learning. Typical single-chip network sizes range from 18 neurons with 846 synapses to 110 neurons with 550 synapses. Larger network sizes can be obtained by concatenation.

The second approach is based on an ASIC processor, which can be programmed for a variety of biologically plausible neuron models with flexible topology (i.e. it possesses programmable and learnable membrane potential, synaptic weights, delays, threshold, and transfer function). Each chip possesses furthermore its own communication hardware to realize global communications through fault-tolerant interconnection multiplexing.

INTRODUCTION

In recent years, a considerable effort has been spent on creating VLSI hardware for artificial neural networks [1]. Up to now, most of the proposed and realized networks are based on very simple models of the biological neuron, such as McCulloch-Pitts model [2]. There have been some attempts to combine the pulse coding scheme of biological neurons with silicon technology [3], [4], [5], but they lack the generality required to probe the multitude of alternative models and topologies.

ASIC prototyping. The obvious way to design a neural network starts with simulations using one of many limited-purpose or single-purpose simulation programs, that are in existence today. They usually model the ideal network and fail to indicate physical limitations and/or manufacturability. Design by simulation requires an amount of computing power, that often exceeds the capabilities of present computers, if such physical detail is called for. As simulations merely exemplify the probable functionality of the network, hardware prototyping is still required.

The practical application of neural processing is seriously restrained by lack of well-established synthesis techniques. Typically, attentive probing is required for learning, fault-sensitivity, component testability, and system parameter sensitivity. To reduce simulation times, two architectural concepts can be distinguished:

--   hardware accelerators, where the parallelism in the network simulation is used to speed up the computation, and

--   neural hardware, where the network itself is casted into one or more silicon chips. This can be subdivided into:

--   -emulators, where general-purpose devices (i.e. microprocessors, DSP's) are programmed to act as one or more neurons [6], [7], and

--   -neural chips, where the network is realized by means of unique chips [8], [9], [10],

and an example of both will be described in detail here.

While hardware accelerators merely aid the simulation, it is not possible to make use of the neural network's built-in asynchronous, massively parallel signal processing. Digital hardware emulators often do not stress biological information processing details, especially when using simple multiply-and-add operations followed by a linear transfer function. Furthermore, they simulate parallel processes by essentially synchronous, sequential hardware.

In this paper, first, a state-of-the-art solution to the design of neural chips is given. In order to provide a smooth transition from pure network simulation to the actual neural realization, construction is based on a sequence of small steps for fast turn-around prototyping with ASIC's [11]. This can be mechanized by the IMS Collective Computation Circuit, the IC^3, where a specialized neural cell library is used to build an adjustable neural network from a semi-custom Sea-Of-Gates template [12]. By this, one can quickly follow technological advantages to build neural networks as large as possible.

Multiprocessing. Neural hardware systems have the capability to implement neural networks for real-time applications (for instance in speech and image recognition), but have not yet reached the performance of human creatures. Whether this is a matter of network size (i.e. of integration volume and density) or a matter of modeling accuracy (are todays models of biological neurons and synapses sufficient?) remains to be debated. Clearly, this question has no easy answer. To gain insight into the basics of biological neural systems as well as to construct advanced neural computers, it is important to build small nets of nerve cell electronic equivalents and to analyze their behavior. The main features of neural systems (especially their inherent massive parallelism, asynchronous operation, and large number of interconnections) require special hardware for real-time implementations and applications of neural networks.

Therefore, the second part of this paper introduces a multi-processing environment based on a programmable neuro-computer chip, which can be used to implement biological neural features in artificial neural networks. This flexible chip can only emulate a few neurons but, although the chips are only connected locally, global communication is possible using a special high-speed on-chip communication processor. Having gained a better understanding of biological signal processing, it should be possible to substitute the general-purpose neuro-processor by special hardware, thus increasing the integration density. Nevertheless, the high-speed communication processor will still be necessary to combine single neural-chips in larger networks.

Technology. While the modern silicon technologies successfully vie with implementation of individual neuron functions, the problem of chosing either digital or analog realization of the neural functions remains undecided [13]. Though biology employs analog neural functions using pulse density coded signals and parallel connections in space, today's VLSI cannot easily realize equivalent high complexity. On the one hand, analog circuits exhibit high functionality but lack long time storage of analog values, unless analog EEPROM's are available. Also, analog circuits are more susceptible to noise, crosstalk, temperature effects, power supply variations, and deviations of technological parameters. Future down-scaling of analog devices will be limited by laws of thermodynamics. On the other hand, the connectivity of VLSI chips is limited internally by chip area and the number of routing layers, and externally by the pin count of the IC carriers. In this paper, the above mentioned problems are tackled by adopting a digital implementation of neural functions with digital weight storage. Where the first concept utilizes a logic-level scheme of time-multiplexing (both on- and off-chip), the second concept is based on an internal bus structure while for inter-chip communication a dedicated communication processor is applied. Thereby, a message passing scheme creates both local and global communication in a network of only locally connected neuro-computer chips.

THE IMS COLLECTIVE COMPUTATION CIRCUITS

The key element in an IC^3 Application-Specific Integrated Neural Circuit (ASINC) is the binary rate multiplier. It consists of (a) a counter and (b) a control register, that selects the counter contents for which the incoming pulse will be passed on. A pulse stream for an n-bits counter will therefore be multiplied by a factor a with value range $0 <= a < 2^n/(2^n-1)$. In the synapse, such a situation is actually occurring; therefore the synapse can be built from a mere binary rate multiplier. In a neuron, the situation is more complicated. Here the dendrite information is collected in an up/down counter, which in turn serves as control register for a counter driven by the neuron clock. The sign of the up/down counter thresholds the rate pulse to reach the axon. The ratio between the neuron clock and the synapse pulses, as collected over the width of the up/down counter, sets the slope of the resulting transfer function. A schematic rendering of the hardware implementation is shown in Figure 1.

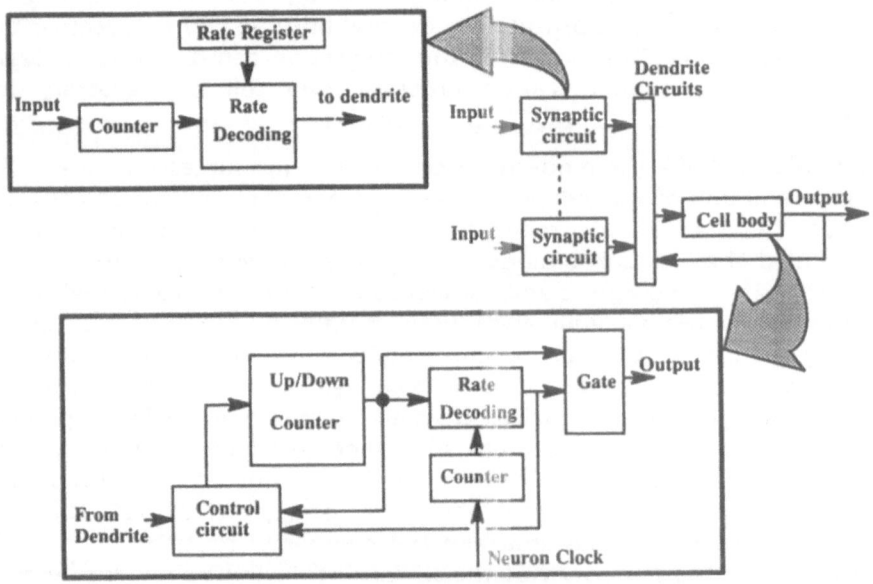

Figure 1   The single neuron and its components

Implementation

Timing. The main architectural problem lies in the relation in time between the internal pulses. A neuron feeds pulses over the synapses to other neurons. If more than one synapse excitates/inhibits a single neuron at a given time, while this neuron is also involved in internal house-keeping affairs, proper functioning of the asynchronous logic can not be guaranteed. Four alternatives have been proposed (see Figure 2):

1.   Introduce time slots so that only one synapse at a time can deliver a pulse to the neuron input. (Time-Division Multiplexing)

2.   Use special hardware so that the addition of simultaneous pulses can be handled, e.g. instead of using one input line several lines

are used, each indicating a particular number of pulses. (Digital Encoding)

3. Accept the possibility that pulses are lost, when two or more pulses arrive at the neuron input at the same time. (Straight Multiplexing)

4. Use several pulse levels, e.g. pulse amplitude will double, when two pulses arrive at the same time. (Analog Encoding)

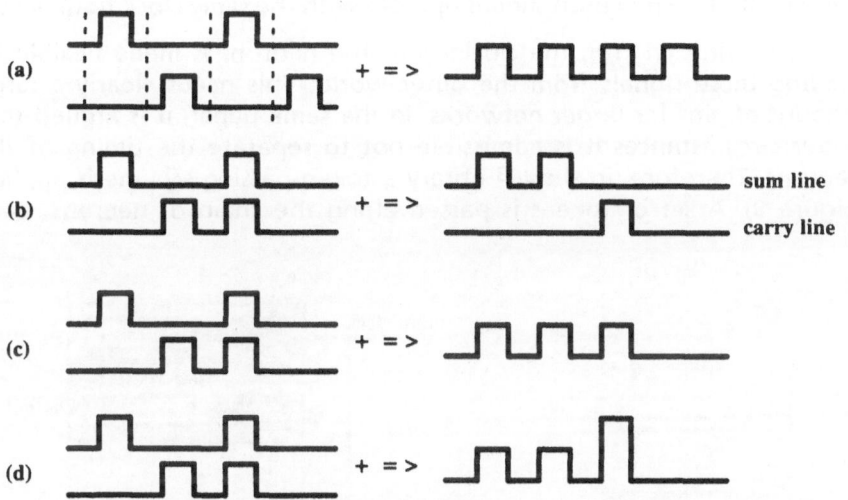

Figure 2   Various dendrite timing schemes
    (a) Separated time slots   (b) Encoded transfer
    (c) Asynchronous transfer   (d) Pusle amplitude modulation

Analog encoding is clearly not relevant for the fully digital realization. Digital encoding, on the other hand, may lead to wide buses for large networks and introduces adder arrangements in the dendrites. For specific topologies this may be bearable [9], but for arbitrary topologies the effect on signal timing can be disastrous. This leaves some way of multiplexing as a reasonable technique [10].

Multiplexing. A synapse from neuron$_i$ to neuron$_j$ can produce an output pulse, when it receives a pulse from neuron$_i$. This means, that simultaneous synapse pulses on the dendrite of neuron$_j$ can only occur, when the output pulses of the neurons, which are connected to neuron j, occur at the same moment. As there exists only one connection between two neurons, simultaneous pulses on the dendrites can be avoided by synchronizing the neurons. Each neuroni will have its own time slot (T_i), in which an output pulse can occur; in other words, all synapses from neuron i will generate pulses on the dendrites only in this time slot T_i.

Compared to a completely asynchronous neural network, whose maximum neuron output pulse frequency equals $f_{neuron,max}$, the maximum output frequency will be equal to $f_{neuron,max}/\#neurons$. This means, that the network becomes slower, when more neurons are used. For this reason, the method is not feasibie for very large neural networks, but for networks with a limited number of neurons the loss in speed could be acceptable. Another disadvantage of neuron synchronization is the fact, that each neuron should operate with the same clock frequency.

Token passing. In [10], the clocking of the neurons is made flexible by feeding these signals from the outer world. This needs clearly a large amount of pins for larger networks. In the same paper, it is argued that under circumstances it is admissible not to separate the timing of the neurons. Therefore, in the IC^3 library a token-passing scheme is applied (Figure 3). A set of tokens is passed along the chain of neurons. Each

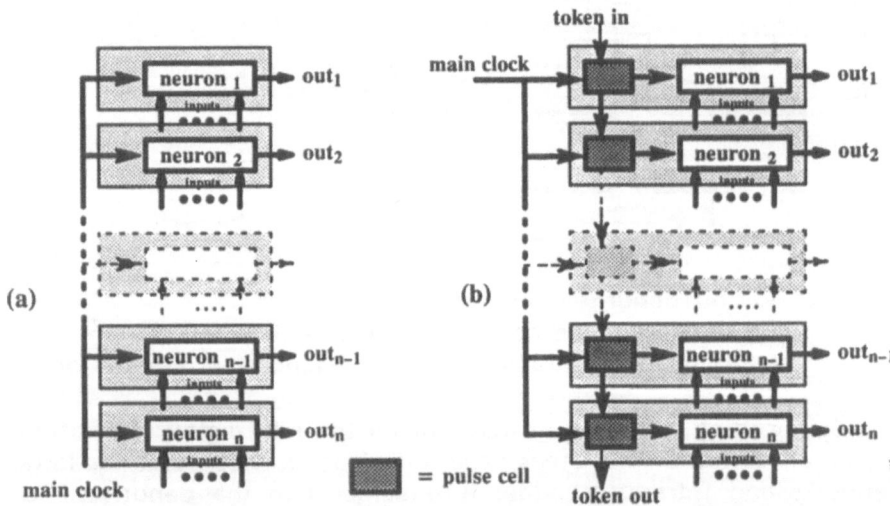

Figure 3   Additional hardware needed to implement neuron synchronization. In (a) all neurons operate in parallel. In (b) each neuron has its separately activated time slot

neuron will potentially fire, if the main clock arrives, and, if an active token is present. If only one token is circulated at a time, the neurons are guaranteed to operate in separate timeslots; if more tokens are present, simultaneous firing of neurons can be accommodated. Note, that whether simultaneous firing leads to pulse contention on the dendrites is foremost given by the topology of the network. A degree of parallelism can be introduced to optimize speed.

Accuracy. From all the differences between analog and digital realizations of neural networks, the most crucial one is probably the data accuracy. It has been shown, that, while learning needs sigmoid transfer functions and highly accurate data, recall of the same quality can be achieved with hardlimited functions and 4% accuracy [14]. Furthermore, there is experimental data, indicating, that this accuracy requirement can be decreased further by using an overmass of hidden neurons.

This sets the number of bits for the synapse weights; the number of bits for the neuron rate multiplier follows from the required transfer function. For a hardlimiter, no extra bits are needed; for a linear transfer the number of bits increases with decreasing slope (see Figure 4). The choice is made selectable through a control bit per neuron.

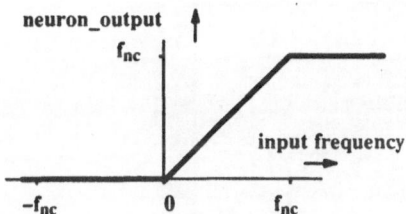

Figure 4   The neuron transfer function (negative input values are represented as a negative pulse frequency)

Mapping on the Gate Forest template

Over the past years, the steeply increasing cost of manufacturing has induced a trend to prototype on semi-custom templates. The Gate Forest template [12] is such a channelless gate array. Implementing a neural network efficiently on such a template implies therefore a library of cells with feedthrough opportunities.

The Gate Forest. The Gate Forest is a Sea-Of-Gates technique, where the transistors are placed under a 45 degrees angle to optimize feedthrough opportunities. This is illustrated in Figure 5a. The core cell consists of a PMOS transistor, 2 NMOS transistors and a small size NMOS transistor. Two of these cells suffice to build a static latch structure. In contrast to standard cells [15], the cells of the neural network library are targeted to use a vertical double cell as unit. This is illustrated in Figure 5, showing the weight storage latch inside a rate multiplier bit cell. The wiring of the standard cell is placed largely underneath the logic, taking one or more core cells. As each core cell offers 13 routing opportunities, of which for the rate multiplier only 8 are used, empty space can not be avoided. In the distributed cell, all vertical area is used in exchange for cell width, thus optimizing the layout.

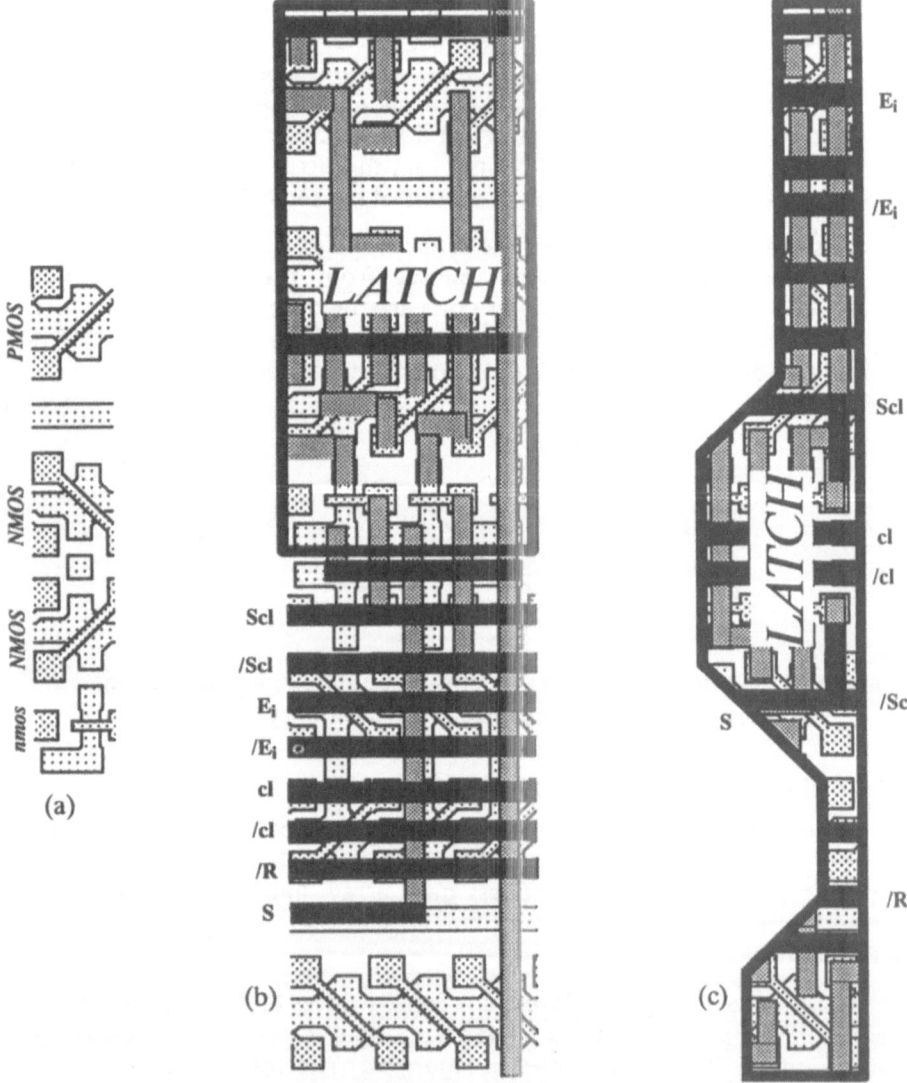

Figure 5   (a) Detail of Gate-Forest, (b) standard-cell latch and
(c) distributed latch as applied in synapse

Synapse and Neuron. The rate multiplier, as required for a synapse as
well as for a neuron, is based of the rate multiplier bit cell, consisting of a
weight bit storage, a bit counter, and a bit output switch. To built an
entire rate multiplier also, a begin cell as well as an end cell are needed.
To program the functionality of an individual synapse within the neural
network, a separate output selector is created. Furthermore, to access the

weight storage as well as the control bits, a random access address decoder is available. From these components, a synapse can be created by abutment as shown in Figure 6.

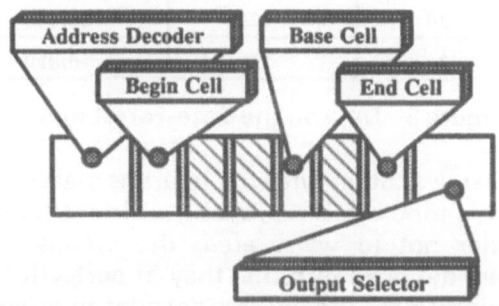

Figure 6   Cell abutment for a 6-bit synapse

Similar considerations are applicable in creating a neuron (Figure 7). It takes two times the width of a synapse, as it has to accommodate roughly twice the amount of rate multiplier bits, and two times the height of a synapse, as it contains next to the rate multiplier also the up/down counter.

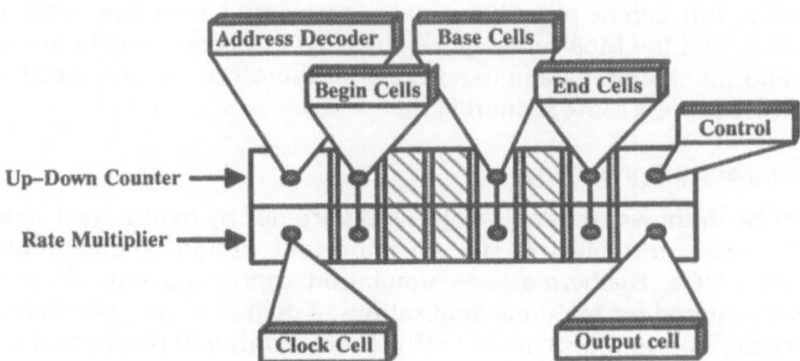

Figure 7   Cell abutment for a 6-bits neuron

Configurations. Gate Forest templates exist in different templates and are therefore of different capacity. The neural network cell library is constructed such, that a network can be easily constructed from the netlist and physical mapping values, obtainable from a neural network simulator such as NNSIM [16]. Different topologies can be selected to support widely differing demands on synapse-to-neuron ratio (Figure 8).

The most popular one, at this moment, is the Matrix-of-Synapses topology (MoS), where all synapses that share a dendrite to a neuron, are

Master	# hor. rows	# vert. rows	#transistors	#IO–pads
GFxx1	102	694	283,152	228
GFxx2	62	476	118,048	134
GFxx4	44	298	52,448	96
GFxx9S	24	170	16,320	56

Figure 8   Data on the Gate-Forest masters

placed on top of one another with the neurons placed at the end of this linear array. Such a topology is reasonably efficient for a fully-connected network. In order not to waist area, the synapse and the neuron conglomerates are designed such that they fit perfectly together by mere abutment. To this purpose the address decoder in a synapse creates the strobe signals for two neighboring synapses simultaneously. The neuron then takes the width of two synapses with shared address decoder. onsequently, a stack of synapses can be serviced from neurons at the bottom side and neurons at the top side [11].

An alternative is the Matrix-of-Blocks (MoB) topology, where the synapses feeding a common neuron, are located orthogonally on both sides of this neuron giving a conglomerate cell of rectangular perimeter, which in turn can be placed in a matrix to create a neural network. Both the MoS and the MoB topology can be used in silicon compilation based on abutment and parametrization and therefore do not need CPU-intensive place & route support [11].

Testing and programming

It has been shown, that neural networks are not by nature fault-tolerant .[17]. Hence, testability is still an important design consideration for neural ASIC's. Furthermore, as simulation cannot go into all physical detail required for real-time applications, a degree of programmability is necessary during prototyping. Lastly, some words will be devoted to the prospect of silicon compilation for neural ASIC synthesis.

Testing. Lets assume an integrated neural network with X neurons and Y n-bit synapses. Each of the synapses houses 2n memory locations and accomplishes a weight resolution of 2^{-n}. The bit length of the up/down counter in a neuron is chosen to be 2^n to guarantee that the neuron rounding errors can be neglected. Therefore each neuron contains 4n memory locations. For the complete structure this sums up to 4nX + 2nY. For example: a fully connected Hopfield network with 20 neurons and 400 6-bits synapses, needs in the order of 5K memory locations, of which more than half are used inside counters.

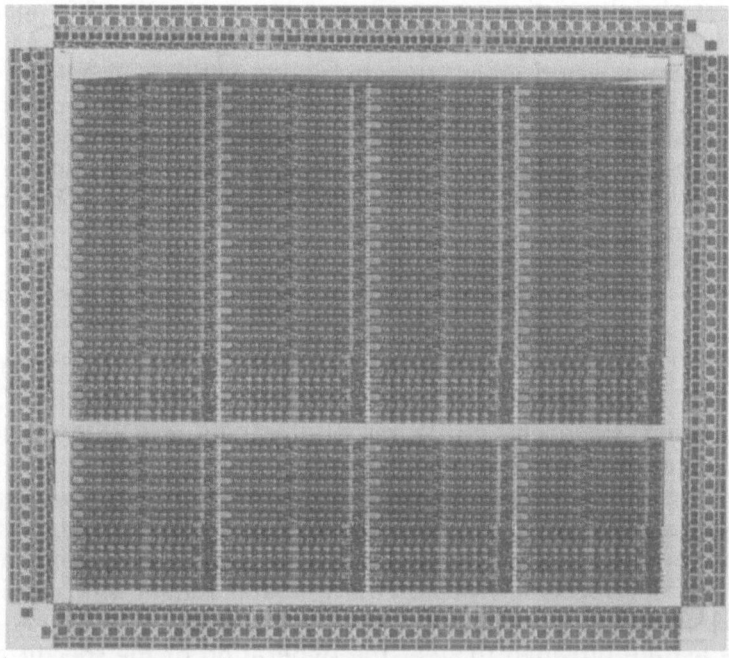

Figure 9   Picture of an IC³ personalized for mobile robot application

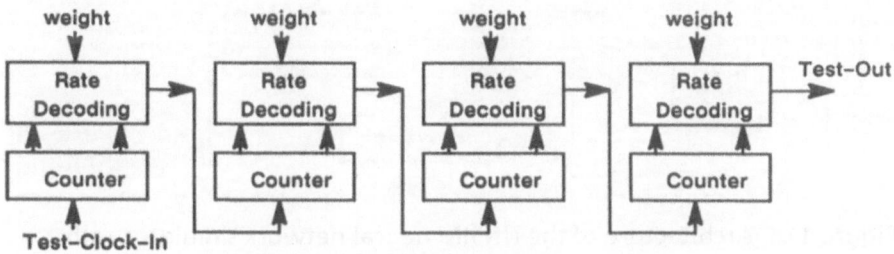

Figure 10   Rate multiplier testing

From the above discussion one finds, that the testing problem is mainly due to the abundance of counters. It is well-known that counters require exponentially increasing test time. Introducing a conventional scanpath inside the counters leads to an unacceptable hardware overhead.

Therefore the alternative arrangement shown in Figure 10 is applied to the counters in the rate multipliers. This testing scheme is based on the fact that with a weight of 1 a rate multiplier will pass all but one of the incoming pulses. So after reset, the incoming pulses will ripple through the series arrangement, till all counters have reached the value 2^{n-1}. Then the first counter will fall back to the value 0 without passing the pulse. Next the first counter is raised to 1, while the second counter falls back to 0 without passing the pulse. And so on. Hence, at the output of the series arrangement one observes 2^{n-1} pulses followed by a gap at size of the number of counters involved. This approach results in a very short test time, that is linearly dependent on the number of counters, rather than exponentially on the number of memory locations in the scanpath, and is therefore suitable for very large neural networks.

Programming. In generating neural networks by means of the IC^3 cell library, three ways of programming can be distinguished. The first way involves the semi-custom technology itself, where a Gate-Forest master is personalized. In simulation, not only the learn strategy and the neural topology are determined but foremost the design dimensions are set, such as synapse accuracy, neural transfer and redundancy required to achieve failure tolerance. At the end of this laborous effort, a netlist and a value list are provided to steer the generation of the specific chip or chip set (Figure 11). The resulting design is used to personalize a Gate-Forest through Direct-Write E-Beam.

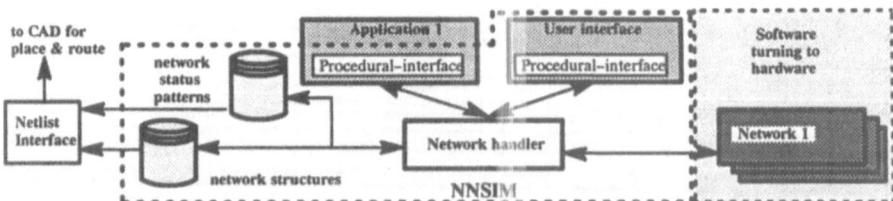

Figure 11   Architecture of the NNSIM neural network simulator with netlist interface

This chip can still be reconfigured using the control bits. Every synapse is equipped with two control bits that influence the output selector, as shown in Figure 6, to direct the output to the dendrite or (for test purposes) to the input of a next synapse. This latter configuration can also be used to extend the accuracy of a synapse by doubling its value range. Each neuron is likewise equipped with two control bits. Furthermore the sign bit, used for a synapse to discriminate between excitatory and inhibitory effects, is used here to select between hardlimited and linear transfer function.

A third way to influence the topology is by inserting multiplexers. In the Matrix-Of-Synapse topology, the synapse input signals are running horizontally through the network. Using multiplexers, these lines can be programmably cut and new synapse signals are inserted at such points. If need be, the resulting design can be re-personalized using non-programmable cells to achieve maximum packing density.

Silicon compilation. The CAD for neural silicon compilation is on the physical level based on conventional commercial software. To link the neural network simulator and the physical design software, additional CAD is required to solve a number of non-standard problems. For instance, the simulated netlist must be partitioned according to the physical capacity of the applied ASIC template (container). Usually, the container capacity is not fully utilized and therefore some neurons and synapses remain unused. These must be either de-activated by properly selecting the control bits, or applied to enhance the fault-tolerance of the network. If the latter option is selected, the network has to be re-simulated.

Another topic is the per chip place & route. Though, using the MoS or MoB topology, much is already fixed, placement is additionally restricted by the token-passing scheme in Figure 2. This dictates, that electrically neighboring neurons should also be physically neighboring, or additional routing is required.

ROBOCAR USING AN IC³ ASINC

In this example, a robot car with ultrasonic sensors is driving autonomously under control of a neural network (Figure 12). The

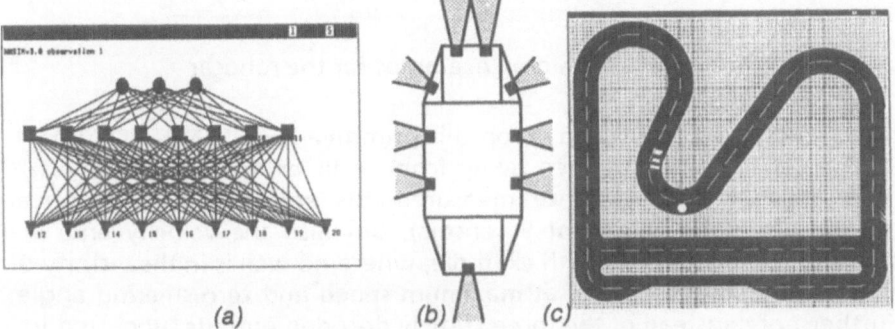

Figure 12   Neural network (a) that can be used to control a car (b) on the road (c)

138

feedforward neural network is trained using the Error Back-Propagation Rule to keep the car on route. In [18] the sensor distance information, supplied by each of the 9 sensors, is logarithmically coded with 4 bits and therefore the network consists of 36 input neurons, 8 hidden neurons and 3 output neurons to represent the modes of movement (stop, fast forward, slow forward, turn right 45°, turn left 45°, turn right 90°, turn left 90°, slow backward). The car is driven on a training track, and the learned information is filtered before actually training the neural hardware.

In our approach, the situations, on which the car should react were analyzed beforehand, and a set of seven basic situations are distinguished (Figure 13). These situations are based on the recognition,

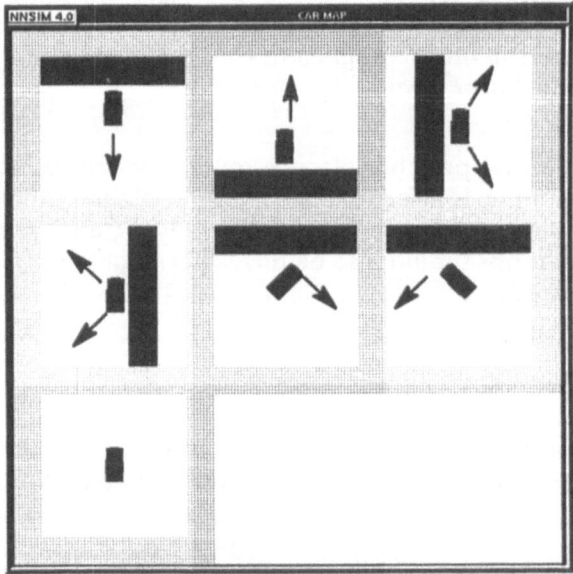

Figure 13   Training examples for the robocar

that all, what a car needs to do on this primitive level, is NOT TO CRASH. So the car is trained to keep away from walls as much as possible. As input, not only the distance measurements are taken (for historical reasons, from 13 instead of 9 sensors), but also the velocity and the steering angle. In the seventh example, where no wall is in the vicinity of the car, we default the car at maximum speed and zero steering angle. Furthermore, instead of the three staticly decoded outputs suggested in , only velocity and steer angle control increments are supplied as outputs. This basic behavior can be used to built more complex robot navigation systems [19].

In the first instance, the above training set was applied to a feedforward network with one hidden layer and a learning rate of 0.9. As the structure of the learning set is also crucial to the success of learning [17], several experiments were made showing, that training the network 500 times with a set, taking all examples once but the first one twice, could do the job. However, on the testing ground, the car came sometimes to a halt being undecisive about the action to take Once a feedforward network was taken with 2 hidden layers, this problem did not show up anymore. However, with a learning rate of 0.7 and a momentum rate of 0.4, it requires 40,000 steps to learn.

Two concepts for providing the stimuli can be applied. The 1-bits per input version as in [18], which leads to a large amount of input neurons and therefore of first-level synapses. Alternatively, one may apply the coded sensor information directly as weights of the incoming synapses. In this scheme only 15 input neurons, 8 + 4 hidden neurons and 2 output neurons are present. A neural network of this size fits on one GFxx2, as shown in Figure 5. This master is available in a 2 um double-metal CMOS technology and, as can be derived from Figure 8, measures 9.57 · 8.69 mm^2 supplying 134 I/O-pads. The design is targetted to run on a 20 MHz main clock, providing a system performance of better than 1.2 million interconnects/second.

DISTRIBUTED PROCESSING NEURAL HARDWARE

In this section, pulse-coding is employed for hardware neurons and asynchronous information processing to implement a fully parallel artificial neural network, using digital signal processing. An analog implementation with discrete electronic components is presented in [20]. Learning rules can be directly embedded into a neural network, using some of the neurons as special learning neurons [21], thus needing no external learning by a host computer. Hence, this concept differs fundamentally from today's neural network implementations.

General-purpose computer. General-purpose computers are not specially designed for neural networks, but can be used for such purposes. If they are arranged into a network, the connection bandwidth is limited because of the lack of special communication hardware. Even transputers designed for an improved communication still use software routing algorithms [22]. SIMD computers, like the Connection Machine ([23], [24]), allow variable bandwidth, but cause problems, when different types of neurons and/or synapse models should be used in a single ANN, because all processors run the same program.

Neuro-computer. The presented neuro-computer is an MIMD computer with asynchronous data distribution and a special hardware routing

140

communication processor. Building large networks, is not just a problem of fabricating neurons and synapses, but also a problem of connecting the chips together and realizing communications. Here it is prefered to separate the switching communication task from the neural processing, but integrate both together on a single chip. As will be shown, this enables not only connection and communication between adjacent chips, but also global communication within the network. A communication processor is introduced, which enables separate processing of communication signals, allowing local and global communication by routing messages through the network of locally interconnected neuro-computer chips.

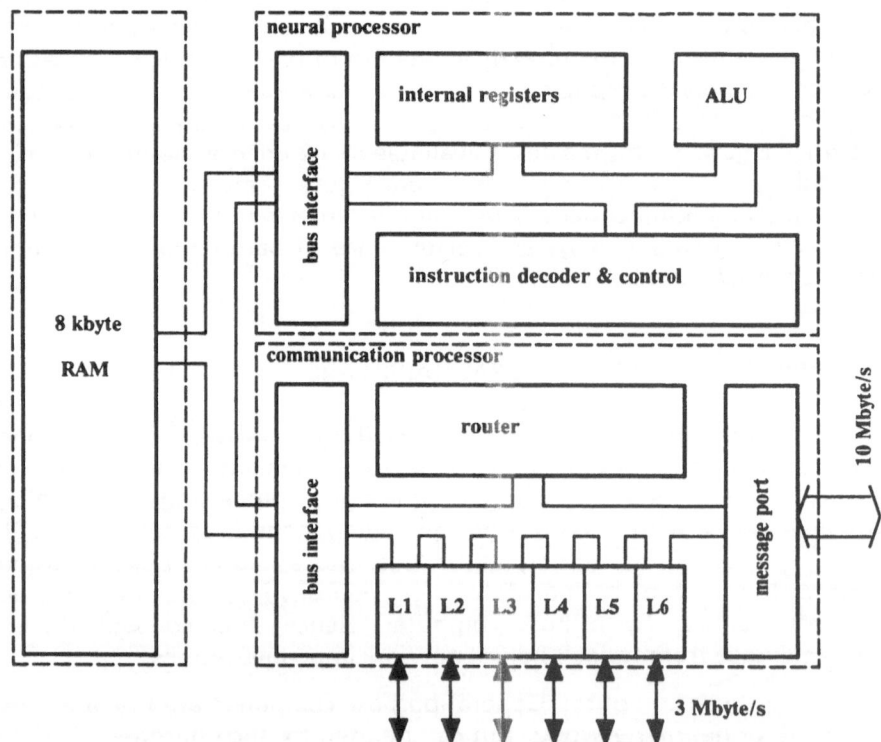

Figure14   Block diagram of the proposed neuro-computer

The block diagram of the proposed neuro-computer is shown in Figure 14. It is apparent, that the actual neural processor is strictly separated from the communication processor, as mentioned above. All received and transmitted messages must pass through the communication processor. Each neuro-computer has its own global address to ensure reception of personal messages. At the same time, the communication processor can pass along messages from other neuro-computers without accessing its internal neural processor and thus increases routing capability. Furthermore, the communication processor can generate addresses for messages to be transmitted by the neuro-computer. Only in case a personal message has been received, the incoming data are sent to the neural processor.

Communication Processor

The implemented routing algorithm is capable of auto-routing using the target addresses and capable of fault-tolerant re-routing by choosing alternative ways towards the receiver, including detours but no returns.

The most challenging part of our neuro-computer chip is the communication processor, which handles the local message routing according to the message target address. Each chip contains its own communication processor, which controls six serial and one parallel link. Its auto-routing capability and extremely high transfer rate of 100 Mbit/s per serial link allows building of a high-speed lossless communication network with a bandwidth of 1 Mmessage/s including all address and control data. The message format (Figure 15) contains a start and control field, an address field that holds the target chip address, and a three byte data field. This data format is common to all message types including membrane potential, synapse weight, delay time, threshold, transfer function and activity pulse messages.

start control	chip address (16–bit)	data 1 (8–bit)	data 2 (8–bit)	data 3 (8–bit)

Figure 15   Message format

Serial and parallel links. Because we use bi-directional links, a token passing algorithm must be implemented for access control to the communication line. The neuro-computers form an rectangular grid, so that only adjacent chips are linked. This allows simplification of the token passing algorithm. The communication processor is to send a single message after having received either a free token or a message token and not only after receiving a free token. Using this protocol, it is guaranteed, that a link cannot be blocked by one processor, while sending a burst of messages. Moreover, an increase of the transmission rate can be expected because free tokens are only sent in the idle state.

The optional parallel message port (Figure 14) is treated in the same way as the serial links, but has a bandwidth of 10 Mbyte/s. It is connected to the internal message bus, which is controlled by the routing unit. This parallel message link is suitable for distributing a parallel video datastream to serial links (see Section).

Message handling. Personal messages, i.e. incoming messages containing the address of the receiving neurocomputer, are sent via the internal port to the neural processor. Other messages are simply passed along through the external ports to other neuro-computers, till they have reached the final destination, i.e. the destination neuro-computer.

The destination of passed messages, i.e. north, south, east, and west, is determined by the address hits scored in the ports. Finally, the communication processor can also generate addresses of neuro-computers, to which the messages including data from the transmitting neuro-computer should be sent to. These addresses define the neuro-computer clusters, to which a single neuro-computer broadcasts identical messages, thus effectively reducing broadcasting transmission requirements. By changing these addresses, we can effectively redefine the topology of the neural network.

The bypassing scheme enables the neuro-computers to communicate globally, though they are connected physically only with their nearest neighbors, either in rectangular or hexagonal arrays. Physically the chips are arranged in a rectangular grid, but virtually they can be connected in any grid, as shown in Figure 18 for a hexagonal grid. Also, the routing

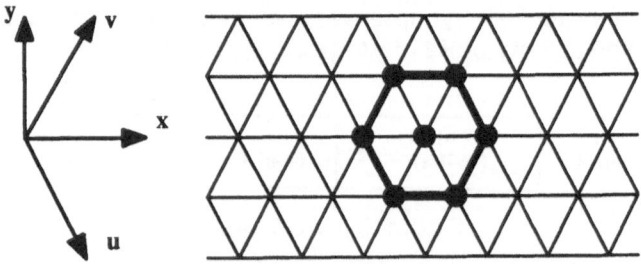

Figure16   Mesh configuration

algorithm becomes simpler, when using a rectangular grid. Each communication processor only has to detect whether the address of the incoming message matches its coordinates in the grid. In this case the data field is transferred to the neural processor by writing the neuron address - which is part of the target address - and the data field into the common memory. If the target address does not match, the router has to determine an outgoing link, on which it can pass the message towards its target chip.

Neural Processor

The neural information in our scheme is contained in the pulse period (i.e. pulse frequency and phase). The pulse frequency (or firing rate) lies in the range from less than 1 Hz up to a maximum of 500 Hz, i.e. much lower than the system clock rate, which is in excess of 10 MHz for a realization in CMOS technology. The propagation and machine delays are thus much shorter than the minimum pulse period and do not affect the neural processing. As the messages run at a very high clock rate, their time constants are much smaller than the time constants of the neurons and thus the neurons appear to operate in asynchronous mode. This allows processing of messages upon arrival, a correspondence to data-flow concepts.

Neural model. The synapse process receives excitatory and inhibitory data signals, which are individually weighted and delayed. All weights and delays are programmed under control of learning algorithms, that can be implemented either externally or internally (i.e. in the neural processor itself). The neuron parameters are membrane potential, transfer function and threshold. Furthermore, the operational mode distinguishes between external or internal learning. During external learning, the learned parameters can be adapted externally (i.e. under control of an external host computer), and then sent to each neuron individually. If internal learning is used, the neuron realizes internally the delta learning rule and the output can be used for weight and delay adaption of synapses connected to other neurons. There is no special mode for associative recall (i.e. operation with fixed parameters), because this can be simply achieved by disabling all learning functions of the neuron.

Neural function. The implemented neural processor is the integrated 8-bit µP FhG-IMS2205 (compatible with the M146805) developed at the FhG-IMS [25]. The neural processor is able to receive incoming messages either as digitally coded parameters or as activity signals of pulse trains (see Figure 17). The activity signals must be weighted and delayed, before a low-pass filter can calculate the membrane potential of the neuron. The output activity can be computed either by applying a non-linear function to the membrane potential or by a special pulse train generator. This generator compares the membrane potential with a time-dependent threshold value and generates an output pulse, if the potential exceeds this variable threshold.

It should be pointed out, that each neural processor can simulate several neurons of different types simultaneously. The number of neurons depends on the complexity of the required neuron model. All neural features mentioned above are programmable.

A CISC microprocessor is used for the realization of the neural processor, because of its low storage requirement, flexibility, and ease of

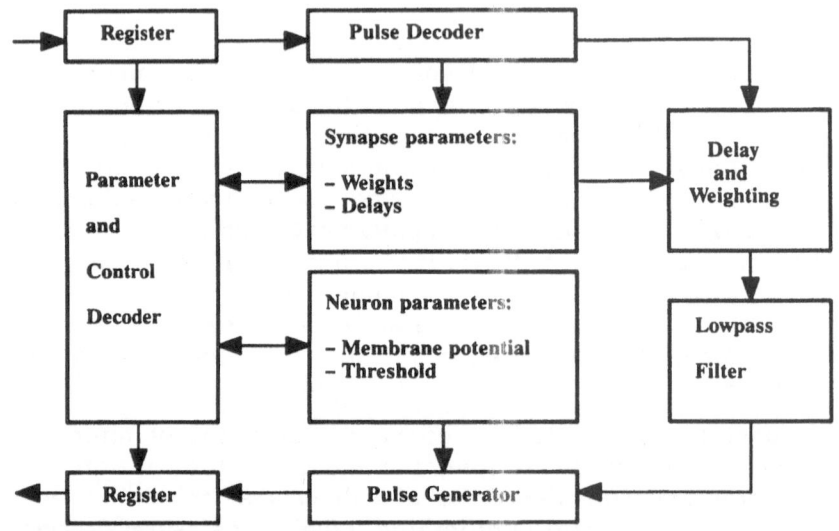

Figure 17   Structure of the neural function

programming. This allows to include the 8 kbyte dynamic RAM on the neuro-computer chip and results in a lower pin count and thereby a smaller chip package. Because of the use of an on-chip RAM, there is no need of directly connecting the neural processor to the communication network. As already mentioned, communications are handled by the communication processor, which exchanges data with the neural processor via the on-chip memory (Figure 14).

Implementation of Neural Networks

The presented VLSI-chip is designed for emulation of 1-4 neurons with up to 40 synapses, freely connected to the neurons. In future versions, it may be possible to connect more than one neural processor to one communication processor on the same chip, thus substantially increasing the number of neurons and synapses per chip. Large numbers of chips can be arranged into an array to build large neural networks. The 2-byte address field of the message format (see Figure 15) essentially determines the maximum array size: there can be up to 64,000 chips connected which yields up to 256,000 neurons and 2,560,000 synapses.

Loading, control, and monitoring of these networks can be done by one or more host computers. The host computer provides interaction between the user and the emulator hardware. It can be used for network design and development using editing tools to set and change the network topology, to initialize the state of the network, to monitor the network dynamics, and to read out the parameters of the adaptive network after a training period. The conceptual design of the emulator control software has been presented in [26]. It uses a virtual network interface; thus it can facilitate the application to other hardware simulator projects.

Silicon realization. The chip is being fabricated using a 2 μm CMOS/SOI process. The designed chip has a chip size of 94 mm² and a total pin count of 64 including internal bus signals. The size of the neural processor is 14 mm², and the size of the communication processor is 10 mm² including 6 mm² for the 6 serial link ports. The speed values, mentioned in this section, are based on simulation and verified by prototype fabrication.

PARALLEL IMAGE PROCESSING BY THE NEURO-COMPUTER

Though the presented concept can be used for various applications, a selected example will be shown in order to clarify the advantages of the proposed approach. Artificial neural networks are often proposed as a new approach to image processing, but only very few hardware implementations of neural nets are actually suitable for image processing applications. In order to point out the salient features of the multi-processing concept in this paper, an application of the neuro-computer for image processing is presented, which is able to load and store image data at video frequencies and to process data at rates that are acceptable for most industrial applications.

System architecture

The proposed architecture consists of several neuro-computer layers (see Figure 18). To process a 512 by 512 pixel grey-tone image, 16 * 16 neuro-computer chips are required in the first network layer, each of it storing at least 1 kbyte segment of image data. At this stage the neural processor is programmed to realize a very simple neuron model. In this way filtering, edge detection, and other preprocessing operations can be carried out. Due to the programmability of the neural processor, classical signal processing operations, usually unsuitable for neural nets (e.g. median filtering), can also be realized. The second layer of the network has to carry out data reduction and feature extraction operations on the preprocessed data. The lower data rate in the following network layers allows the use of more sophisticated neuron models or a reduction of the array size. The third lower layer of neuro-computers is used to implement trainable and self-organizing classifiers. So it should also be possible to experiment with neural structures observed in brain [27].

Video bus interface

The data distribution network, shown in Figure 18, contains a 10 Mbyte/s video bus interface and a frame buffer. Only 17 neuro-computers and a very simple additional address generator are needed. The address generator has to compute a target chip address for each three byte video data packet (see message format in Figure 15). These addresses must be generated in such a way, that the messages are equally distributed between the six serial links of the chip. This is necessary, because the serial links have a useful data bandwidth of only 3 Mbyte/s. An arrangement of the sixteen target chips, shown in Figure 19, allows use of a simple 4-bit ring counter to build the address generator. Because the counter is incremented for each message, and, because the communication processor of the topmost interface chip determines the shortest connection to the target chip, the messages are shared equally between its six links.

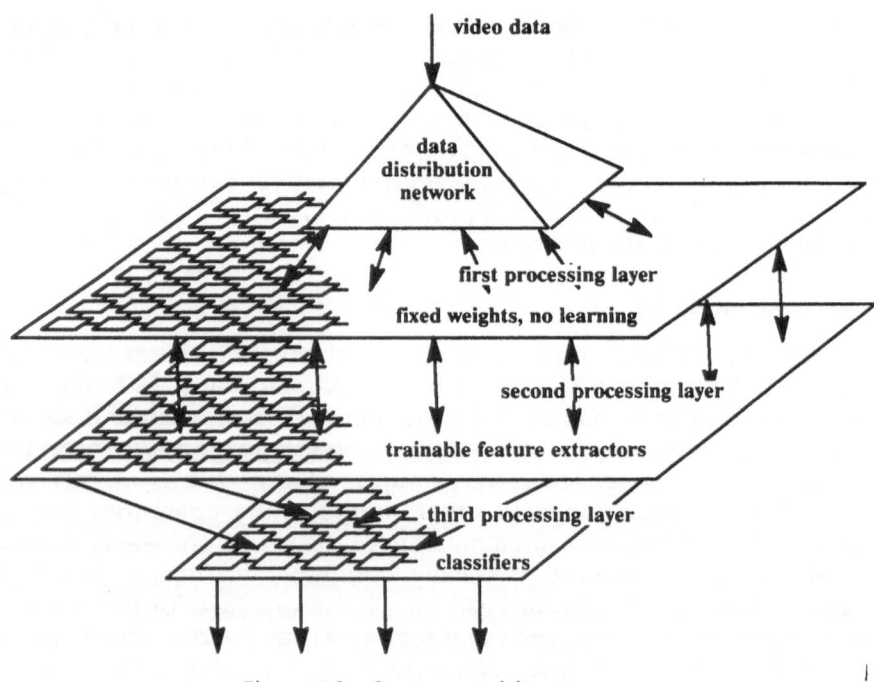

Figure18   System architecture

The remaining sixteen neuro-computers, thus far only serving as a frame buffer, now can be used to redistribute the image data to the first image processing layer (Figure 18). Normally each of the neuro-computers in the first layer should hold an equal sized (32*32) segment of the original image. So, for the distribution of the buffered data into the processing layer, the frame buffer computers first have to compute the destination pixel and target computer address. This calculation of addresses is not restricted to performing a memory load operation, also any point transformation can be realized within the distribution process.

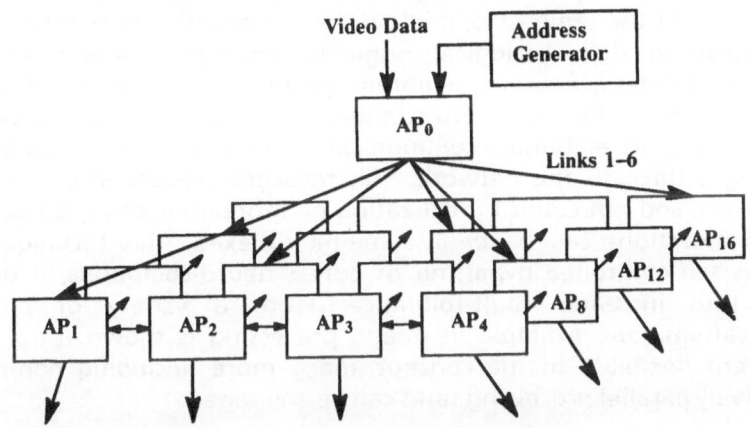

Figure 19   Data distribution network

Image processing layers

The 2-dimensional mesh of neuro-computers, that form the first image processing layer of our network, holds an entire image and the code for the algorithms, that should be applied (Figure 18). Since the neuro-computers can be re-programmed in less than 2 ms, it is possible to compute even sequences of different image processing operations, like convolving or rankvalue-filtering in a single preprocessing layer. If higher performance is required, additional layers can be attached. The use of bidirectional links also between the different network layers allows us to build arbitrary network structures; e.g. feedback loops useful for space-variant operations in contrast enhancement, edge detection and many other applications. All these operations can be executed using a very simple neuron model. For feature extraction and classification we will implement more sophisticated models, including learning or optimization algorithms like back-propagation [28].

CONCLUSIONS

Two concepts of neural hardware based on pulse-density modulation are presented. The first concept is highly suitable for fast prototyping of application specific neural systems. It is aimed to provide a direct migration path from abstract simulation model to the real neural hardware. Its successive refinements are based on silicon modeling in a semi-custom technology using E-Beam Direct-Write for quick turn-around. An example of application in mobile robot navigation shows that realistic problems can be solved.

The second concept places high emphasis on solving communication problems. At the same time, it allows implementation of neurons, which are close to their biological originals, including variable synapses, synaptical delays, neuron membrane potential, transfer function, and neuron threshold. The interconnection problem was solved by introducing an elaborate communication system, that allows passing messages through the network. This measure reduces interconnection hardware and pin count in realization on monolithic chips, because the interconnections can be easily time-multiplexed. This technique also allows self-controlled bypassing of defect neuro-computers, if desired, and thus increases fault-tolerance. From a variety of potential applications, one example in image processing is shown. Due to the inherent flexibility of the concept, many more (including non-neural) massively parallel processing tasks can be envisaged.

Both presented concepts increase the capability of artificial neural networks and should help to answer the question raised in the introduction: "Is the performance of artificial neural networks hampered by network size or is it a matter of modeling accuracy?"

ACKNOWLEDGEMENTS

The authors gratefully acknowledge the cooperation with B.Hosticka, G.Hess, and W.Renker of the FhG-IMS Duisburg, and with A.J.Beltman, J.Leenstra, S.Neusser, and A.Reitsma of the IMS Stuttgart. Furthermore, thanks are due to Allan Murray, not only for pioneering pulse-density modulated circuits, but also for promoting Scottish songs.

The IC^3 project is supported by Daimler-Benz AG and the BMFT under contract "TV 8296 3". The contribution of A.J.Siggelkow is funded through ESPRIT-BRA project 3049 "NERVES". The neuro-computer project is supported by the Ministery of Science and Research of North-Rhine-Westfalia under contract "IV A 6 - 102 007 89".

REFERENCES

[1]   C. Mead and M.Ismail, Analog VLSI implementation of neural systems (Kluwer, Boston, 1989).

[2]   W.S.McCulloch and W.H.Pitts, A logical calculus of the ideas immanent in neuron activity, Bulletin Mathematical Biophysiks 5, pp.115-133.

[3]   N.E.Cotter, K.Smith and M.Gasper, A pulse width modilation design approach and programmable logic for artificial neural networks, in: Proceedings 5th MIT Conference on Advanced Research in VLSI (MIT Press, Cambridge, 1988) pp.1-15.

[4]   R.Eckmiller, Electronic simulation of the vertebrate retina, IEEE Transactions on BME 22 (1975) pp.305-311.

[5]   A.Hamilton, A.F.Murray and L.Tarassenko, Programmable analog pulse-firing neural networks, in: D.S.Touretzky (Ed.), Advances in Neural Information Processing Systems 1 (Morgan Kaufmann, San Mateo, 1989) pp.671-677.

[6]   M.Duranton and J.A.Sirat, Learning on VLSI: A general-purpose digital neuro-chip, Digest IJCNN'89 (Washington D.C., June 1989) pp. II.613-II.620.

[7]   U.Ramacher and J.Beichter, Systolic architectures for fast emulation of artificial neural networks, in: J.McCanny et al. (eds.), Systolic array processors (Prentice-Hall, New-York, 1988) pp.277-286.

[8]   A.F.Murray and A.V.W.Smith, Asynchronous VLSI neural network using pulse-stream arithmetic, IEEE Journal on Solid-State Circuits 23, No.3 (June 1988) pp.688-697.

[9]   J.Tomberg et al., Fully digital neural network implementation based on pulse density modulation, Digest CICC'89 (San Diego, May 1989) pp. 12.7.1-12.7.4.

[10]   Y.Hirai et al., A digital neuro-chip with unlimited connectibility for large-scale neural networks, Digest IJCNN'89 (Washington D.C., July 1989) pp. II.163-II.169.

[11]   L.Spaanenburg, A.J.Beltman and J.A.G.Nijhuis, The IMS Collective Computation Circuits, Technical Report IMS-TM-07/90 (IMS, Stuttgart, July 1990).

[12]   M.Beunder et al., The CMOS Gate Forest: an efficient and flexible high-performance ASIC design environment, IEEE Journal of Solid-State Circuits 23, No.2 (April 1988) pp.387-399.

[13]   H.P.Graf and D.Henderson, A reconfigurable CMOS neural network, Digest ISSCC'90 (San Fransisco, February 1990) pp.144-145.

[14]   L.Spaanenburg, Basics of analog synapse memory, Technical Report IMS-TM-06/90 (IMS, Stuttgart, June 1990), also available as ESPRIT/BRA-3049 deliverable DR1-F3.

150

[15]   H.Yoshimura, M.Nagatami and S.Horiguchi, 32-bit microprocessor design, in: S.Goto (Ed.), Design methodologies, Advances in CAD for VLSI 6 (North-Holland, Amsterdam, 1986) pp.357-398.

[16]   J.A.G.Nijhuis, L.Spaanenburg and F.Warkowski, Structure and application of NNSIM: a general-purpose neural network simulator, Microprocessing and Microprogramming 27, No.1-5 (August 1989) pp.189-194.

[17]   J.A.G.Nijhuis, B.Hfflinger, F.A.vanSchaik and L.Spaanenburg, Limits to the fault-tolerance of a feedforward neural network with learning, Digest 20th Fault-Tolerant Computing Symposium (Newcastle, England, June 1990) pp.228-235.

[18]   R.Opitz, private communication, 1990.

[19]   E.Sorouchyari, Mobile robot navigation: a neural network approach, Digest Journées D'léctronique (Lausannne, Switzerland, October 1989) pp.159-176.

[20]   J.R.Beerhold, M.Jansen and R.Eckmiller, Pulse-processing neural net hardware with selectable topology and adaptibe weights and delays, to be presented at IJCNN'90 (San Diego, July 1990).

[21]   S.Candit and R.Eckmiller, Pulse coding hardware neurons that learn boolean functions, Proceedings IEEE Int. Conference on Neural Networks (IEEE Press, January 1990) pp. II.102-II.105.

[22]   B.Lang, Ein paralleles Transputersystem zur digitalen Bildverarbeitung mit schneller Pipelinekopplung, Proceedings of the 11th DAGM Symposium on Pattern Recognition (Hamburg, October 1989) pp.372-379.

[23]   S.Shams and K.W.Przytula, Mapping of neural networks onto programmable parallel machines, Proceedings of the IEEE Int. Symposium in Circuits and Systems (New-Orleans, May 1990) pp.2613-2617.

[24]   D.Hillis, The Connection Machine (MIT-Press, 1986).

[25]   R.Lerch, Y.Manoli, A.Both, H.G.Hck and T.Neumann, A flexible embedded-core processor system for automotive applications, Proceedings Int. Symposium on Automotive Technology and Automation (Florence, Italy, May 1990) pp.623-628.

[26]   B.Kreimeier, Hardware-independent control software for pulse-processing neural network simulators, Poster presentation at ICNC'90 (Düsseldorf, W.-Germany, March 1990).

[27]   P.Richert, G.Hess, B.Hosticka, M.Kesper and M.Schwarz, Distributed
        processing hardware for realization of artificial neural nets, in:
        R.Eckmiller, G.Hartmann and G.Hauske (Eds.), Parallel Processing in
        neural systems and computers (North-Holland, Amsterdam, March
        1990) pp.311-314.

[28]   D.E.Rumelhart and J.L.McClelland, Parallel Distributed Processing 1
        and 2 (Bradford Books, 1986).

VLSI DESIGN OF AN ASSOCIATIVE MEMORY
BASED ON DISTRIBUTED STORAGE OF INFORMATION

U. RÜCKERT

INTRODUCTION

In the last decade there has been an increasing interest in the use of artificial neural networks in various applications. One of these application areas, where the analysis of the performance of a neural network approach is comparatively advanced, is associative memory. Artificial neural networks (ANNs) are well suited for the implementation of associative memories at least because the processing elements (artificial neurons) in an ANN operate in a highly parallel way and thus a considerable gain in speed is to be expected. This parallelism is one of the major reasons for investigating new computational models inspired by neurophysiological processing principles. The idea is that information is stored in terms of synaptic connectivities between artificial neurons while the activities of the neurons represent the stored patterns. Thus the information is stored distributed over many units and not anywhere in particular. Each unit participates in the encoding of several informations (patterns). ANNs combine both storing and processing information whereas conventional computers are based on sequential processors operating on the contents of a passive memory in which information is accessed by finding the right place in memory.

Many different models have been discussed in literature under such names as "Lernmatrix", "Correlation Matrix", "Associative Memory" etc. [1,2]. A bottleneck in both theory and applications of ANNs seems to be the current lack of suitable parallel hardware exploiting the system inherent parallelism. Thus, development of parallel computing architectures is a fundamental requirement in making some of these models a realistic alternative to more conventional forms of information processing. On the other hand, there are purely technological reasons for studying parallel systems, because they may be the best way to increase the speed and the power of computation in the future. In this respect, ANNs are well adapted for VLSI system design, especially those models using simple processing units and regular interconnection schemes. The highly regular and modular architecture of certain ANN models is an attractive property that circuit designers want to transfer to parallel VLSI hardware. In system design this aspect tends to play a more important part, as Computer Aided Manufactoring enhances VLSI process to higher integration levels and Computer Aided Design leads to easy-to-handle design of smart chips.

154

There are two different approaches for supporting these models on parallel VLSI hardware [3]: *General-Purpose Neurocomputers* for emulating a wide range of neural network models and *Special-Purpose VLSI-Systems* which are dedicated to a specific neural network model. This paper is devoted to a special-purpose hardware implementation of a very simple associative memory based on neural networks. The memory has a simple matrix structure with binary elements (connections, synapses) and performs a pattern mapping or completion of binary input/output vectors. To the authors knowledge, this comparatively simple model of a distributed associative memory was first discussed by Willshaw et. al. in 1969 [4]. However, similar structures have been more generally discussed, e.g. by Kohonen and Palm [1,2]. The characteristics of the implemented model are described in the 2nd section.

The important aspect for VLSI implementation of this simple memory model is the close relationship to conventional memory structures. Hence, it can be densely integrated and large scale memories with several thousands of columns (model neurons) can be realized with current technologies already. Furthermore, the regular topology results in a rigorous modularization of the system indespensible for a successful management of the design and test complexity of VLSI systems. In this respect a pure digital and a hybrid analog/digital VLSI architecture will be described in the 3rd section and discussed in the 4th section.

THE ASSOCIATIVE MEMORY CONCEPT

In general the basic operation of an associative memory is a certain mapping between two finite sets X and Y. In a more abstract sense these two sets may be regarded as questions and answers or stimuli and responses, both coded as vectors of numbers (Figure 1).

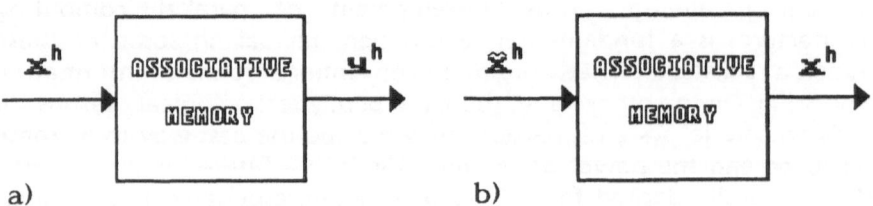

a)   b)

Figure 1   Basic operations of an associative memory:
heteroassociation (a) and autoassociation (b).

The associative memory should respond with y^h to the input x^h for every pair (x^h, y^h) stored in the associative memory. The paired associates can be selected freely, independently of each other. This operation is often called *pattern mapping* or *heteroassociative recall* [1,2]. Further it would be convenient if the associative memory responds with y^h not only to the complete input x^h but also to sufficiently large parts of it. In other words the mapping should be fault-tolerant to incomplete or noisy versions of the input pattern. A special case of this functionality is the *autoassociative memory* where the stored pairs look like (x^h , x^h). Given a sufficiently large part of x^h the memory responds with the whole pattern (*pattern completion*). Besides the discussion of the fuzzy term "sufficiently large" the input can be any part of the stored pattern and it even can be a noisy version of x^h. This operation is called the best match search in terms of pattern recognition.

Among the many different implementations of an associative memory in the field of neural networks the following simplest type is very attractive as well as effective in regard to VLSI implementation. The *Associative Matrix (AM)* is a nxm matrix of binary storage elements w_{ij}, the connection weights (Figure 2). The input vectors x^h as well as the output vectors y^h take a binary form.

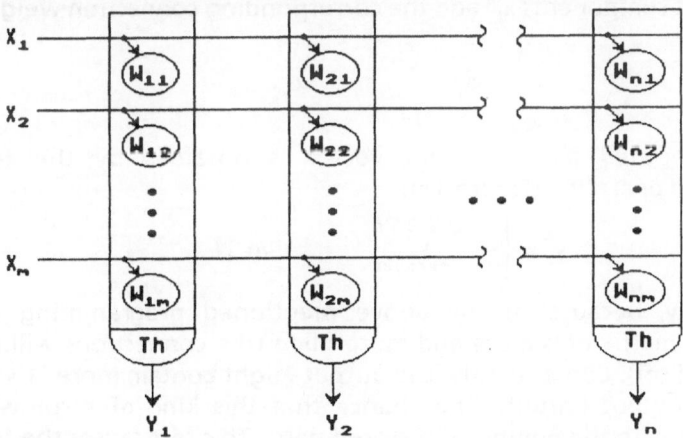

Figure 2   Structure of an Associative Matrix (AM)

The mapping is build up in the following way: The input vector x^h as well as the output vector y^h of every pair which should be stored in the AM (h = 1, .. ,z) are applied to the matrix simultaneously. At the beginning all storage elements in the matrix are zero. Each storage element at the crosspoint of an activated row and column (x^h = y^h = 1) will be switched on, whereas all the other storage elements remain unchanged. This clipped Hebb-like rule [2] programs the connection matrix and the information is stored in a distributed way (Figure 3 a,b):

$$w_{ij}^h = w_{ij}^{h-1} \vee (x_i^h \wedge y_i^h) \ , w_{ij}^0 = 0, \ h = 1..z \ . \tag{1}$$

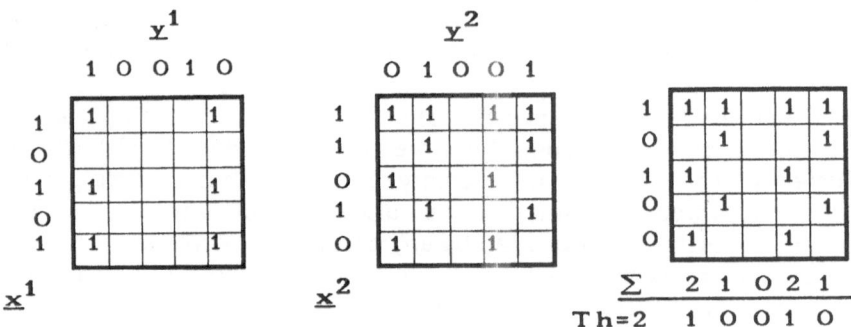

Figure 3: Programming two pattern pairs into an Associative Matrix (a,b) and pattern mapping of an incomplete input pattern (c).

The recall of the constructed mapping is done by applying an input vector to the rows of the matrix. For each column i we sum up the products of the input components x_i^h and the corresponding connection weights w_{ij}:

$$S_i = \sum_{j=1}^{m} x_j^h \cdot w_{ij} \qquad (2)$$

The associated binary output vector is obtained by the following threshold operation (Figure 3 c):

$$y_i^h = \begin{cases} 1, & if\, S_i \geq Th \\ 0, & otherwise \end{cases} \quad , \quad Th \in N \,. \qquad (3)$$

Obviously, because of the above mentioned programming rule the memory matrix gets more and more filled (the connections will never be switched off). Consequently, the output might contain more '1's than the desired output pattern. The chance that this kind of error will occur increases with the number z of stored pairs. This fact causes the following quantitative questions:

i) How many patterns can be stored in an AM?
ii) How many bits of information can be stored in an AM?

Both questions were answered, e.g. by Palm 1980 [5]. Summarizing his results, an AM has its optimal storage capacity C for sparsely coded input/output patterns.

Therefore, the number l(k) of active components ('1') in the input (output) patterns should be logarithmically related to the pattern length (n,m). Asymptotically, the optimal storage capacity for hetero-association is given in [5]:

$$C(m,n,k,l,z) \rightarrow ln\,2 \cdot m \cdot n$$

$$for\; n,m \rightarrow \infty\; and\; parameters: k = log\,n\;\;,\;\; l = log\,m\;\;,\;\; z < ln\,2\frac{m \cdot n}{k \cdot l} \qquad (4)$$

Hence, the storage capacity C is proportional to the number of storage elements n ·m and the number of patterns z that can be stored is much larger than the number of columns (artificial neurons). For example, an optimum of C = 593,000 bits for m, n = 1000, z = 34,780 (k = 2,l = 9) can be stored in the AM under the constraint that on the average 90% of the information of the output vector of each pair is stored [5]. Figure 4 shows the storage capacity of an AM as a function of the number of stored patterns (z) and Table 1 as a function of the number of activated components in the input (l) and output (k) patterns, respectively.

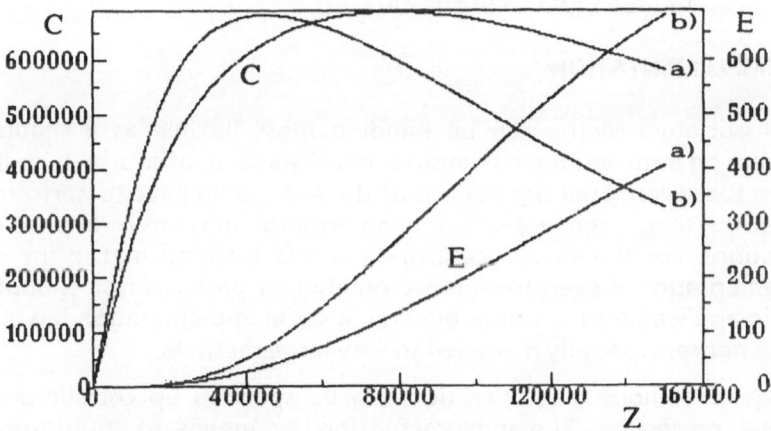

Figure 4   Storage capacity C in bits and the expected number E of additional '1's in the output pattern as a function of the number of stored patterns z (m, n = 1000): a) k = l = 3; b) k = l = 4.

Furthermore, it turns out that the AM works for pattern mapping applications in a more economic way compared to conventional methods (e.g. hashing) and other neural network models, if the number of patterns is large and their individual information content small [6].

These results encourage a hardware implementation in VLSI of this simple associative memory model in situations where such a mapping is a more natural way of storing information than a listing. Especially, because the AM works the more effective the larger the matrix is.

m = n	l	z	C/mn
1024	10	23313	0.565
2048	11	96012	0.591
4096	12	318353	0.604
8192	13	1182378	0.613
16384	14	4407707	0.621

Table 1: Number of patterns z that can be stored with low error probability and the corresponding storage capacity C as a function
of parameters n,m,l (k = 3)

VLSI IMPLEMENTATION

The Associative Matrix can be handled most flexible as a simulation
program on a conventional computer (workstation), of course. It could be
shown that even serial simulations of the AM would have to perform less
operations than a conventional implementation in terms of bitwise mask
operations. For the special case $m < < n$ and a sparse matrix the serial
implementation is even for a large number of patterns fast enough for
certain applications [7]. But in general, a serial implemetation has a poor
performance, especially if applied to very large matrices.

It is quite obvious that operation can be speeded up considerably by
parallel processing. The implementation by means of multiprocessor
architectures (SIMD machines) is a promising compromise between
flexible modelling - the system is still program controlled - and a
complete parallel processing of large matrices. In fact, at least two
research groups have already designed a parallel associative computer
(SIMD) based on a set of conventional microprocessors communicating
via a common bus [8,9].

Consequently, the highest degree of parallelism is achieved by task-
dedicated VLSI systems. It is well in the range of current technologies to
implement an AM effectly on VLSI chips. So far two different special-
purpose VLSI architectures have been designed for an AM at the
University of Dortmund: a digital and a hybrid analog/digital
implementation. Both of them will be discussed in this paper.

The system architecture in both cases is split up vertically into "slices";
each slice manages an equal number of columns. The slices are controlled
by a conventional microprocessor (system control, Figure 5), distributing
input data in an appropriate way to the slices and collecting output data

from the slices. In consequence of the sparsely coded input/output patterns the microprocessor transfers and collects the patterns optimally by means of the addresses of the activated components. Hence, a transfer operation of a m-bit pattern takes only log(m) cycles and address lines in the serial case.

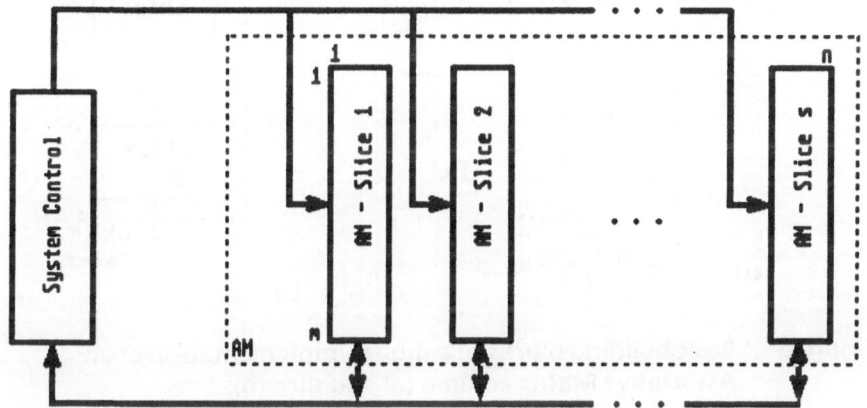

Figure 5   Partition of an m×n Associative Matrix into slices.

Digital Implementation

In the case of digital implementation the columns of an AM are controlled by a special slice chip comprising several very simple processing units (PUs). Each PU controls one column of the matrix and computes bit-serial the weighted sum (2) of the input pattern and the respective column. Because the input/output signals as well as the connection elements are binary, the basic building blocks of a PU are a counter and a comparator (Figure 6a).

The programming algorithm (1) for the connection matrix is realized by a simple OR-logic-Block and is incorporated on the chip, too. The connection matrix can be build up by conventional RAMs (Figure 6b).

In order to transfer the output pattern to the system control the addresses of the '1's in the pattern are generated locally in the slice chips by an additional priority-encoder. All slice chips are connected to a common bus and the access to the bus can be controlled by daisy-chaining or by an additional priority- encoder-logic (Figure 7). In case of the daisy-chaining method the time for transferring the output pattern is proportional to the number of slice chips. In the other case the time is proportional to the number of '1's in the output pattern. In both cases the transfer of the associated output pattern and the calculation of the weighted sum of inputs can be pipelined.

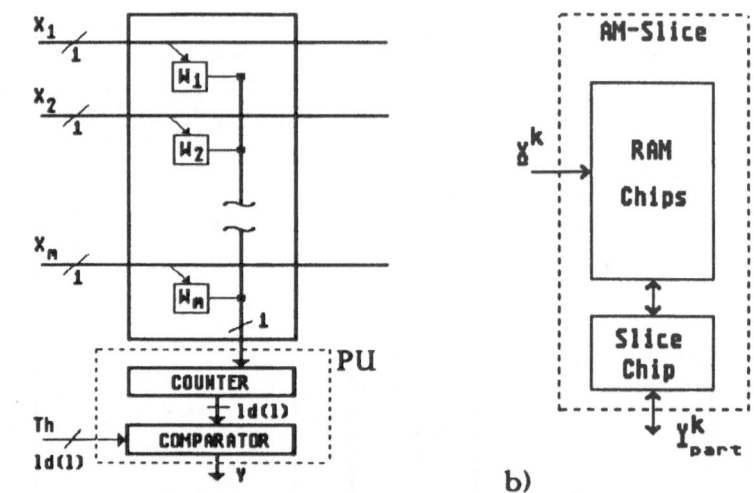

Figure 6   Basic building blocks of a digital implementation of an
Associative Matrix column (a) and slice (b).

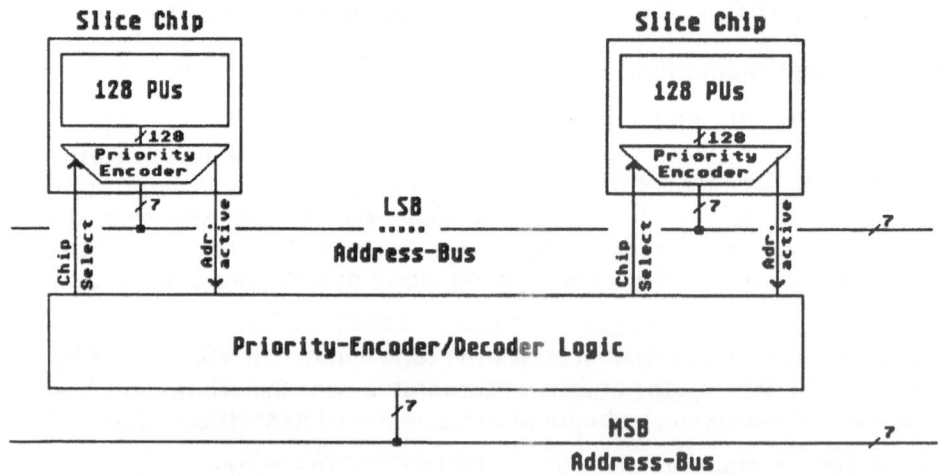

Figure 7   Transfer scheme for the generated addresses of '1's in the
output pattern.

Up to now several standard-cell-designs comprising 32 PUs (e.g. 2μm
CMOS, 38mm^2, ≈ 20,000 transistors, 53 pads, 10MHz) and a full-custom-
design comprising 128 PUs (2μm CMOS, 1cm^2, 48,000 transistors, Figure 8)
of a slice chip have been finished. Instead of realizing a whole chip in
silicon we have first fabricated and tested successfully a single PU of the

full-custom-design at the University of Dortmund (Figure 9). The tested PU is able to perform the calculations for the equations (1) - (3) at least at a clock rate of 12 MHz. Optimization in respect to speed hasn't been done yet. Instead, we have concentrated on a modular and testable design. As

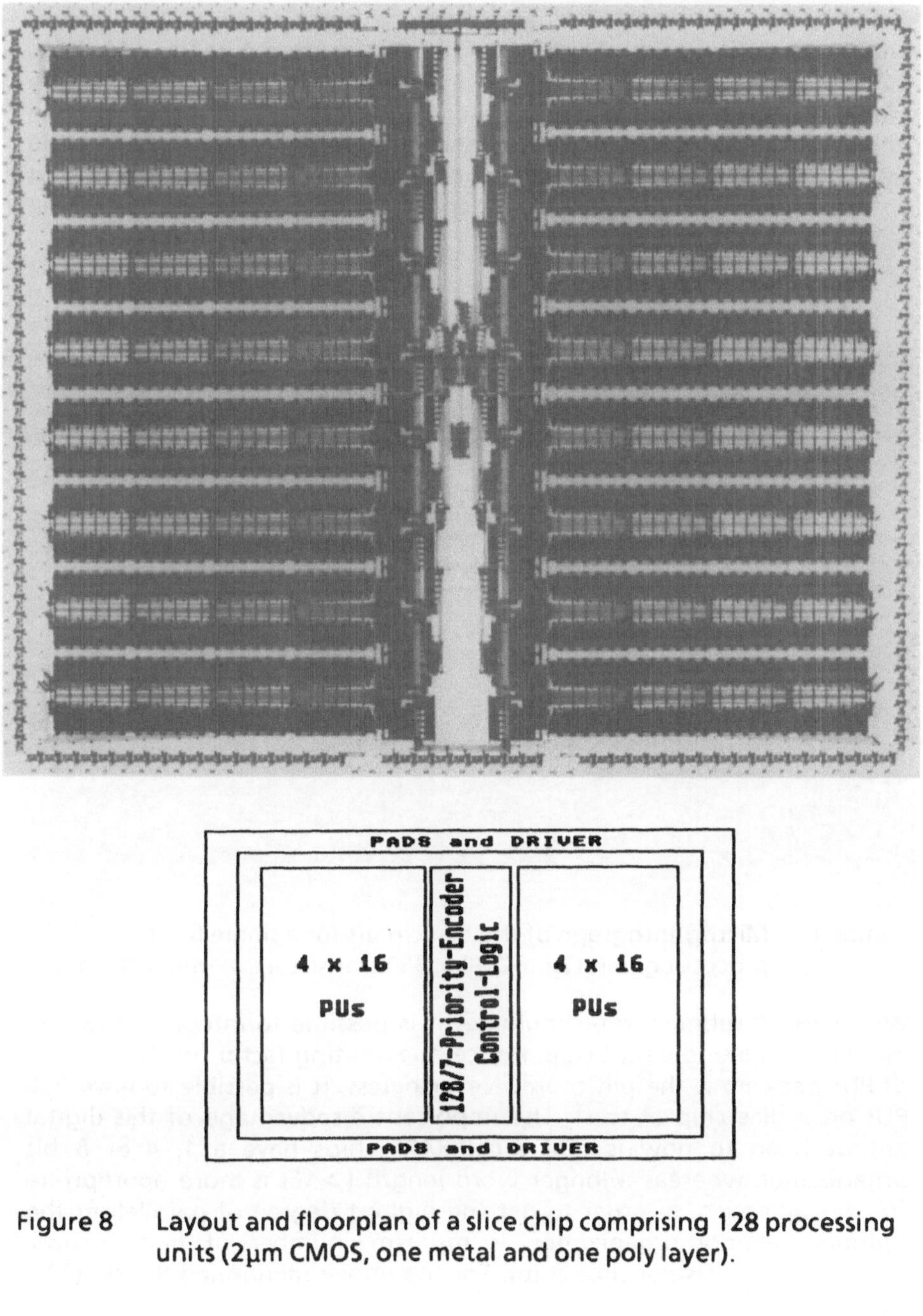

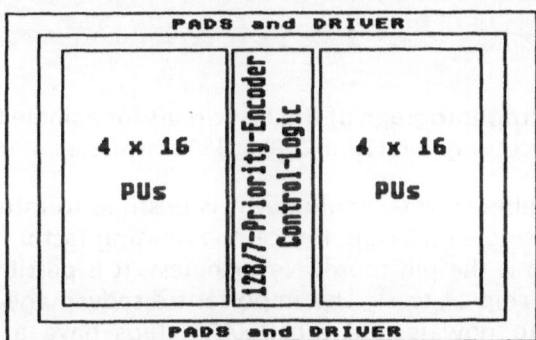

Figure 8   Layout and floorplan of a slice chip comprising 128 processing units (2μm CMOS, one metal and one poly layer).

can be seen from the microphotograph in Figure 9, the PU is built up by six bit-slices. The bit-slices can be configured to a scan-path for test reasons and an extension to a 8 or 16 bit PU is easily achieveable by duplicating the bit-slices.

Based on this facts, a 8192x8192-AM can be build up by a 64 MBit-RAM and 64 slice chips, each comprising 128 PUs, for example. This AM stores more than one million sparsely coded patterns with low error probability (Table 1), which corresponds to a storage capacity of 40 Mbits. Such an implementation performs a pattern mapping within 10µs. The association time is proportional to the number of '1's in the input/output patterns $(\log(m) + \log(n))$, hence independent of the number z of stored pairs.

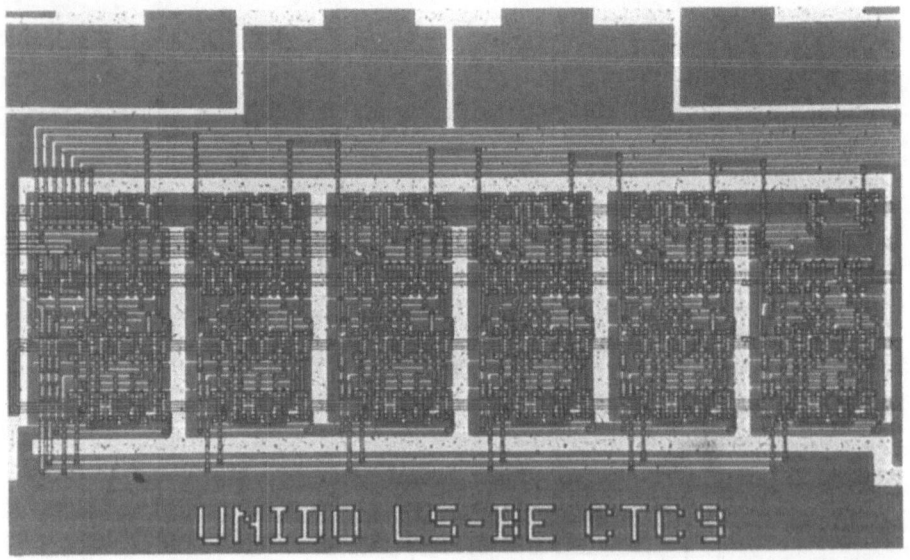

Figure 9   Microphotograph of the test circuit for a single 6-bit processing unit (2µm CMOS, 350 transistors, 1.1mm x 0.4mm).

With current submicron technologies it is possible to integrate 512 and more PUs on a single slice chip. Hence, the limiting factor for the number of PUs per chip is the pin count. Nevertheless, it is possible to have 256 PUs on a slice chip at least. The important disadvantage of this digital approach up to now is that most RAM chips have a 1, 4 or 8 bit organization whereas a longer word length (>16) is more appropriate for this approach. In order to get the highest degree of parallelism the optimal memory organization is m×u (m = number of matrix rows, u = number of PUs per slice chip). For the above mentioned 8kx8k-RAM,

1MBit-RAM chips with a 8kx128 organization are required, for example. Therefore, the system architecture has to be slightly modified in order to make effective use of currently available memory chips. Work on this topic is carried out in the moment.

Digital / Analog Implementation

The largest computational load implementing an AM is incurred by the weighted sum of input signals (2). Using analog circuit techniques [10], this sum can be effectively computed by summing analog currents or charge packets, for example. In Figure 10 a simple circuit concept is proposed in CMOS technology. The matrix operation is calculated by current summing and the threshold operation is done by an analog voltage comparator.

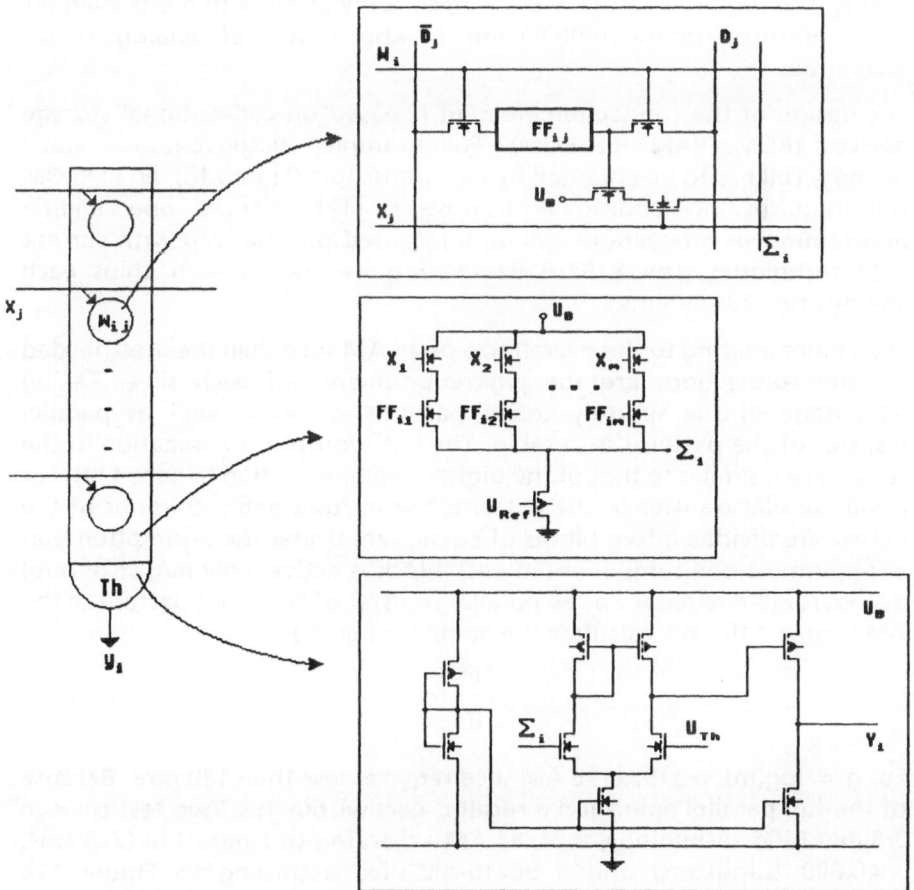

Figure 10   Analog implementation of an Associative Matrix column.

The accuracy of analog circuits is not as high as for digital circuits, but they can be build much more compactly and are more appropriate for the highly parallel signal transfer operations immanent in neural networks. Nonlinearities or parasitic effects of the devices allow us to realize complex functions, as an exponential or square-root function [11]. Note, however, that analog circuits are not so densely integrated as it may seem at first glance. They demand large area transistors to assure an acceptable precision and to provide good matching of functional transistor pairs, as used in current mirrors or differential stages. Furthermore, analog circuits are influenced by device mismatches due to the fabrication process and it is very difficult to control offset voltages, for example. Consequently, analog implementations should be applied to artificial neural networks requiring only modest precision. One example for such a network is the AM because there are only log(m) terms contributing to the weighted sum S_i. The required accuracy of an AM is only about 4 to 5 bits even for large matrices (m,n > 10,000) and in the range of analog circuit techniques.

The design of the connection element is based on conventional storage devices (ROM, RAM, EEPROM). For example, a conventional static memory cell has to be enlarged by two transistors (Figure 10), an EEPROM cell requires no additional transistors [12]. Hence, one million programmable connections can be integrated on one chip with current VLSI techniques. The 8192x8192-AM requires 64 of such chips each comprising 128 columns.

Even more limiting to the overall size of an AM slice than the area needed for the connections are the pin requirements of each slice. Taking advantage of the sparsely coded patterns, serial as well as parallel transfer of the patterns is suitable. The input/output organization in the serial case is similar to that of the digital implementation (Figure 11a). For a full parallel transfer of the patterns the m rows and n columns of the matrix are divided into g blocks of equal size. Under the assumption that at uppermost one component in each block is active, only m/g (1 of m/g) decoders are needed for a full parallel transfer of the input pattern to the AM (Figure 11b). We calculate the number of pins as:

$$p = log_2 \left(\frac{m}{g}\right) \cdot g \qquad (5)$$

For g ≈ log(m), a 8192x128 AM slice requires less than 130 pins. Because of the full parallel operation a recall occurs within 1μs. Two test chips in 2.5μm CMOS technology, a 64x64 AM according to Figure 11a (7x8mm^2, ≈ 40,000 transistors) and a 96x16-AM-slice according to Figure 11b (7x5mm^2, ≈ 20,000 transistors) [10], with programmable connections have been fabricated at the University of Dortmund. With both test chips the above mentioned functionality has been tested and verified. A

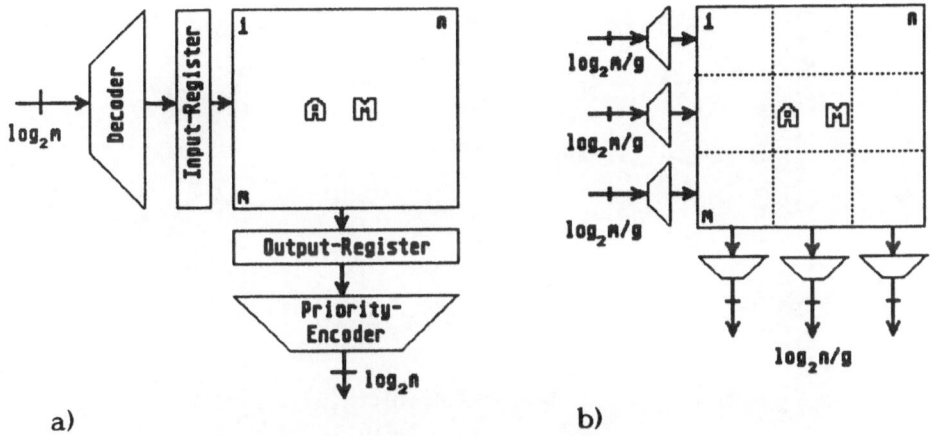

Figure 11   Input/output organization of a semi-parallel (a) and full
parallel (b) analog implementation of an Associative Matrix.

microphotograph of the 64x64 AM test chip in CMOS technology is
shown in Figure 12. In this case, data input is done through a buffer in
order to reduce the pin count, whereas for test reasons the output of
each column is attached to a separate pad. The accuracy of the proposed
analog cuircuits turned out to be at least 4 bit. In other words, sparsely
coded binary patterns with up to 16 activated components can be
handled correcly by this simple implementation.

Because of the modular and regular structure of the architecture, the
implementation of large AMs (n,m > 10000) is feasible. A further
attractive feature of the AM is its fault tolerance to defective
connections. Even in the presence of 5 % defects, up to 20,000 sparsely
coded patterns ($l = 13$, $k = 3$) can be stored in a 1000x1000 AM with low
error probability. Therefore, the AM will also be well adapted for the
envolving wafer-scale-integration technique.

DISCUSSION

Though the AM system concept is comparatively simple, it has very
attractive features in regard to other associative or neural network VLSI
implementations:

--   the asymptotic storage capacity is $0.69 \cdot n \cdot m$ bits

--   the number of sparsely coded patterns that can be stored in an AM is
much larger than the number of columns (artificial neurons)

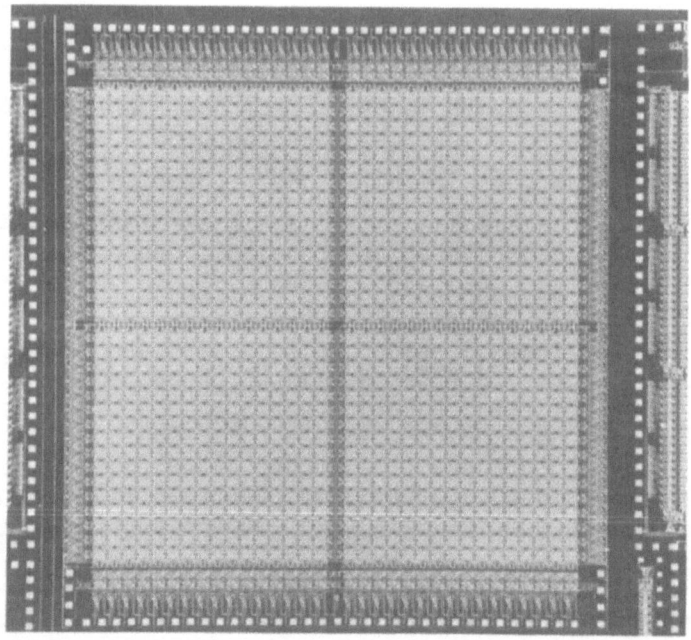

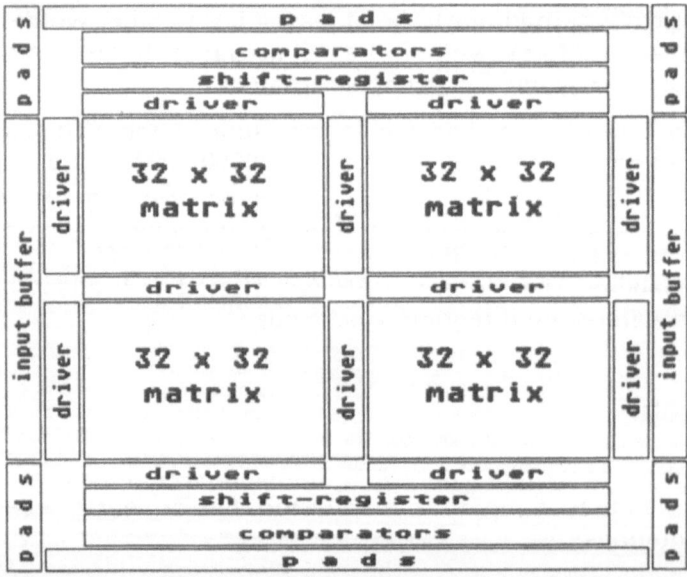

Figure 12   Microphotograph and floorplan of the analog/digital
implementation of the 64x64 Associative Matrix test chip.

-- the number of operations during association is only O(log(n) ·m) instead of O(m ·n)

-- the simpler circuit design requires less silicon area.

Because of the modular and regular structure of the proposed architectures, the implementation of very large AMs (n,m > 10000) is feasible. This aspect is very important for practical applications where the AM has to be extended to a useful number of storage elements. Work on possible applications of an associative memory of this type is done at the moment by differnt research groups, e.g. in the field of speech recognition, scene analysis and information retrieval.

Comparing both VLSI approaches presented above, we can call on efficient software tools for a fast, reliable and even complex digital system design. For the memory matrix we can use standard RAM chips employing the highest density in devices. In general, the matrix dimensions (n,m) can be extended by using additional RAM chips. An important disadvantage up to now is that most RAM chips have a 1 or 4 bit organization whereas a longer word length (> 16) is more appropriate for the digital implementation.

On the contrary, the design of analog circuits demands much more time, good theoretical knowledge about transistor physics and a heuristic experience of layout. Only a few process lines are characterized by analog circuits. The noise immunity and precision is low compared to digital circuits. The fixed matrix dimension in case of a special-purpose implementation is a further disadvantage. In their favour, we point out that analog circuits can be build much more compactly and are more appropriate for the highly parallel signal transfer operations immanent in neural networks. For example, a 1000x1000 AM can be integrated on one chip, whereas the digital concept requires several slice chips and at least one RAM chip. In conclusion, both approaches have their advantages and it remains to be seen which type of implementation will be more effective in certain applications.

REFERENCES

[1] T.Kohonen, "Associative Memory: A System-Theoretical Approach", Springer Verlag, Berlin, 1977.

[2] G.Palm, "Neural Assemblies", Springer Verlag, Berlin, 1982.

[3] P.C.Treleaven, "Neurocomputers", Int. Journal of Neurocomputing, Vol. 89/1, pp. 4-31, 1989.

[4] D.J.Willshaw, O.P. Bunemann, H.C. Longuet-Higgins, "A Non-Holographic Model of Associative Memory", Nature 222, No. 5197, pp. 960-962, 1969.

[5] G.Palm, "On Associative Memory", Biol. Cybernetics 36, pp. 19-31, 1980.

[6] G.Palm, "On Associative Memories", in: Physics of Cognitive Processes, E.R. Caianiello, Ed., World Science, pp. 380-420, 1986.

[7] H.J.Bentz, M. Hagstroem, G. Palm, "Information Storageand Effective Data Retrieval in Sparse Matrices", Neural Networks, Vol. 2, No. 4, pp. 289-293, 1989.

[8] G.Palm, T.Bonhoeffer, "Parallel Processing forAssociative and Neural Networks", Biol. Cybern. 51,pp. 201-204, 1984.

[9] U.Rueckert, J.Moschner, "A SIMD Architecture for Parallel Simulation of Associative Networks", to be published, 1990.

[10] K. Goser, U. Hilleringmann, U. Rueckert, K. Schumacher, "VLSI Technologies For Artificial Neural Networks", IEEE Micro, Dec. 1989, pp. 28-44, 1989.

[11] C. Mead, "Analog VLSI and Neural Systems",Addison-Wesley, 1989.

[12] U. Rueckert, I. Kreuzer, K. Goser, "A VLSI Concept for an Adaptive Associative Matrix based on Neural Networks", Proc. COMPEURO Hamburg, May 1987, pp. 31-34, 1987.

SILICON INTEGRATION OF LEARNING ALGORITHMS AND OTHER AUTO-ADAPTIVE PROPERTIES IN A DIGITAL FEEDBACK NEURAL NETWORK

P.Y.ALLA, G.DREYFUS, J.D.GASCUEL, A.JOHANNET,
L.PERSONNAZ, J.ROMAN, M.WEINFELD

INTRODUCTION

In the past few years, a lot of efforts has been devoted to the integration of neural networks [1,2], with much emphasis on the implementation of the network itself, leaving the burden of learning to a host, possibly parallel computer [3]. However, the idea of implementing training on the chip itself is attractive for two reasons : (i) the learning phase is usually very time-consuming ; (ii) on-chip learning makes the network more autonomous and opens the way to building elaborate assemblies of networks. The present paper discusses the capabilities of a neural network chip, fully connected with feedback, using binary neurons with parallel synchronous dynamics, intended to be used as an associative memory [4] ; the chip integrates a learning algorithm and also some additional, potentially useful features such as self identification of correct relaxation on a stored vector (a prototype), and to the discussion of the main silicon implementation issues.

The main architectural and technological choices

The first integrated neural networks were based on analog technology, with the advantage of speed and compactness, but with severe limitations in terms of programmability and precision of the synaptic coefficients, as well as some sensitivity to crosstalk and spurious noise [5-14]. Nowadays, many innovative efforts are dedicated to various techniques to overcome these deficiencies, and the issue of analog versus digital technology is far from being settled. Nevertheless, digital circuitry has theoretically none of the aforementioned drawbacks: it allows the implementation of any function (for network relaxation as well as for in-circuit learning), with arbitrary precision, is strongly immune to noise, and has no specific interfacing problems. The real difficulties arise from the connectivity issues and the larger number of transistors per function (memory or arithmetic units), leading to rather large silicon areas. Even though less numerous, there are or have been several attempts to design digital networks [15-20].

A solution to the connectivity problem lies in trading off parallelism with computation time, using some kind of multiplexing scheme. We have chosen a linear systolic architecture [21,22], in which each neuron has its own synaptic coefficients stored locally, in a circular shift register (see figure 1). This choice limits and simplifies the data lines connecting each neuron to the rest of the network, but leads to a large area for each neuron, which is mainly devoted to memory. Taking into account the possibilities of present technologies, we decided to focus our efforts towards a network of 64 fully-connected binary neurons with feedback, with a chip surface of about one square centimeter, which is conservative enough to give reasonable yields. Designing such a circuit consists mainly in designing the elementary neuron, which can be duplicated as a macrocell with automatic connection of the signal ports.

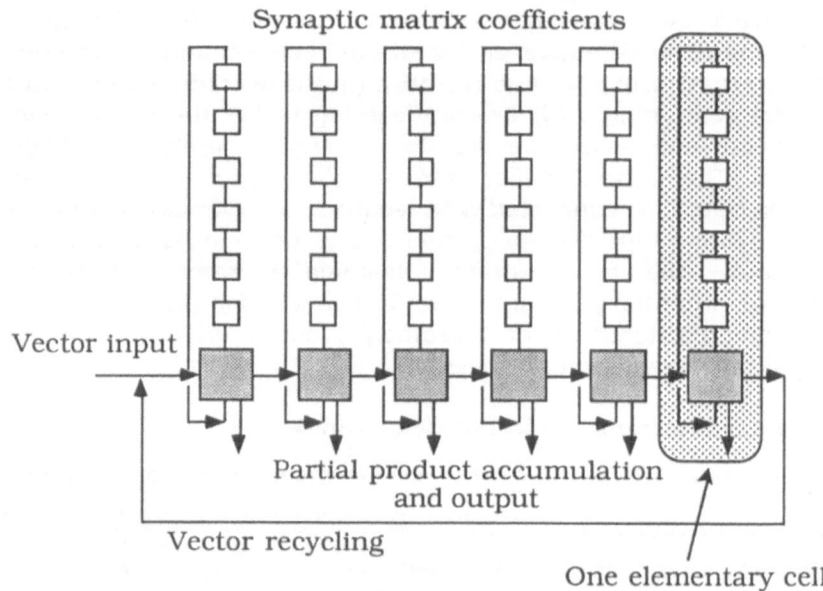

Figure 1

One of the advantages of digital architecture lies in the fact that training can be integrated on the chip, at no great expense in terms of additional complexity or silicon area, provided an appropriate learning rule is used. As we shall see later, other features can be rather easily added to the basic architecture, providing supplementary possibilities in view of multi-network architecture implementation.

In addition, it should be noticed that the same architecture, with minor alterations, can be used to implement multilayer networks of binary or multivalued neurons [17], which are becoming increasingly popular with the emergence of constructive methods allowing a rational use of such networks for classification [23].

Towards multi-network architectures : some additional properties

A common approach to applications of neural networks has been the use of one single network to perform the whole task. It can be argued that a hierarchical approach might be more efficient : therefore, one should try to find ways of assembling relatively small networks, strongly connected, into higher-level architectures, with sparser connections, as biology would suggest it. However, this strategy requires that the basic building blocks have the ability of making decisions about information processing and routing autonomously, including learning, thereby alleviating the need for control from an external superviser [24]. Any external supervision, downloading synaptic coefficients or assessing the results of a relaxation, for instance, may induce excessive overhead and impair drastically the interest of these specific integrated components.

Implementing a learning rule

The choice of a learning rule depends strongly on the task that the network is supposed to perform, and on the architecture of the network. In the present case, we want to take advantage of the storage and retrieval properties of a Hopfield network of binary neurons. Therefore, we need a learning rule which has the ability of storing binary patterns as attractors of the dynamics of the network. In addition, the learning rule should be easily integrable, which means that it has to be local (contained in each neuron), not needing additional signal routing or global external calculations to be broadcast to each cell. [25]

The projection rule [26] was shown to allow the storage and retrieval of any set of prototype patterns, but its direct implementation on an integrated network is very difficult. It has been proved [27] that the Widrow-Hoff iterative learning rule can be used to train a fully connected network intended for use as an associative memory : if the number of prototypes is smaller than the number of neurons, the synaptic matrix obtained by the Widrow-Hoff rule is identical to the projection matrix. In addition, the procedure is iterative and local : the computation of the variation of the weight of the connection between neurons i and j, in neuron i, requires a multibit information (the potential) which is locally available, in addition to the state of neuron j. Thus, it is a good candidate for integration on the same chip as the network itself.

Since the use of standard floating-point arithmetic in each neuron is ruled out because of its excessive complexity, the main issues in the integration of the training algorithm on the network itself are (i) the accuracy of the synaptic coefficients used in the retrieval phase, and (ii) the accuracy required for the computation of the coefficients. The first issue is related to the fault tolerance of the network : Hopfield-type networks exhibit some degree of fault tolerance, in the sense that the retrieval properties of the network degrade gracefully if the synaptic coefficients are expressed on a small number of bits [28]. Thus, one can capitalize on this property to use integer arithmetic with limited accuracy. To the best of our knowledge, the second issue, namely the precision required for the training phase itself, has never been investigated in depth.

We first recall the Widrow-Hoff rule :

$$\Delta C_{ij} = \frac{1}{N} \left(\sigma_i^k - v_i^k \right) \sigma_j^k$$

where C_{ij} is the coefficient of the synapse transmitting information from neuron j to neuron i, N is the number of neurons, σ_i^k is the i-th component of prototype number k, and v_i^k is the potential of neuron i when the state of the network is the k-th prototype ; v_i^k is given by :

$$v_i^k = \sum_{j=1}^{N} C_{ij} \sigma_j^k .$$

It is easy to notice that this rule is local, since each neuron only needs the knowledge of the states of the others, coded on one bit, as would be the case with Hebb's rule.

The matrix obtained by the Widrow-Hoff rule converges to the projection matrix provided it is initialized as the zero matrix.

Speed of convergence of the learning procedure

The first issue to be discussed is the speed of convergence of the iterative rule. When the prototypes are orthogonal, the Widrow-Hoff procedure reduces to Hebb's rule, and it converges after a single presentation of the prototypes. Thus, it may be conjectured that the speed of convergence will decrease with increasing correlations between prototypes, and with a larger number of neurons. Indeed, the number of presentations of the prototypes set increases quadratically in the range of N investigated. Figure 2 shows the number of presentations of the training set required in order to have

$$|1 - \sigma_i^k v_i^k| < \frac{1}{N} \quad \forall i, \forall k$$

as a function of the number of neurons. Clearly, the number of presentations for a small number of neurons is manageable. For random uncorrelated patterns, the number of presentations is roughly independent of the number of neurons. For random correlated patterns, the number of presentations increases quadratically in the range investigated. The graphs shown on Figure 2 were obtained for a ratio a = p/N (where p is the number of prototypes) equal to 0.25, and for average overlaps of 0 (uncorrelated patterns) and 0.16 (correlated patterns).

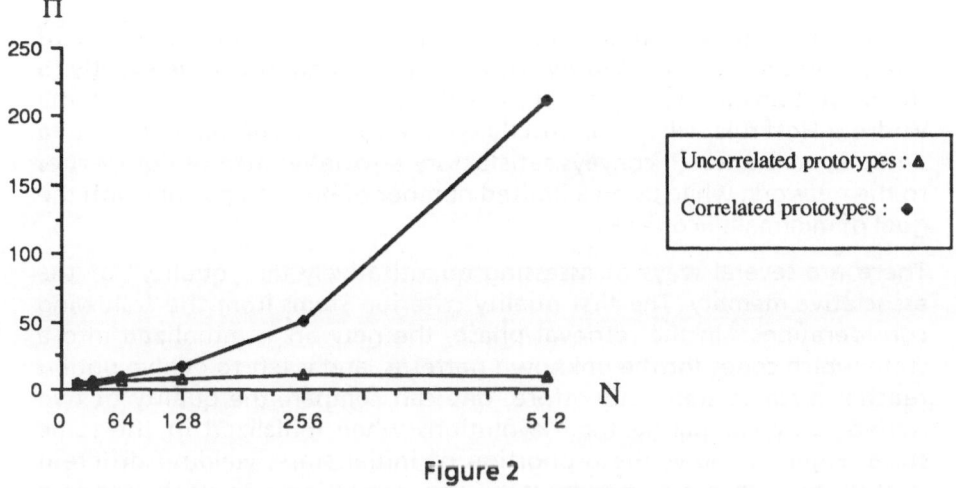

Figure 2

The accuracy required in fixed point arithmetics

We now turn to the influence of the computation with limited accuracy on the learning phase. We assume that the synaptic weights are coded on b = log2M + 1 bits (including the sign bit), and, for simplicity, we assume that M is a multiple of N. Thus, the Widrow-Hoff rule can be written as :

$$\Delta J_{ij}^k = \frac{1}{N} \left[M\sigma_i^k - \sum_{m=1}^{N} J_{im}^{k-1} \sigma_m^k \right] \sigma_j^k \ ,$$

where $J_{ij} = M\,C_{ij}$.

The first term within brackets is an integer. The summation itself is also an integer, but the division by N introduces a truncation error. Thus, the Widrow-Hoff procedure with integer arithmetic can be expressed as :

$$\Delta J_{ij}^k = \frac{1}{N} \overline{\left[M\sigma_i^k - \sum_{m=1}^{N} J_{im}^{k-1} \sigma_m^k \right]} \sigma_j^k$$

where the bar over an expression means the integer part of it.

We define the integer potential as :

$$u_i = \frac{1}{N} \sum_{j=1}^{N} \overline{J_{ij} \sigma_j} \quad .$$

After convergence of the algorithm, all ΔJ_{ij} are equal to zero, and all potentials are equal to $\pm M/N$.

Given the discretization and truncation errors introduced by the use of integer arithmetics, the Widrow-Hoff matrix will not converge exactly to the projection matrix, in general ; specifically, the matrix obtained by the Widrow-Hoff rule will not be strictly symmetrical. The problem is to get a synaptic matrix which conveys satisfactory associative memory properties to the network, while using a limited number of bits, compatible with the goal of minimal silicon area.

There are several ways of assessing quantitatively the "quality" of the associative memory. The first quality criterion stems from the following considerations : in the retrieval phase, the network is initialized into a state which codes for the unknown patterns, and is left to evolve until it reaches a stable state ; therefore, one can compare the quality of two networks by comparing their evolutions when initialized in the same state. Figure 3 shows the proportion of initial states yielding different evolutions with a network built by the projection rule with standard floating-point accuracy, and for a network built by the Widrow-Hoff rule, with coefficients coded on b bits. Each point is an average on 10000 random states and 20 sets of prototypes.

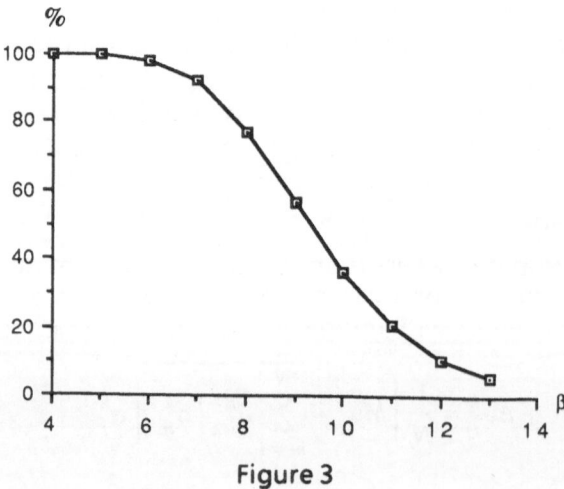

Figure 3

The diagram shows that a coding on 13 bits gives less than 10 % difference between the behaviour of a network of 64 neurons, with a synaptic matrix computed by the projection rule, and the behaviour of the network with a synaptic matrix computed by the Widrow-Hoff rule. (16 uncorrelated prototypes).

It can be argued that the really important issue is not the behavior of the network when initialized in any random state, since the relevant parts of state space are the basins of attraction of the prototypes. Thus, one can also assess the quality of a learning rule, for the task of auto-association, by investigating the size of the basins of attraction of the stored prototypes. In order to determine this size, the following procedure is used : after completion of the learning phase, the state of the network is initialized at a Hamming distance H_i of one of the prototypes, and the network is left to evolve until it reaches a stable sate which is at a Hamming distance H_f of that prototype. A distance H_f equal to zero corresponds to a perfect retrieval. Figure 4 shows the results obtained with the projection rule, on a network of 64 neurons ; 16 prototypes, with equally probable random components + 1 or -1, were stored ; the histograms show the distribution of the final normalized distances $h_f = H_f/N$, for two values of the normalized initial Hamming distance $h_i = H_i/N$, equal to 0.125 and 0.25.

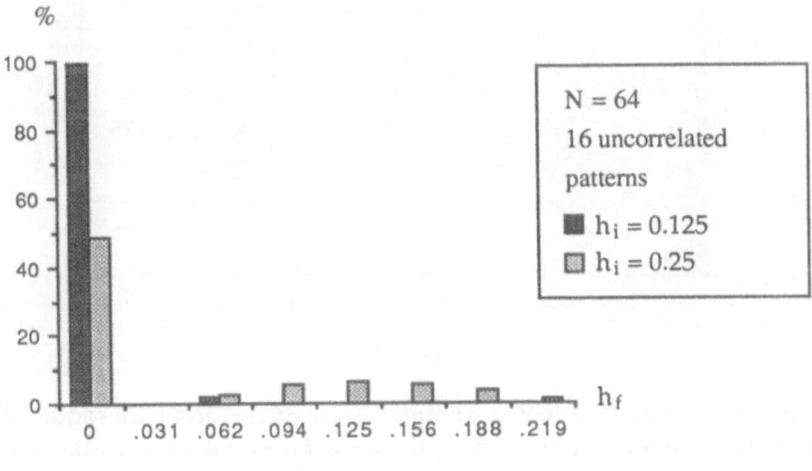

Figure 4

Figure 5 shows the same results obtained with the Widrow-Hoff procedure on integers with b = 7 bits and b = 9 bits respectively. Clearly, the auto-associative properties of the synaptic matrices coded on 9 bits are quite similar to those obtained when making use of the projection rule with floating-point accuracy. Although, as stated above, the synaptic matrix is not exactly symmetrical, no cycles have been observed with this choice of parameters.

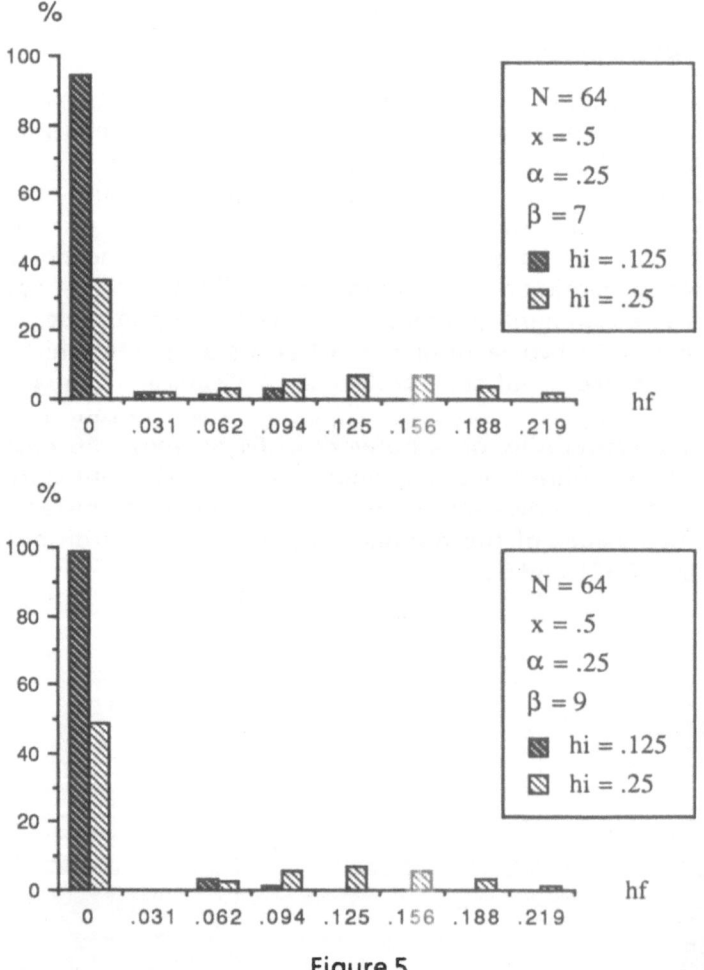

Figure 5

The present study is the first systematic study of the accuracy required to express the synaptic coefficients, for a given network architecture, a given learning rule and a given task. Much effort is still required in order to get a general framework for tackling such problems, which will be of central importance if artificial neural networks ever come into actual industrial use. We now turn to other important implementation issues.

Self identification of successful retrieval

In the context of auto-associative memory applications, it seems desirable to give to the network the ability of signalling by itself whether it has or not succeeded in recognizing a pattern, since a network gives always an output (whether it is a feedforward or a feedback network) for every input stimulus. This intrinsic property of neural networks in general, whatever their architecture, is nearly always overlooked, but it seems clear that it must be taken into account when real applications must be implemented.

We have devised a mechanism which allows the automatic recognition of "good" convergence, producing a binary signal indicating whether the final state reached is a prototype or a spurious state. The vectors presented for storing by the network are divided into two fields: the longer contains the prototype itself, and the other one, called the label, contains the result of the coding of the first through a cyclic error correcting code [29]. This coding uses the primitive polynomial $x^6 + x + 1$, which is well adapted to hardware generation by linear feedback shift registers, in good agreement with the general architecture of the network. The whole vector (prototype concatenated to its code) is stored. In the retrieval phase, the attractor on which the network has relaxed is submitted to the same coding: if the information-carrying field is coherent with the label-carrying field, there is a strong probability that this attractor corresponds to a prototype, and vice-versa. Figure 6 shows schematically how this mechanism operates.

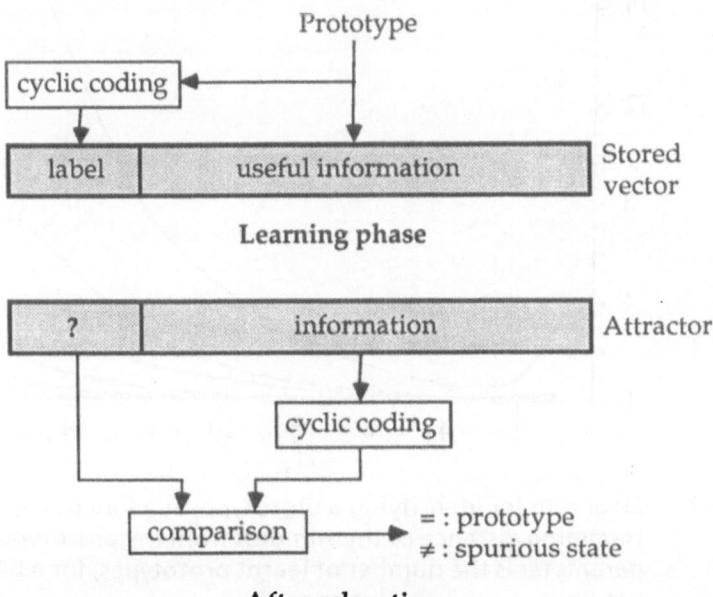

Figure 6   Schematic diagram of attractor labeling, during the learning phase, and after retrieval

Simulations with random uncorrelated or partially correlated prototypes have shown that the probability of correct identification increases with the length of the label, and that a six-bit label is sufficient to give a good rate of success with a network size of 64 cells. The following diagram (figure 7) illustrates the efficiency of the labeling scheme for identifying prototype convergence, as a function of the number of stored prototypes. As can be expected, the failure rate of the labeling scheme increases both with the initial distance from the studied attractor and with the learning ratio of the network. However, staying within realistic conditions, that is to say an initial distance such that the initial stimulus lies in the basin of attraction of the prototype, the diagram shows that an identification accuracy of about 98% or more can be guaranteed. In addition, simulations showed that most spurious states, wich are known to be thresholded linear combinations of the prototypes, are well identified as such. A similar approach, using a less elaborate coding, was suggested earlier for classification tasks using similar networks [30].

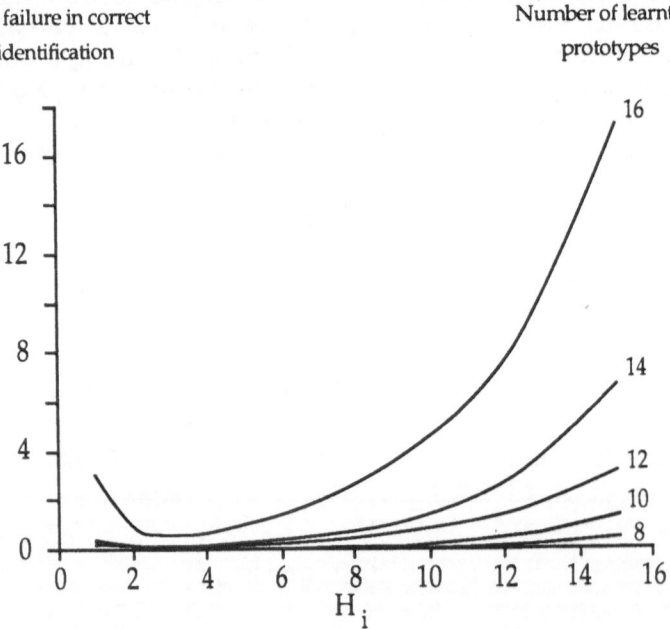

Figure 7   Error rate for identifying a prototype, as a function of initial Hamming distance of the stimulus from the prototype. The parameter is the number of learnt prototypes, for a 64 cells network

A simple but efficient pseudo-annealing mechanism

Another interesting feature would be the use of a probabilistic rule for the neuron decision. It is known that this improves the attractivity of the stored prototypes, but it is also known that it slows down the relaxation of the network [31]. Implementing a real probabilistic threshold function inside each neuron would have added an unacceptable complexity. We chose to implement a very crude mechanism, consisting in flipping a fixed number of neuron states (this number being the remote equivalent of a temperature parameter), the flipped neurons being drawn at random at each updating cycle of the network. The particular random generator required must be able to place a fixed number of bits at random locations in a 64-bit wide vector. It works during the updating cycle of the network, generating a bit-string which is serially xored with the network state vector. This simple mechanism is easy to implement into digital circuitry, and represents only a slight additional overhead to the basic architecture.

Although the added noise is independent of the state of the neuron, simulations have shown that it does lead to a noticeable improvement in prototype attractivity, as shown later, but the number of perturbed neurons must stay low enough. The decision of annealing must be taken only in case of unsuccessful "cold"relaxation of the network, since this situation provides the fastest convergence, and should be tried first. For this purpose, the self-identification mechanism is very useful, since it can be used without supervision to trigger this annealing.

We have made several simulations with various values of a (8, 16 or 24 prototypes for 64 neurons), and measured the percentage of convergence on a prototype, starting from various Hamming distances from this prototype. This is a measure of the prototype attractivity. The experiments have been performed first on cold networks, with the self-identification being used to trigger annealed retries (no more than three) in case of convergence on a spurious attractor. The annealing consisted in flipping randomly 2 or 4 neuron states at each parallel iteration of the network. The results are shown on the following table. The figures are the overall success rate, in percent, of exact retrieval of the prototype (this is a rather stringent criterion), under various learning conditions, and as a function of the pseudo-temperature. It is easy to see that the pseudo-annealing, however coarse, can enhance the performances significantly, two bits being enough in almost all cases, and even a maximum for heavily loaded networks (high values of a). Increasing the number of perturbed neuron states is apt to lead the state of the network out of the basin of attraction of the prototype.

Using it all together

Even with a network alone, a certain amount of auto-adaptivity is made possible by the self-identification property. For instance, some pattern

p	H_i	t = 0	t = 2	t = 4
8	16	78.8	93.4	94.2
8	20	36.5	63.8	65.0
16	10	76.3	88.9	89.0
16	14	28.2	53.9	52.6
24	6	38.3	60.6	56.3

Table 1   Effect of pseudo-annealing on the overall retrieval success. p and H_i are respectively the number of learnt prototypes and the initial Hamming distance, and t indicates the number of flipped neurons. The results are the percentage of exact convergence on a prototype

that is not recognized can be automatically stored for later learning. This property may open the way to associations of networks. For instance, if several networks are wired in parallel (i.e., have the same inputs), each one having learnt various representations of a given symbol (for example the same letter with various positions and shapes), the whole device may operate as a classifier, the output of which is using only the self-identification bit of one network to signal the recognized letter; the particular prototype on which the network has converged being of no interest, is not used as an output. One potential advantage is that learning or relearning is local to each network, and does not involve the whole classifier. Another advantage would be that it might be possible to deal with some invariance difficulties (translation or size), not having to be treated by the learning rules themselves. Simulations are underway on a Connection Machine to assess the properties of various multiple network couplings.

The silicon implementation

Figure 8 shows the fundamentals of the network architecture. We have made conservative choices, using a state of the art industrial technology, namely two-metal 1.2μm CMOS. The main efforts have been devoted to the design of the synaptic memory, since it represents the main part of the neuron. For a low number of stored words (64 times 9 bits in each neuron memory), using a standard RAM would have been wasteful because of the relatively important amount of ancillary circuitry (read-write, addressing, etc...). Since the synaptic coefficients are always stored or read in a serial fashion, we designed a custom 9 bit wide circular shift

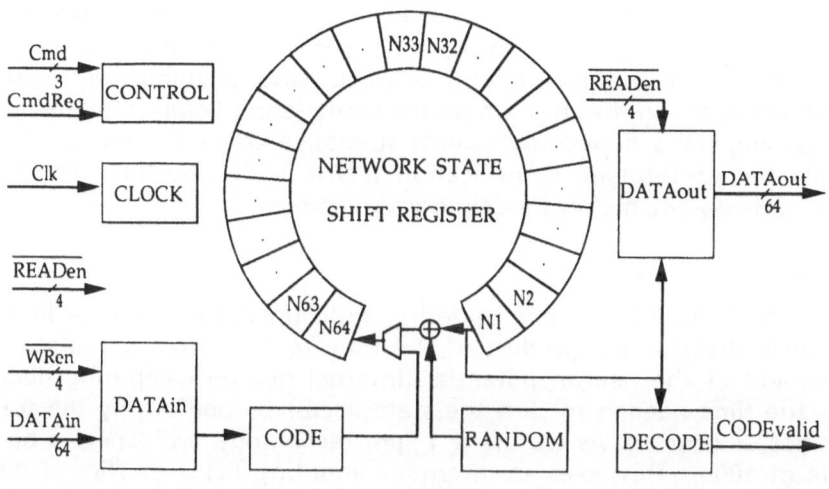

Figure 8

register, with a power and area lower than that of conventional silicon-compiler generated RAM, using semi-static tristate circuitry. On the other hand, the other elements of the neuron (control, arithmetic and logic unit) have been designed using a normal standard cell approach, which simplifies the design, and allows parametrization to some extent. It uses fixed-point arithmetic, twelve-bit wide including the sign bit. Provision is made to take into account over- or underflows, which are transformed into saturations, so as to avoid errors.

Convergence detection is rather straightforward. During the relaxation phase, each neuron stores its previous state in a one bit local register, which is xored with the new state bit at the end of an updating cycle. The local results in each neuron are just combinatorially ored serially, giving a low true signal when the whole network is stable. Thus, an extra updating cycle is needed, in which no updating actually takes place, just for asserting the convergence. During the learning phase, since each neuron calculates a synaptic coefficient increment, the six most significant bits are ored to give a local convergence signal (null increment), which is in turn ored serially through all neurons, the same way as in the relaxation stage. Since a subsequent division trims the lower bits, there is no point in making use of these less significant bits for convergence detection.

Input/output operations are performed in parallel, the necessary serialization of the data taking place inside the chip. It is possible, for use with 16 bits buses such as the MC68000's, to multiplex the 64 bits in four 16 bits blocks : this is the reason why the WRen (write enable) and the READen (read enable) are four-bit wide, each bit enabling the corresponding block while the three others are kept at high impedance. In addition to these I/O control signals, the main external control signals are three command bits (Cmd), a clock, and a command request (CmdReq) entering the chip, a network ready (end of relaxation), an end of learning (all synaptic coefficients stable), a command accept, and finally the prototype identification (CODEvalid), one bit each, plus some state signals, mainly for debugging purposes.

Performances

The overall duration of one updating cycle (in the learning as in the retrieval phase) is the product of the time required to accumulate an increment of the neuron potential (internal neuron computing delay) plus the time needed to shift the state vector by one bit, by the total number of neurons. As we use a 1.2μm technology, we expect a basic cycle of 100ns. This gives an internal computing delay of 75ns, a shift delay (including various housekeeping tasks) of 25ns ; thus, for a complete updating cycle we expect a duration of 6.4 ms, as already shown by functional simulations. In the retrieval phase, if we suppose that for instance four cycles are needed, counting the extra cycle during which convergence is assessed, pattern recognition is performed in approximately 30ms. In the learning phase, the speed is lower for two reasons. The first reason is that the calculation of the coefficients increments needs two cycles : one for the calculation of the potentials, and the second for the calculation of the increments themselves. The second reason is that one has to reach the convergence of the coefficients to stable values corresponding to the exact learning rule : the numbers of iterations needed is roughly proportional to the square of the number of prototypes, as seen earlier. Thus, in the same conditions as above, learning 15 prototypes may take approximately 5 ms, which is probably an absolute maximum, corresponding to a rather high learning ratio. Expressed in the so-called exotic unit of "connections updates per second" (or "cups"), this is equivalent to 2.108 cups in our circuit. Whatever unit is used, the overall speed of the circuit is certainly high enough for research purposes, and probably also for most real-life applications.

Using the networks in a real environment

We are currently designing a testbed for the chip, and also for forthcoming multichip structures. It is based on a special prototyping card called MCP", supplied by Apple Computer for interfacing in a Macintosh II NuBus. This card includes a real-time multitasking system using a local MC68000 processor, 1 Mbyte of RAM and auxiliary circuitry. This setup provides a high level interface between any particular proprietary VLSI chip sitting on the card and the host computer operating system. We intend to use the well known ergonomy of applications running in the

Macintosh to drive the network, that is to say supplying global control signals and data, and asynchronously reading the results from the network. The only function of the host computer is that of an intelligent input/output device for the chip, with no "neural computations" to perform. Several prototyping cards (up to 5) may be inserted in the bus slots, and can work in parallel, allowing a flexible multi-chip architecture. In order to implement larger multi-network architectures, we intend to design a whole external card, able to host more chips, with mixed hard-soft reconfiguration capabilities.

Future integrated networks with many more neurons : will this be possible ?

Sixty four neurons are not enough to cope with some real-life problems, such as image processing. Networks four or even sixteen times larger would be probably more interesting in the future. A 256 cells network is clearly out of reach at present times if we want to keep the same architecture, the same technology, and still design the network in one chip. The area scales roughly like the square of the numbers of neurons, because the synaptic coefficients must be coded on a greater width (two more bits), and the arithmetic unit must also be made wider acordingly. We can design cascadable chips, each containing 16 neurons arranged in an open loop, but each with a 256 word 11-bit wide memory, occupying approximately the same area as our present 64-neuron chip. Sixteen such chips could be arranged using "silicon hybridation" in a four by four array using a surface of about 20cm^2, which is manageable. The main drawback is the following : since the updating cycle is longer, all the characteristic times (learning as well as relaxation) are increased proportionally to the number of neurons. Apart from finding more efficient architectures, if they exist, the only solution that we see at present is to use smaller technologies, thus increasing the clock rates. Roughly, if the technology is shrinked by a quantity k (k > 1), the clock rate should be increased by this factor. For a 0.6 µm process, the loss in speed would then be only by a factor of two for the 256 cell network with respect to the 64 cell network, with the important advantage of approximately dividing the silicon area by four, possibly allowing a monolithic implementation. Of course, some important issues such as the connection delays between chips have not yet been addressed in detail. Finally, it is not unlikely that a 1024 cell network could be designed, having an acceptable speed, using technologies below 0.5 µm and multichip silicon hybridation, by the mid of the decade.

184

REFERENCES

[1]   L.Personnaz, A.Johannet, G.Dreyfus : Problems and trends in integrated neural networks; in Connectionism in Perspective, R. Pfeifer, Z. Schreter, F. Fogelman-Soulié, eds (Elsevier, 1989).

[2]   M.Weinfeld : Neural networks as specialized integrated circuits : an academic exercise or the promise of new machines ? ; International Conference "Neural networks : biological computers or electronic brains", Lyon, 6-8 march 1990.

[3]   A.Petrowski, L.Personnaz, G.Dreyfus, C.Girault : Parallel implementations of neural network simulations; in Hypercube and Parallel Computers, A. Verjus, F. André, eds. (North Holland, 1989).

[4]   J. J.Hopfield: Neural networks and physical systems with emergent collective computational abilities; Proc. Natl. Acad. Sci. USA, vol. 79, pp. 2554-2558, 1982

[5]   J.Alspector, R.B.Allen, V.Hu, S.Satyanarayana : Stochastic learning networks and their electronic implementation. IEEE Conf. on Neural Information Processing Systems, Natural and Synthetic, Denver (USA), 1987; in: Neural Information Processing Systems, Natural and Synthetic, D.Z.Anderson, ed., (American Institute of Physics, 1988).

[6]   M.Holler, S.Tam, H.Castro, R.Benson: An electrically trainable artificial neural network with 10240 "floating gate" synapses; Proceedings of International Joint Conference on Neural Networks, Washington D.C., June 1989

[7]   A.Moopenn, H.Langenbacher, A.P.Thakoor, S.K.Khanna: A programmable binary synaptic matrix chip for electronic neural networks. IEEE Conf. on Neural Information Processing Systems, Natural and Synthetic, Denver (USA), 1987; in: Neural Information Processing Systems, Natural and Synthetic, D.Z.Anderson, ed., (American Institute of Physics, 1988).

[8]   M.Sivilotti,M.R.Emerling, C.Mead: A novel associative memory implemented using collective computation; in: Proc. Chapel Hill Conf. on VLSI, pp. 329-342, (Computer Science Press 1985).

[9]   S.Tam, M.Holler, G.Canepa: Neural networks synaptic connections using floating gate non-volatile elements; Neural Networks for Computing, Snowbird, 1988.

[10]   D.B.Schwartz, R.E.Howard, J.S.Denker, R.W.Epworth, H.P.Graf, W.Hubbard, L.D.Jackel, B.Straughn, D. M.Tennant : Dynamics of microfabricated electronic neural networks; Appl. Phys. Lett., vol. 50, pp. 1110-1112, 1987.

[11]   J.P.Sage, K.Thompson, R.S.Withers : An artificial network integrated circuit based on MNOS/CCD principles; in: Neural networks for computing, J.S.Denker ed., (American Institute of Physics, 1986).

[12]   D. B.Schwartz, R. E.Howard : Analog VLSI for adaptive learning; Neural Networks for Computing, Snowbird, 1988.

[13]   Y.P.Tsividis, D.Anastassiou : Switched capacitor neural network; Electronic Lett. vol. 23, pp. 958-959, 1987.

[14]   U.Rueckert, K.Goser : VLSI design of associative networks. International Workshop on VLSI for artificial intelligence, Oxford (GB), July 1988; in : VLSI for artificial intelligence, J.G. Delgado-Frias and W.Moore eds., (Kluwer Academic, 1989).

[15]   A.F.Murray, V.W.Smith, Z.F.Butler : Bit-serial neural networks , IEEE Conf. on Neural Information Processing Systems, Natural and Synthetic, Denver (USA), 1987; in: Neural Information Processing Systems, Natural and Synthetic, D.Z.Anderson, ed., American Institute of Physics, 1988

[16]   S.Jones, M.Thomaz, K.Sammut : Linear systolic neural network machine; IFIP Workshop on Parallel Architectures on Silicon, Grenoble, 1989.

[17]   J.Ouali, G.Saucier, J.Trilhe : A flexible wafer scale network; ICCD Conference, Rye Brook, (USA), September 1989

[18]   V.Peiris, G.Columberg, B.Hochet, G.van Ruymbeeke, M.Declercq : A versatile digital building block for fast simulation of neural networks; IFIP Workshop on Parallel Architectures on Silicon, Grenoble, 1989

[19]   U.Ramacher : Systolic architectures for neurocomputing; IFIP Workshop on Parallel Architectures on Silicon, Grenoble, 1989

[20]   M.Duranton, J.Gobert, N.Mauduit, : A digital VLSI module for neural networks. nEuro'88 conference, Paris, June 1988; in: Neural networks, from models to applications. L.Personnaz and G.Dreyfus, eds., (IDSET, 1989).

[21]   S.Y.Kung, J.N.Hwang : Parallel architectures for artificial neural nets; Proc. IEEE International Conf. on Neural Networks, vol. II, pp. 165-172, 1988.

[22]   M. Weinfeld : A fully digital CMOS integrated Hopfield network including the learning algorithm; International Workshop on VLSI for Artificial Intelligence, Oxford, 1988, in VLSI for artificial intelligence, J.G. Delgado-Frias and W.Moore eds. (Kluwer Academic, 1989).

[23]   S. Knerr, L. Personnaz, G. Dreyfus : Single-layer learning revisited :
        a stepwise procedure for building and training a neural network;
        Proc. NATO Advanced Workshop on Neurocomputing, F.
        Fogelman, J. Hérault, (Springer Verlag, 1990).

[24]   M.Weinfeld : Integrated artificial neural networks : components
        for higher level architectures with new properties; Proc. NATO
        Advanced Workshop on Neurocomputing, F. Fogelman, J. Hérault,
        (Springer Verlag, 1990).

[25]   L.Personnaz, A.Johannet, G.Dreyfus, M.Weinfeld : Towards a
        neural network chip: a performance assessment and a simple
        example; in Neural networks, from Models to Applications,
        L.Personnaz and G.Dreyfus, eds.(IDSET, Paris, 1989).

[26]   L.Personnaz, I.Guyon, G.Dreyfus : Collective computational
        properties of neural networks: new learning mechanisms; Phys.
        Rev. A, vol. 34, pp. 4217-4228, 1986.

[27]   S.Diederich, M.Opper : Learning of correlated patterns in spin-
        glass networks by local learning rules; Phys. Rev. Lett., vol. 58, no.
        9, pp. 949-952, 1987.

[28]   H.Sompolinsky : Neural networks with non linear synapses and a
        static noise ;Phys.Rev. A, vol. 34, no. 3, pp. 2571-2574, 1986.

[29]   W.W.Peterson, E.J.Weldon : Error correcting codes; MIT Press, 1972

[30]   I.Guyon, L. Personnaz, P. Siarry, G.Dreyfus : Engineering
        applications of spin-glass concepts; Lecture Notes in Physics, vol.
        275, J.L.van Hemmen, I. Morgenstern, eds (Springer, 1986)

[31]   P.Peretto, J. J.Niez : Stochastic dynamics of neural networks; IEEE
        Trans. on Systems, Man and Cybernetics, vol. SMC-16, pp. 73-83,
        1986.

FAST DESIGN OF DIGITAL DEDICATED NEURO CHIPS

J. OUALI, G. SAUCIER, J. TRILHE

ABSTRACT

This paper proposes a distributed, synchronous architecture for artificial neural networks. A basic processor is associated to a neuron and is able to perform autonomously all the steps of the learning and the relaxation phases. Data circulation is implemented by shifting techniques. Customization of the network is done by setting identification data in dedicated memory elements. A neuron processor which performs the relaxation phase has been implemented on silicon. It is shown that in a silicon compiler environment dedicated networks can be easily generated by cascading these elementary blocks.

INTRODUCTION

Neural computing relying on non algorithmic information processing has led in the recent last years to a tremendous research effort all over the world [1,2,3]. It appears clearly that neural computing will be practical and efficient if dedicated, fast architectures exist. Presently, neural computing is simulated on existing Von Neumann or parallel computers (CRAY, GAPP, Transputer). Attempts have been made to build dedicated architectures with commonly available components and some other studies concentrate on the design of either analog [4] [5] or digital [6] to [9] specific components.

This paper focuses on the design of a flexible neural architecture. The goal is to provide an implementation frame for a large variety of architectures (Hopfield, layered network) including both learning and relaxation phases. Some other parameters such as precision are also taken into account as design parameters. A silicon compilation approach allows to provide architectures with a given number of bits for state and coefficient encoding without additional design efforts.

THE BASIC ARCHITECTURAL CHOICES

The neural processor

In our approach, a physical entity is associated with each neuron. It is an application specific processor, in a previous paper [8] we have described the controller which implements the relaxation phase, we add here also the learning phase. We suppose that the neural network is made up of one layer (Hopfield network) or several layers (multilayered network). Each neuron is connected to all the neurons of the previous layer in a layered network or to all other neurons in a Hopfield network. Symbolically, during the relaxation phase, the neural processor implements the classical operations shown in figure 1.

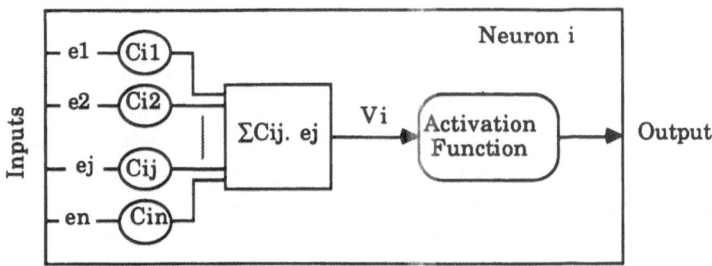

Figure 1   Function performed by the neural processor during the relaxation phase

Connections between neurons and global architecture

It has been pointed out frequently, that a large number of connections between the neurons is a severe drawback for their implementation on silicon. Practically, virtual connections are implemented to replace spatial connections. This means that if the processor cannot be connected to a large number of neighbors, the state of these neighbors will circulate in a shift register and the processor will pick up at the right instant the values of the state of its predecessors.

The global architecture is made up of elementary blocks which are 2-D arrays of neuron processors. A block corresponds either to a layer or a portion of a layer in a layered network or to a portion of a whole Hopfield network. In a block, the processors are organized in rows between two buses. These buses are made up of bus segments (figure2) which can be connected by soft switches (software programmable switches) when required during the phases of neuro-computing.

We will detail this global architecture using the example of 8 processors realized on a 4 x 2 array. Complete architectures will be built later on by assembling such basic blocks. These 8 processors either constitute a layer or a portion in a multilayered network, or in a Hopfield network.

First, let us show how the computation of potentials is performed; the equations of potentials V_i are given in figure 3.

In a first computation step, bus segments are connected as shown in figure 4 (a). The switches of the second column are open. The state of neuron N1 is available on the bus segment of the second row for both

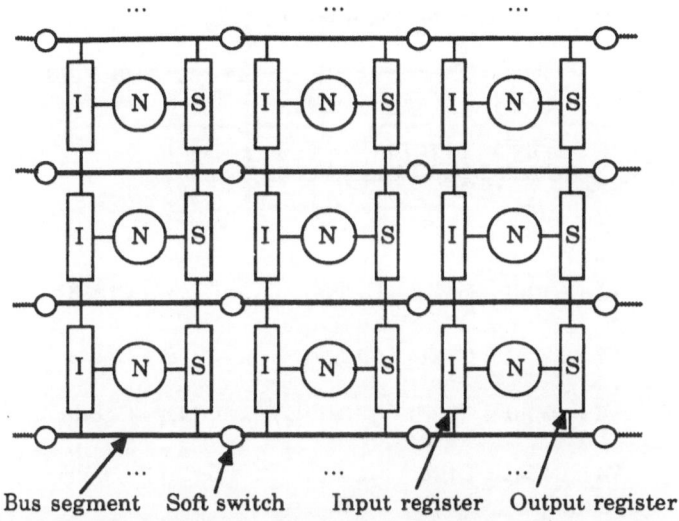

Bus segment   Soft switch   Input register   Output register

Figure 2   Global architecture (partial view)

$$V1 = C1,1e1 + C1,2e2 + ... + C1,7e7 + C1,7e7$$

$$V2 = C2,1e1 + C2,2e2 + ... + C2,7e7 + C2,8e8$$

$$V7 = C7,1e1 + C7,2e2 + ... + C7,7e7 + C7,8e8$$

$$V8 = C8,1e1 + C8,2e2 + ... + C8,7e7 + C8,8e8$$

Figure 3   Operations required for evaluating the potentials of the
artificial neurons

neurons N1 and N2, the state of neuron N3 is available similarly on the
next bus segment and the same for neurons N5, N7. The computations
performed during this step are shown in figure 4 (b).

In the second step of the computation, a circular shifting is performed in
the first layer as indicated in figure 4 (c). The state of neuron N1 has been
shifted on the next bus segment and is now available for both neurons N5
and N6.

It is clear that 4 elementary shiftings - a complete circular shifting - will be
necessary to perform half of the complete computation. Two complete
circular shiftings (8 elementary shiftings) will be required to end up the
computation. Once this computation is over, the controller initializes the
computation of the activation function.

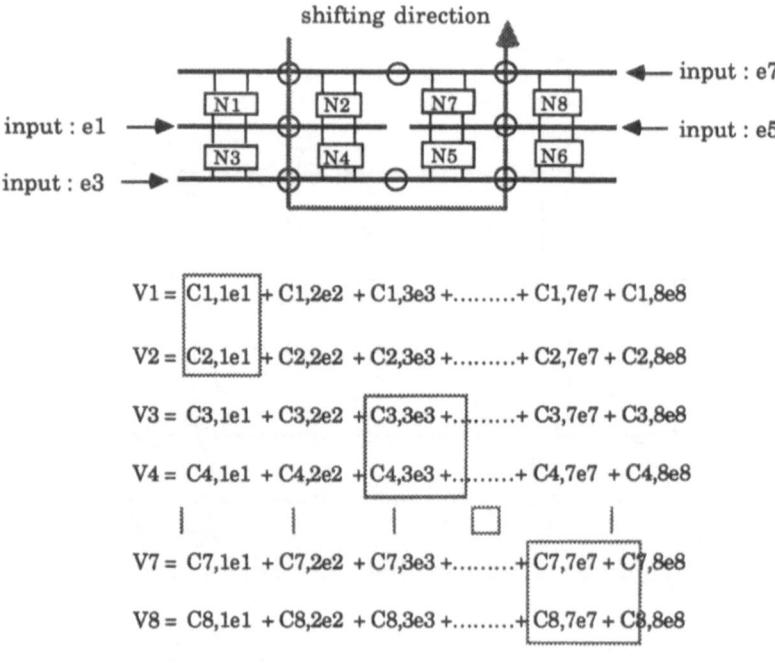

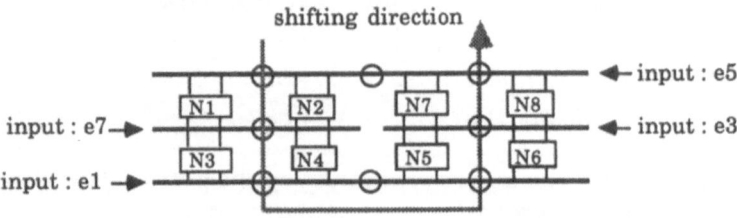

Figure 4   (a) A block of 8 neurons
(b) Operations performed in the first computation step
(c) Second computation step

Generally speaking a row of N rows will be implemented on a (N/k x k) array. In any case N elementary shiftings are required and k/N circular shiftings.

DETAILED ARCHITECTURE

The architecture is a distributed synchronous architecture. Each neural processor has autonomous control and is able to perform all the phases of neuro-computing (potential and activation function computations, learning and relaxation phases). The neuron processor contains a set of identification registers which contain the necessary information about the global architecture of the network. An initial loading of these registers will therefore customize the network. The local memory is a RAM for a chip implementing both learning and relaxation phases and a EPROM for an optimized chip dedicated to the relaxation phase of chosen application. For an optimized ASIC used for relaxation, the ROM technology should be used for both the identification registers and memory coefficients. The first prototype realized presently uses a 64x8 bits RAM to simplify the design problem. External control is just required to provide the feeding of the data. This approach is quite different from other ones based on powerful multiplication devices leaving outside any of the other functions

Neuron structure

The neural processor (figure 5) is made up of :

1. seven identification registers (ID__R):
    - N1 containing the number of bits encoding the state
    - N2 containing the number of rows in the block multiplied by two
    - N3 containing the number of coefficients stored in the memory
    - N4 containing the shifting direction of the inputs in the layer.

    for example: if N4 = 0 the shifting of the inputs is performed down
                    if N4 = 1 the shifting of the inputs is performed up
    - N5, N6 ,N7 containing each the necessary number of shiftings to format the result of each multiplication.

2. a local memory storing the coefficients $\{C_{ij}\}$.

3. a data path able to perform mutiplications and additions/subtractions.

4. a controller controlling the computation of the potential, the state of the neuron and the data transfer with the other neurons.

5. an input register storing and shift the input value.

6. an output register storing the state of the neuron.

7. three counters C1, C2, C3.

    C1 is initially loaded with the value of the register N1 and is decremented during the multiplication.

    C2 is used on each shifting state where its is initially loaded with the value of one of the register N2, N5, N6, N7 and is decremented during the shifting.

    C3 is initially loaded with the value of the register N3 and is decremented after each read cycle in the memory.

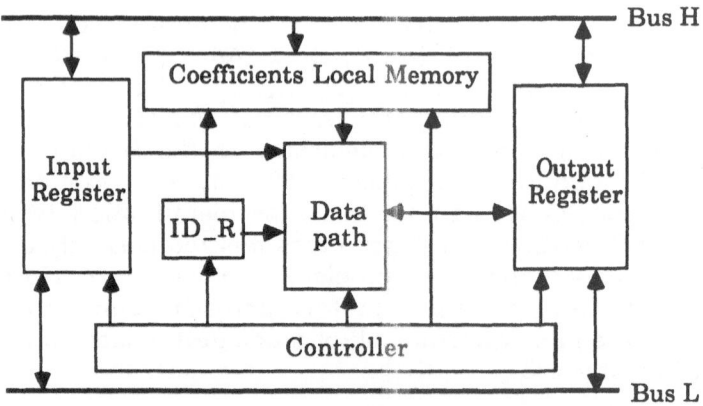

Figure 5   Neural processor

Algorithm used for the potential computation

The data path used in the relaxation phase (figure 6), computes the potential of the neuron $\Sigma C_{ij}.e_j$ where the C_{ij} are the coefficients and e_j the input states. Both of the C_{ij} and e_j are represented in 2's complement and encoded on N1 bits. The data path is made up of 2 operators, the multiplier and the adder/substractor.

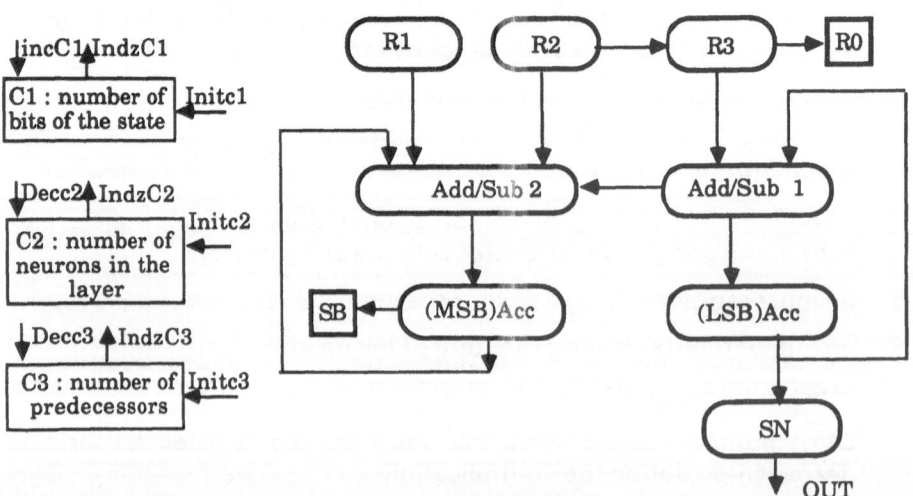

Figure 6   The data path structure

The multiplication performs a Booth's algorithm for 2's complement fixed point representation ; it is based on additions substractions and shiftings.

Algorithm used for the activation function

The sigmoidal activation function used in our demonstrator could be either stored as a table of values or hardwired. The first solution is simple and straightforward but the memory size becomes important if a high precision is required. We have chosen the second solution, and the first problem to solve is to approximate the theoretical function by a simple polynomial which can be easily implemented on silicon.

Two classical approximation methods are generaly used, the interpolation and the least squares approximation. The drawback of the polynomial approximation is the cost of the computation which increases when the degree of the polynomial is high. We have chosen the least square approximation and if we impose certain constraints to the polynomial, we obtain the following equations :

$$O(V) = a(V_s - V)^2 + 1 \qquad \text{if } 0 \leq V < V_s$$

$$O(V) = +1 \qquad \text{if } V \geq V_s$$

$$O(V) = -1 \qquad \text{if } V \leq -V_s$$

$$O(V) = -a(V_s + V)^2 - 1 \qquad \text{if } -V_s < V \leq 0$$

where a is a negative real and V_s the saturation potential

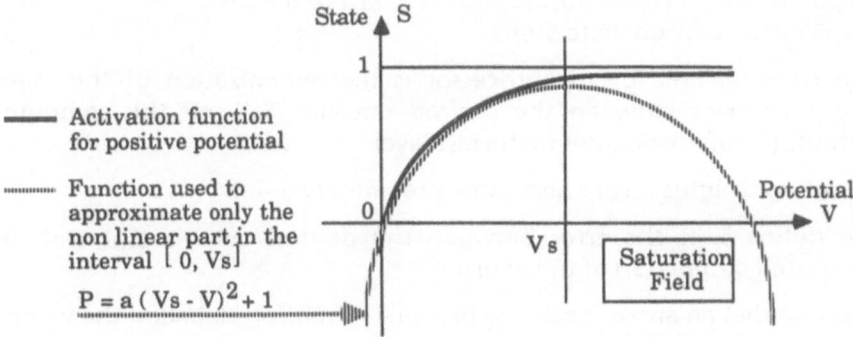

Figure7   Approximation of the positive part of the activation function

As previously said, the coefficients and the states are represented in fixed point 2's complement. t_1 bits encode the integer part of the state, t_2 bits encode the integer part of the coefficient and t_3 bits encode the integer part of the saturation potential. The integer part of the neural potential will be encoded on $(t_1 + t_2)$ bits. The test of the saturation will be done as follows on the $t(t = t_1 + t_2 - t_3)$ MSB of the potential :

--   if all the t bits are the same (0 or 1), the S_t bit will be equal to zero, then the potential is between $-V_s$ and V_s and the state will be computed

--   if the t bits are not identical, the S_t bit will be equal to one. The potential V is out of the interval $[-V_s , V_s]$ and the neuron will be saturated. The sign bit stored in the register S_B will determine the state value : -1 or + 1.

Algorithm used for the learning phase

There are many different approaches for the learning phase of artificial neural networks ; it can be used an external supervisor or not, the network can expand or not. We propose to implement one of the most popular, the backpropagation (constant size network and supervised learning), for which the neural processor elaborates its successive synaptic weights by comparing its output for a specific training set to outputs wished by the external supervisor for this set. The backpropagation can be described as follows :

Step1 : Initialization of the weigths
Step2 : Presentation of the inputs and desired outputs
Step3 : Calculation of actual outputs
Step4 : Verification of the minimization criterion
        if this criterion is satisfied the learning is finished
        if not the learning is not completed and go to Step5
Step5 : Update weigths starting from the last to the first layer
Step6 : Repeat by going to Step2.

The criterion used by the processor is the minimization of the mean quadratic error between the desired outputs (S_d) and the computed outputs (S_c) of the neurons in the last layer

Updating weigths . Let's give some preliminary definitions :

We define δ_j as the error between the desired outputs (S_d) and the computed outputs (S_c) of the neuron j.

Suppose that an arc connects the neuron i to neuron j labelled the weight w_{ij}.

We call weighted error for the neuron j induced on neuron i, the expression $w_{ij} \times \delta_j$.

If we consider that the neuron i has k successors in the next layer, we call global weighted error

$$\Delta_j = \sum_{l=1}^{k} w_{jl} \times \delta_l$$

The updating of the weights is done according to the following formula :

$$w_{ij}(t+1) = w_{ij}(t) + \eta * \Delta_j * x_i * f'(x_i).$$

where η is a constant, x_i is the output of the neuron i and f' is the derivate of the activation function f which is approximated by a polynomial.

Implementation of the algorithm. We will detail how the mean quadratic error and the global weighted error can be calculated on the architecture previously presented.

Suppose that we have a network made up of L layers. The last layer is made up of NI neurons, folded in m columns (from C_1 to C_m) and R rows.

-- Each neuron in the last layer calculates its mean quadratic error δ_j^2

-- Two columns, C_1 and $C_{(m/2)+1}$, will respectively accumulate the quadratic errors of the columns from C_2 to $C_{m/2}$ and from $C_{(m/2)+1}$ to C_m.

-- An arbitrary neuron, for example neuron N_1, will accumulate the final mean quadratic error. Therefore, partial errors circulate from column C_1 and column $C_{(m/2)+1}$ to neuron N_1.

-- The neuron N_1 will compare the result of this sum J to a value V fixed by the supervisor.

If J is less or equal to V, the prototype presented is learned by the network

If J is greater than V, a new update of weights is necessary.

We illustrate this by an example of a network having 12 neurons in the last layer organized in 4 columns and 3 rows.

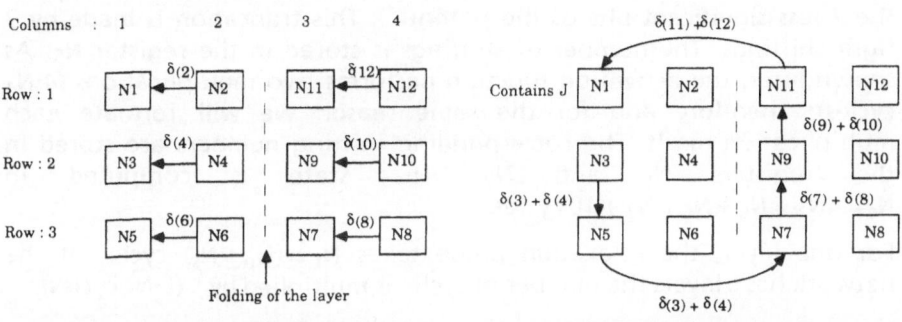

Figure 8a   Sum of the errors   Figure 8b   Sum of the errors
of C_1 and C_2, C_3 and C_4 of C_1 and C_3

Computation of the global weighted error : Suppose that the errors of the neuron in layer i are known. The layer i will calculate the global error of each neuron in layer i-1 before updating its weigths. Each neuron in layer i will contribute by calculating a weighted error term for a neuron N_i in layer i-1 and we will adopt the same principle of data circulation, used for calculating the mean quadratic error, to accumulate the weighted errors.

EVALUATION AND FIRST RESULTS

Global flowchart for the relaxation phase and performance evaluation

In the relaxation phase, each neuron computes its potential and its state as shown in (figure 9). If one column of neurons is selected, a first initialization step done in 4 cycles loads the identification values of the registers $N_1...N_7$. A second initialization step loads the coefficients in the memory in N3 cycles. To multiply all the memory coefficients by all the inputs states we need N_3 cycles.

In each cycle we load R_1 and R_3 with the coefficient value with a 4 bit integer and 4 bit decimal part. The multiplication needs N_1 cycles. The result will be encoded on 16 bits - with 5 bits integer part - and stored in R_2&R_3 registers. In order to accumulate the partial potential, after each multiplication, we use the two adder/substractor operators (Add/Sub1 and 2) to add the contents of R_2 & R_3 to the contents of the accumulator. The result of the partial sum of the potential and the final potential will be stored in the accumulator. Consequently, the computation of the potential is performed in $Np = N_3 \cdot (2N_1 + 4)$ cycles.

Once the potential is computed, the neuron calculates its state using the activation function. Only the 8 most significant bits of the potential result are used for the computation of the state. Therefore, we must suppress the 7 less significant bits of the potential. This truncation is made by 7 right shiftings. The number of shiftings is stored in the register N_5. As shown later, the activation function performs two multiplications ($4 \cdot N_1$ cycles); therefore and for the same reason we will formate each multiplication result. The corresponding shifting numbers are stored in the registers N_6 and N_7. The state is computed in $N_s = 4N_1 + N_5 + N_6 + N_7 + 10$ cycles.

For one layer, the relaxation phase takes $N_L = (N_p + N_s)$ cycles. If the network has L layers the number of cycles is multiplied by L ($L \cdot N_L$ cycles).

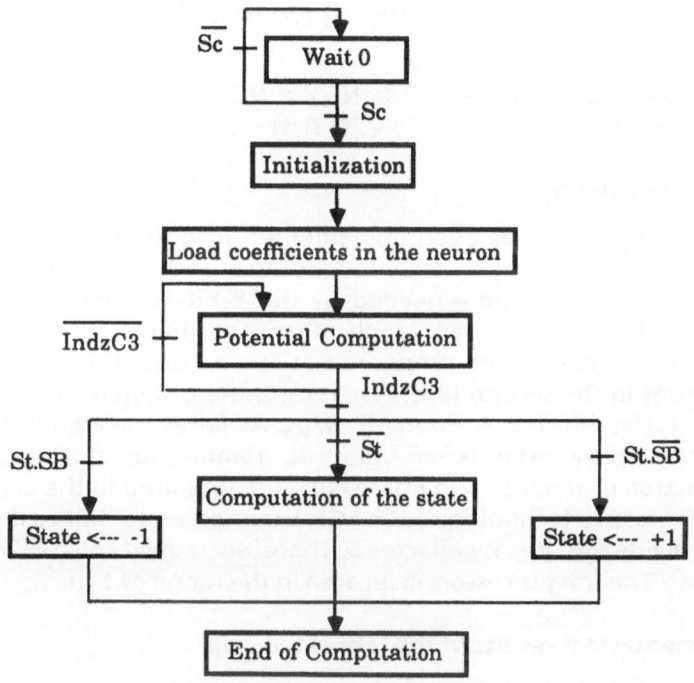

Figure 9   Global control algorithm

Evaluation of performance for the learning phase

Suppose that we have a network made up of three layers which have respectively $N(1)$, $N(2)$ and $N(3)$ neurons. Neurons are organized as follows :

- $N(1)$ in R rows and $C(1)$ columns
- $N(2)$ in R rows and $C(2)$ columns
- $N(3)$ in R rows and $C(3)$ columns

We need $(C(3) - 2)*2R + A$ cycles to calculate the mean quadratic error ; A corresponds to the number of cycles needed for a substraction $(S_{di} - S_{ci})$ and a multiplication $(S_{di} - S_{ci})^2$.

To calculate the global error D_i, we need $(C(3) - 2)*2R + B$ cycles ; where B is the number of cycles needed to multiply the weight by the error δ_j. In the step, the second layer will calculate its errors δ_j in B cycles and calculate the global error of the neurons in the first layer in $(C(2) - 2)*2R + B$ cycles. Simultaneously, the third layer updates its weights in $N(2)(B + 1)$ cycles. Finaly the first layer will calculate its errors δ_j in B cycles and update its weights $N(0)(B + 1)$ cycles. Simultaneously, the second layer will update its weights in $N(1)*(B + 1)$ cycles. $N(0)$ is the number of inputs in the first layer.

If we consider that $N(0) > N(1) > N(2) > N(3)$, then the backpropagation will require $N(0)(B + 1) + 2R(C(2) + 2C(3)-4) + 5B$ cycles.

Practical evaluation

The first prototype has been implemented in a $2\mu m$ CMOS technology using the block generation techniques of VLSI Technology. The basic cycle clock is fixed by the time needed by the 8 bit addition which is 50ns. Consider for instance an application for hand written numbers recognition. The network proposed has 45 neurons in the first layer and 10 neurons in the second layer. Therefore, the example will be analyzed in 1100 cycles which correspond to $57\mu s$. As far as the area is concerned, the first layout exhibits an area of $15mm^2$ for the neuron. The introduction of the learning phase will add 10% area to the controller. A more aggressive technology from SGS-Thomson and a full custom design should divide this area by a factor 5. Therefore, a next chip will contain as much as 32 neural processors in an area in the range of 1 cm^2.

Experimental test results of the fabricated chip

The neuron can have three states :

-- the positive saturation case state equal to + 1.
-- the negative saturation case state equal to -1.
-- the non saturation case state between -1 and + 1.

The first part of the test has consisted in verifying the correctness of the computation performed by the neuron in these three cases. For this, we have sent several patterns with the following shifting parameters $N5 = N6 = 5$ and $N7 = 3$.

In the first step, we have tested the positive saturation case. The saturation potential value is equal to + or - 3.96875.

The input control signals are :
-- a reset control of the neuron (R),
-- a transfer control sign (T) opening and closing the neuron switches,
-- N4 performing the data shifting ; if $N4 = 1$ a shifting up is performed otherwise a shifting down,
-- S_c selecting the neuron.

These control signals are synchronized by the clock (C_k). The number of cycles needed for the computation is C.

C, C_k,R,T,N4,Sc	BusH	BusL	OutputBusH	OutputBusL	Operation performed
1: 1 00 1 1	00000000	00000001	xxxxxxxx	xxxxxxxx	N3 = 1
2: 1 00 1 1	00000000	01010001	xxxxxxxx	xxxxxxxx	N5 = 5, N2 = 1

3: 1 00 1 1	00000000	00110101	xxxxxxxx	xxxxxxxx	N7 = 3, N6 = 5
4: 1 00 1 1	00000000	01111111	xxxxxxxx	xxxxxxxx	Coefficient. = 7.9375
5: 1 00 1 1	00000000	01111111	xxxxxxxx	xxxxxxxx	Input state = 0.99216
29:1 00 1 0	00000000	01111111	xxxxxxxx	01111111	Neuron state = 0.99216

A similar negative saturated case has been tested by changing the input state value to the value - 0.99216.

As soon as the saturation case has been tested, two sequences have been sent testing respectively the positive and the negative non saturated states.

C, C_k,R,T,N4,Sc	BusH	BusL	OutputBusH	OutputBusL	Operation performed
1: 1 00 1 1	00000000	00000001	xxxxxxxx	xxxxxxxx	
2: 1 00 1 1	00000000	01010001	xxxxxxxx	xxxxxxxx	
3: 1 00 1 1	00000000	00110101	xxxxxxxx	xxxxxxxx	
4: 1 00 1 1	00000000	00100000	xxxxxxxx	xxxxxxxx	Coefficient = + 2
5: 1 00 1 1	00000000	01101101	xxxxxxxx	xxxxxxxx	Input state = + 0.8515
72: 1 00 1 0	00000000	01101101	xxxxxxxx	01010101	Neuron state = + 0.664

C, Ck,R,T,N4,Sc	BusH	BusL	OutputBusH	OutputBusL	Operation performed
1: 1 00 1 1	00000000	00000001	xxxxxxxx	xxxxxxxx	
2: 1 00 1 1	00000000	01010001	xxxxxxxx	xxxxxxxx	
3: 1 00 1 1	00000000	00110101	xxxxxxxx	xxxxxxxx	
4: 1 00 1 1	00000000	00100111	xxxxxxxx	xxxxxxxx	Coefficient = + 2.437
5: 1 00 1 1	00000000	10101000	xxxxxxxx	xxxxxxxx	Input state = - 0.6875
72: 1 00 1 0	00000000	10101000	xxxxxxxx	10101010	Neuron state = - 0.6718

In the second part, we suppose that the neuron is connected to 4 others and we will test both the computation and the data shifting. The values of the identification registers, the coefficients and the input state values are loaded on the input bus low (Bus L). The data shifting is performed up. Therefore the shifted data will be on the output bus high (OBus H)

The coefficients are successively equal to 2.25 ; 3.0 ; 3.125 and 1.3125. The respectively input states are 0.625 ; 0.125 ; 0.078 and 0.32. The neuron state will be equal to 0.5.

C, C_k,R,T,N4,Sc	BusH	BusL	OutputBusH	OutputBusL	Operation performed
1: 1 10 1 0	00000000	00000000	xxxxxxxx	xxxxxxxx	
2: 1 00 1 1	00000000	00000100	xxxxxxxx	xxxxxxxx	N3 = 4
3: 1 00 1 0	00000000	01010100	xxxxxxxx	xxxxxxxx..	N5 = 5 , N2 = 4
4: 1 00 1 0	00000000	00110101	xxxxxxxx	xxxxxxxx..	N7 = 3 , N6 = 5
5: 1 00 1 0	00000000	00100100	xxxxxxxx	xxxxxxxx	1st Coefficient
6: 1 00 1 0	00000000	00110000	xxxxxxxx	xxxxxxxx	2nd Coefficient
7: 1 00 1 0	00000000	00110010	xxxxxxxx	xxxxxxxx	3rd Coefficient
8: 1 00 1 0	00000000	00010101	xxxxxxxx	xxxxxxxx	4th Coefficient
18: 1 00 1 1	00000000	00001000	xxxxxxxx	xxxxxxxx	1st input state
70: 1 00 1 1	00000000	00010000	00001000	xxxxxxxx	2nd input state

122: 1 00 1 1	00000000	00001010	00010000	xxxxxxxx	3rd input state
174: 1 00 1 1	00000000	00101001	00001010	xxxxxxxx	4th input state
283: 1 00 1 0	00000000	00101001	xxxxxxxx	01000000	Neuron state

In a last example tested, the data shifting is performed down. The identification values and the coefficients are always loaded on the Bus L. The coefficients values are equal to 3.0625, 2, 0.1875 and 1.5625 and the respective input states values are 0.0781, 0.1875, 0.0281 and 0.3984. The neuron state included shown to be equal to 0.64 which is the right value.

$C, C_k, R, T, N4, Sc$	BusH	BusL	OutputBusH	OutputBusL	Operation performed
1: 1 00 1 1	00000000	00000000	xxxxxxxx	xxxxxxxx	
2: 1 00 0 1	00000000	00000100	xxxxxxxx	xxxxxxxx	N3 = 4
3: 1 00 0 0	00000000	01010100	xxxxxxxx	xxxxxxxx	N5 = 5 , N2 = 5
4: 1 00 0 0	00000000	00110101	xxxxxxxx	xxxxxxxx	N7 = 3 , N6 = 5
5: 1 00 0 0	00000000	00110010	xxxxxxxx	xxxxxxxx	1st Coefficient
6: 1 00 0 0	00000000	00100000	xxxxxxxx	xxxxxxxx	2nd Coefficient
7: 1 00 0 0	00000000	00000011	xxxxxxxx	xxxxxxxx	3rd Coefficient
8: 1 00 0 0	00000000	00011001	xxxxxxxx	xxxxxxxx	4th Coefficient
18: 1 00 0 1	00001010	00011001	xxxxxxxx	xxxxxxxx	1st input state
70: 1 00 0 1	00011000	00011001	xxxxxxxx	00001010	2nd input state
122: 1 00 0 1	00000110	00011001	xxxxxxxx	00011000	3rd input state
174: 1 00 0 1	00110011	00011001	xxxxxxxx	00000110	4th input state
277: 1 00 0 0	00110011	00011001	xxxxxxxx	01010010	Neuron state

SILICON COMPILER FOR NEURO-ASICS

Starting from the previous experience, a fully automatic neuro-ASICs silicon compiler can be defined.

Parameters from a high level description language

Starting from a high level language description [10] , the customization data can be fed to the customization registers. These information concern :

1. the number of predecessors of the neuron in the preceding layer.
2. the number of bits encoding the states and the coefficients.
3. the number of input waves for a layer if the number of inputs is greater than the number of elements in the layer.
4. the position of the neural processor in the physical and logical layer as a logical layer may be implemented in several physical layers.
5. the shifting direction of the input values in the layer. For example, neuron from N1 to N4 (figure 4) perform the shifting of the data down and from N5 to N8 perform the shifting data up
6. the number of shiftings required to format the result of each multiplication.

Block generation and blocks interconnection

The local memory is generated by a ROM or a RAM if the relaxation phase only is implemented.

The data path is automatically generated from an architectural description in terms of registers, ALU, bus connections and bit slice structure. Extracted from a high level description tool or given by the user, the following control flowgraphs have to be available

- control flowgraph of the potential computation
- control flowgraph of the activation function
- global flowgraph implementing the data feeding and including the previous ones.

A FSM (finite state machine) compiler generates then an automatic state assigment and a standard cell layout for the global controller. A chip compiler connects all the previous blocks and adds the pads.

Customization of the switches

Two types of switches implement the connection between processors. The first ones are definitely fixed as they are always closed for a given target chip. They will be PROM switches or aluminium segments. The second type switches are electrically programmed from the neuron processors as they are closed during the data loading and the transfer of the neurons states from the first to the second layer steps, or open during the different steps of the potential and the state computations. We give an example of a 2 layers network, with 6 neurons in the first layer and 4 neurons in the second one (figure 10).

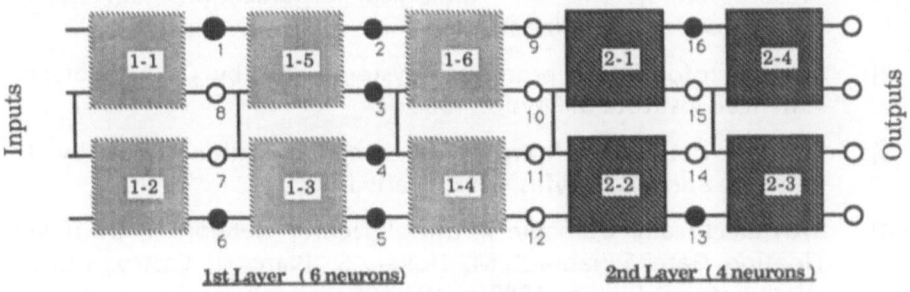

Figure 10   Two layer network (o electrically programmed,
          ● always closed)

The switches from 9 to 12 are used to separate physically the two layers. They are always disconnected during the potential and state computations steps. All the switches are closed during the data loading and the transfer of the neurons states from the first to the second layer

steps. The folding of the first and the second layers are respectively made
by the switches 7, 8 and 14, 15.

CONCLUSION

A distributed architecture for digital implementation of Artificial Neural
Networks, easily cascadable and highly flexible has been proposed and
demonstrated on silicon. It can implement different target architectures
and is customized by using soft switches and a silicon compilation
approach. The silicon compilation approach allows to translate any
control flowcharts on silicon using finite state machine compilers. Thus if
the designer defines his learning algorithm or different variations of
relaxation or activation functions through a control flowchart, no new
design effort is required for generating the controller. Of course the data
path has to contain the adequate resources and floorplanning
adjustment of these processors allows to implement arrays of any size
unless yield limits are reached.

ACKNOWLEDGEMENT

This work has been suported by the Commission of the European
Communities under grant ESPRIT 2059.

REFERENCES

[1]   "Disordered systems and biological organization" Ed. by E.
       Bienenstock et al., Springer, 1986.

[2]   "Neural information processing systems" Ed. by D.Z. Anderson,
       American Institute of Physics, 1988.

[3]   "Neural networks: from models to applications" Ed. by L.
       Personnaz and G. Dreyfus, IDSET, Paris 1989.

[4]   "An Electrically trainable Artificial Neural Network with 10240
       Floating Gate Synapses" M. Holler, S. Pam; H. Castro. IJCNN,
       Washington D.C. ,June 1989, p. 181-196

[5]   "VLSI Implementation of NN Model" H. Graf, L.Jackel and
       Hubbard, Computer, March 1988, p.41

[6]   "A digital VLSI module for neural networks" M. Duranton, J.
       Gobert, N. Mauduit, Neural networks: from models to
       applications, Ed. by L. Personnaz and G. Dreyfus, IDSET, Paris 1989.

[7]   "Architectures Neuroniques" J. Ouali , G. Saucier , J.Trilhe. Journée
       d'Etudes "Architectures Parallèles et Applications aux traitement
       d'Images". ENSERG, Grenoble, France, Mai 1990.

[8]   "Réseau neuromimetique sur silicium: mythe ou réalite ?"J. Ouali,
       G. Saucier J. Trilhe. Int. Workshop Neuro-Nîmes'88, Nimes, France,
       November 1988.

[9]   "Customizable artificial neural network on silicon". J.Ouali ,
       J.Trilhe. Technical revue of Thomson, to be published October 1990

[10]   "ESPRIT II Conference Pygmalion Session". Brussells, Belgium.
       November 89

DIGITAL NEURAL NETWORK ARCHITECTURE AND IMPLEMENTATION

J. A. VLONTZOS, S. Y. KUNG

INTRODUCTION

Digital implementations of neural networks represent a mature and well understood technology, which offers greater flexibility, scalability, and accuracy than the analog implementations. For example, using digital logic and memory it is quite easy to partition a large problem so that it can be solved by a smaller (in terms of hardware) implementation. Using the same logic and memory, a digital implementation can realize more than one network and combine the results in a hierarchical fashion to solve large problems. The main drawbacks of digital VLSI implementations are however their larger silicon area, relatively slower speed and the great cost of interconnecting processing units. These problems are addressed in this paper and solutions to alleviate them are proposed. First, the theoretical foundations for digital VLSI implementations are developed. Based on those, the design of digital systems is proposed that offer great speed and at the same time do not require a large number of interconnections.

SYSTOLIC/WAVEFRONT ARCHITECTURE

In this section we will derive systolic arrays for the implementation of Recurrent Back Propagation networks (RBPs) and Hidden Markov Models (HMMs). The methods presented and the arrays resulting from this exposition are applicable to a great variety of neural network models.

It is shown that operations in both the retrieving and learning phases of RBPs and HMMs can be formulated as consecutive *matrix-vector multiplication* (MVM), consecutive *outer-product updating* (OPU), and consecutive *vector-matrix multiplication* (VMM) problems. In terms of the array structure, all these formulations lead to a universal ring systolic array architectures [7, 8]. In terms of the functional operations, all these formulations call for a MAC (multiply and accumulation) processor[8].

Ring Systolic Design for the Retrieving Phase

Consecutive MVM in the Retrieving Phase The system dynamics in the retrieving phase can be formulated as a consecutive MVM problem interleaved with the nonlinear activation function [7].

$$\mathbf{u}(l+1) = \mathbf{Wa}(l)$$

$$\mathbf{a}(l+1) = f[\mathbf{u}(l+1),\theta(l+1)] \qquad (1)$$

where $\mathbf{u}(l) = [u_1(l),...,u_N(l)]^T$, $\mathbf{a}(l) = [a_1(l),...,a_N(l)]^T$, $\theta = [\theta_1(l),...,\theta_N(l)]^T$ and $\mathbf{W} = \{w_{ij}\}$. The superscript notation T denotes the vector transposition, and the $f[x]$ operator performs the nonlinear activation f_i on each element of the vector x.

A special feature of consecutive MVM problems is that the outputs at one iteration will be used as the inputs for the next iteration. Therefore, careful arrangement of the processor assignment and schedule assignment in the parallel algorithm design is neccessary, and this leads to a ring systolic architecture as shown in Figure 1 [8].

Systolic Processing of Consecutive MVM At the (l + 1)-th iteration of the retrieving phase, each PE, say the i-th PE, can be treated as a network unit, and the corresponding incoming weights (w_{i1}, w_{i2}, ..., w_{iN}) are stored in the memory of the i-th PE in a circularly-shift-up order (by i-1 positions). The operations at the (l + 1)-th iteration can be described as follows (see Figure 1):

1. Each unit activation value $a_i(l)$, created at i-th PE, is multiplied with w_{ii}. The product is added to the residing net-input accumulator $u_i(l + 1)$, which has a zero initial value for the HMMs (or set to be equal to θ_i for the RBPs). After the MAC operation, $a_i(l)$ will move counterclockwise across the ring array and visit each of the other PEs once in N clock units.
2. When $a_j(l)$ arrives at the i-th PE, it is multiplied with w_{ij} and the product is added to $u_i(l + 1)$.
3. After N clock units, the accumulator $u_i(l + 1)$ collects all the necessary products.
4. One more clock is needed for $a_i(l)$ to returns to the i-th PE and the processor is ready for the nonlinear activation operation f_i to create $a_i(l + 1)$ for the next iteration.

The above procedure can be executed in a fully pipelined fashion. Moreover, it can be recursively executed (with increased l) until L iterations are completed.

The nonlinear sigmoid function used in RBPs can be well approximated by a piecewise linear function with 8-16 segments [10]. This again calls for a MAC operation. The nonlinear activation function used in HMMs is

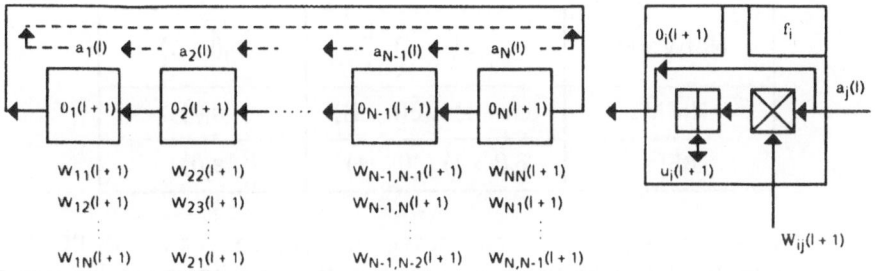

Figure1   The ring systolic architecture for the consecutive MVM of
retrieving phase at (l + 1)-th iteration

simply multiplication. A table-look-up mechanism is provided for fast
retrieval of the slope and intercepts in RBPs and symbol occurrence
probability $f_i(v_k)$ in HMMs.

Ring Systolic Design for the Learning Phase

There are two types of operations required in the learning phase of RBPs
and HMMs: (1) the consecutive *outer-product updating* (OPU), and (2)
the consecutive *vector-matrix multiplication* (VMM). Both types can be
efficiently implemented by the ring systolic ANN based on the same array
configuration, memory storage, and processing hardware in the
retrieving phase.

Given the weight value w_{ij} of the previous recursion, (1) the incremental
accumulation of weight updating can be calculated based on the
consecutive OPU; (2) and at the same time the back-propagated
corrective signal $\delta_i(l)$ of l-th iteration can also be computed based on the
consecutive VMM.

Consecutive OPU in the Learning Phase   In general, the (additive or
multiplicative) updating formulation of the weights at (l + 1)-th iteration
of the RBPs and HMMs can be summarized by the following consecutive
OPU equation (for $0 \leq l \leq L\text{-}1$):

$$\Delta w_{ij} \leftarrow \Delta w_{ij} + g_i(l+1) \times h_j(l+1) \ or$$

$$\Delta \mathbf{W} \leftarrow \Delta \mathbf{W} + \mathbf{g}(l+1) \ \mathbf{h}^T(l+1)$$

where $\mathbf{g}(l+1) = [g_1(l+1), \ g_2(l+1), \ ..., \ g_N(l+1)]^T$, $\mathbf{h}(l+1) = [h_1(l+1),$
$h_2(l+1), ..., h_N(l+1)]^T$. To be more specific:

Systolic Processing for the OPU The operations of the consecutive OPU in
the ring systolic array can be briefly described as follows (see Figure 2):

Networks	$g_i(l + 1)$	$h_j(l + 1)$
HMMs	$\delta_i(l + 1) f_i(\theta(l + 1))$	$a_j(l)$
RBPs	$\delta_i(l + 1) f_i'(l + 1)$	$a_j(l)$

1. The value of $g_i(l + 1)$ is computed and stored in the i-th PE. The value $h_j(l + 1)$ produced at the j-th PE will be cyclically piped (leftward) to all other PEs in the ring systolic ANN during the N clocks.

2. When $h_j(l + 1)$ arriaves at the i-th PE, it is multiplied with the stored value $g_i(l + 1)$, the product will be added to the weight accumulator Δw_{ij}, which is initially set to zero at L-th iteration. Note that the accumulator $\{\Delta w_{ij}\}$ are accessed in a cirularly-shift-up sequence, just like the access ordering of $\{w_{ij}\}$.

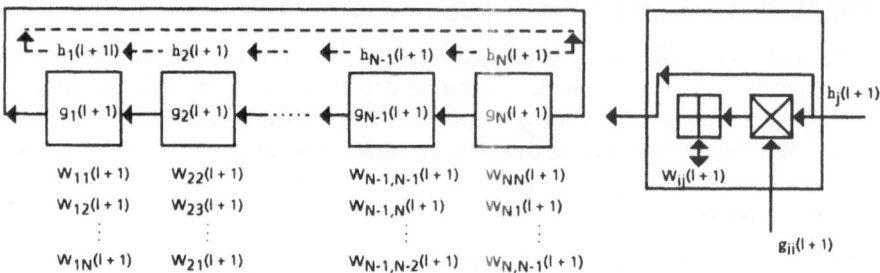

Figure 2   The ring systolic architecture for the consecutive OPU of learning phase at (l + 1)-th iteration

3. After N clocks, all the N x N weight accumulators are incremented for the (l + 1)-th iteration. The ring array is now ready for the consecutive OPU accumulation of the next l-th iteration.

After all the L-layer incremental accumulations of the $\{\Delta w_{ij}\}$, the ring array is ready to perform the local (additive or multiplicative) updating at each PE, this again requires N time units;

$$w_{ij} \leftarrow w_{ij} \oplus \eta \, \Delta w_{ij}$$

where $\oplus$ is a minus operation in RBPs and a muliplication operation in HMMs.

Note that the η value in RBPs is fixed, while the η value in HMMs should be calculated for each recursion, this can be easily done by accumulating $\delta_j(l)a_j(l)$ over all the l iterations inside each individual PE, and one inversion via division operation is required in the end of the accumulation. The same situation happened in the case of modifying the nonlinear activation function $\{f_i(v_k)\forall i,k\}$.

Consecutive VMM in the Learning Phase The backpropagation rule for both RBPs and HMMs can be formulated as a consecutive VMM operations [8];

$$\delta_i(l) = \sum_{j=1}^{N} g_j(l+1)w_{ji} \quad or \quad \mathbf{d}^T(l) = \mathbf{g}^T(l+1)\,\mathbf{W} \tag{2}$$

where $\mathbf{d}(l) = [\delta_1(l), \delta_2(l), ..., \delta_N(l)]^T$.

Systolic Processing for the Consecutive VMM The operations of the consecutive VMM in the ring systolic array can be briefly described as follows (see Figure 3):

1. The signal $g_j(l + 1)$ and the value w_{ji} are available in the j-th PE at $(l + 1)$-th iteration. The value w_{ji} is then multiplied with $g_j(l + 1)$ at j-th PE.

2. The product is added to the newly arrived accumulator acc_i. (The parameter acc_i is initiated at the i-th PE with zero initial value and circularly shifted leftward across the ring array.)

3. After N such accumulation operations, the accumulator acc_i will return to i-th PE after accumulating all the products $\delta_i(l) = \Sigma^N_{j=1} g_j(l + 1)\, w_{ji}$.

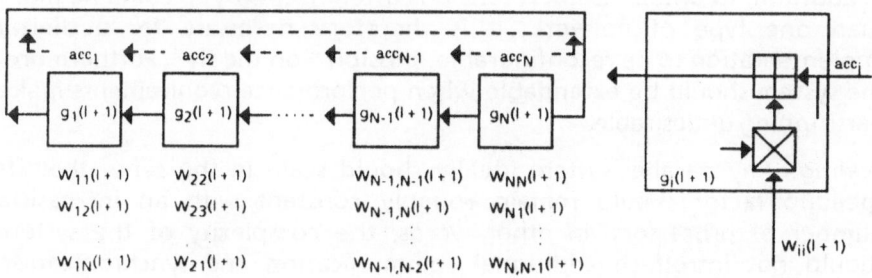

Figure 3   The ring systolic architecture for the consecutive VMM of learning phase at $(l + 1)$-th iteration

DIGITAL IMPLEMENTATIONS

Architectural Considerations

The goal of a digital implementation of ANNs is to provide a high speed, accurate yet flexible platform for neural network algorithm implementation. In the following we will briefly discuss the requirements for a digital ANN implementation and the architectural features they dictate.

High Performance: In order to conduct meaningful ANN research and to meet real world requirements [1], a very large number of units must be used. Given the need for multiple sweeps of the training patterns in networks with units numbering in the thousands, a very high processing rate is needed in order to keep training time in a reasonable range. On the other hand, during recognition, although the operation is much simpler, real time requirements again dictate very high processing rates. Given today's algorithms and hardware technologies, parallel execution is the only way of achieving the required processing rates.

Reconfigurability: Research in ANNs produces a great variety of new algorithms. Programmability is therefore necessary for our machine. It is well known that programmable systems are less efficient than highly optimized special purpose machines. In the ANN case however, as well as for a variety of signal processing and numerical analysis applications, most operations are repetitive and can be "reused" to increase efficiency. In addition, to obtain useful results it is often necessary to combine more than one type of network. It is therefore necessary for a digital implementation to be reconfigurable, possibly "on the fly". Furthermore, the system should be extendable when performance requirements make partitioning undesirable.

Scaling: Any parallel system ideally should scale in the sense that its speedup factor should remain roughly constant with an increasing number of processors. In other words, the complexity of the system should not introduce additional communication and synchronization overhead. Systolic arrays with their regular and local interconnection patterns and their concurrency of computation and communication, are ideal in this respect. It is obvious that the physical size of the parallel digital implementation of ANNs will seldom be adequate for a straightforward mapping of the algorithm to the machine. It is therefore necessary to provide the means (software) to map and partition large problems and at the same time provide large amounts of memory to store weights and intermediate results.

Convenience: In neural network research, there is a need of visualiszing the results of training a network. It is therefore necessary to provide an easy to use graphical user interface. In addition, the user should be able to concentrate on the problem at hand instead of worrying about managing the hardware resources of a parallel system. Some operating system support is therefore necessary (e.g. languages, memory management, communications). More importantly, a library of basic matrix type operations is needed to permit rapid prototyping of algorithms.

Cost Effectiveness: A machine that meets all of the above requirements but is prohibitively expensive will not help ANN research. This dictates the use of commercially available components when possible, and for the custom components simple architecture and small size to enhance yield therefore limiting cost.

Performance requirements dictate the use of a parallel architecture. Neural network algorithms have been shown to correspond to a series of inner and outer product type of operations [11]. In such operations, the limiting factor is the speed one can perform a series of additions since all multiplications can be performed in parallel. It is well known that a ring array is very well suited for such operations although shuffle exchange and tree networks have a performance edge. The ring array requires far less complicated routing and a smaller number of processors thus satisfying the economy requirements as well.

The next decision to be made, given that the overall architecture will be a ring array, is on the number of processing elements and their method of synchroniszation. The use of a global clocking signal greatly simplifies the processing element control but it creates synchronization problems in large arrays. Given the limited propagation speed of the signals both on chip and on a board, processors that are far away from the clock source will receive the clock pulses with a few nanoseconds delay relative to processors closer to the source (clock skew) [6]. In reality, a global clock can drive approximately 16 processors. Above that number clock skew becomes a problem. Our machine, in order to provide the necessary performance, needs more than 16 processors and in addition, it must be extendable. The above considerations therefore dictate the use of an asynchronous clocking scheme where each processor has its own clock and is synchronized with the others via *handshaking signals*.

Convenience and economy considserations dictate a system that will be closely coupled with a workstation in order to take advantage of its graphics and operating system capabilites. This in turn means that our system should have small size and low power requirements. In order to realize such a system, we should either use single chip processors or processors consisting of very few components.

In the following sections we will describe some ANN system implementations based on the above considerations and we will discuss in more detail the architectural tradeoffs needed to obtain our goals.

A Transputer-based CPU Architecture

To provide a performance baseline and an example of an ANN system using single chip processors, in this section we present a design based on the Inmos Transputer. It must be noted that although such a system is low cost, it does not meet the performance requirements for large ANNs.

The Transputer (T800) [9] is a general purpose RISC-like microprocessor. In addition to an ALU, it contains a floating point unit capable of operating on 64bit numbers at a 1MFLOP rate, 4K of internal high-speed RAM and four bidirectional hardware links that can be used for communications between Transputers at 20Mbits/sec. The Transputer is optimized to execute Occam, a parallel processing language based on CSP [2]. The Transputer and Occam provide a natural way to overcome the problem of partitioning, that is, mapping a network of large logical size to processor network of small physical size. An Occam process runs identically whether it is assigned one processor or many processes share the same processor since the hardware links have software counterparts operating in the exact same way.

Implementation of ANNs and HMMs: Currently, our array consists of four Transputer Modules (see Figure 4) [5,11]. Each module, in addition to a T800 Transputer contains 2Mbytes of external dynamic RAM. The modules are attached to a motherboard containing an INMOS C004 link interface and a T212 Transputer that is dedicated to control the C004. the motherboard is connected to the backplane of the host, an IBM PC-XT. The links of each Transputer are connected to the C004 and can be reconfigured in any network configuration in 1 μsec under the control of the T212.

One of the transputer modules is assigned the role of Ring Master and is responsible for running the user interface and accessing the host memory to load the program and data to all processors. After the initialization phase, the Master module executes the same program as the other processors and after the algorithm is finished, it collects the output and stores it in the host memory.

The physical size of our array (4 processors) is much smaller than the size of the model it has to imlplement. In terms of software, our program will be identical but more than one state (neuron) will be allocated to a single processing element by the configuration program. The question of how many processes to allocate on each processor is easily solved in our case since all processes are the same. To have a perfectly balanced load throughout our system, it is sufficient to allocate an equal number of processes to each processor.

The only remaining problems in the above implementation is the calculation of the sigmoid function. Direct evaluation using an

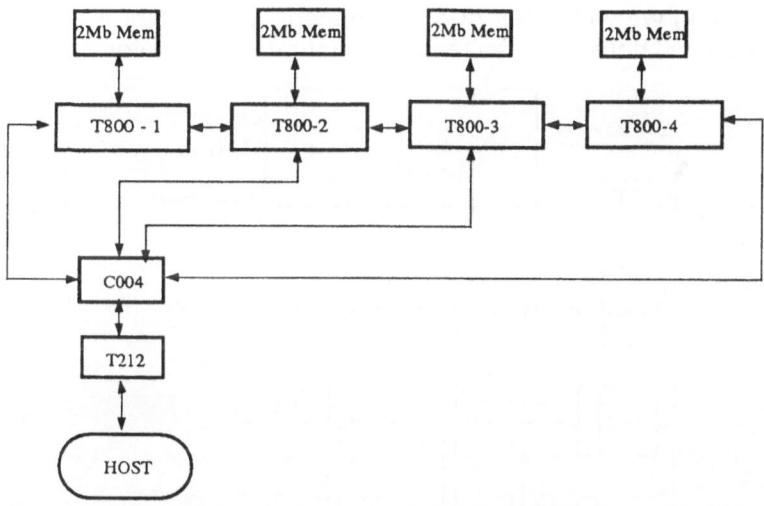

Figure 4   Transputer Array System

exponentation library function is 64 times slower than a multiplication.
We can however use a piecewise linear approximation of the sigmoid
[10]. The slopes and intercepts of the approximating lines are stored in a
small lookup table and the evaluation of the sigmoid requires accessing
the table and then performing a multiply/accumulate operation. Using
the above procedures on our Transputer ring array we were able to
achieve a processing rate of 10M connection evaluations per second in
the retrieving phase and 2.5M connections per second in the learning
phase. In contrast, in the straightforward realization of back-propagation
we achieved only 1.1M connections per second in the retrieving phase
and .6M connections per second in the learning phase.

Multichip Architectures

Architecture and Instruction Set

In this section we will describe the instruction set and a generic
architecture for the processing element of our system which will
subsequently be modified depending on the constraints of different
implementations.

Processing element block diagram: A generic block diagram for the
processing element is depicted in Fig. 5 [4]. It consists of the controller,
address generator, two memory banks, the I/O subsystem and the
floating point unit containing a multiplier and an adder. Each block is
accompanied with a set of registers. In order to provide fast execution,
we provide a number of busses that permit simultaneous data
movements from a variety of sources to a large number of destinations. In

214

the following we will detail how this generic architecture is to be used
and speciliazed depending on the implementation technology.

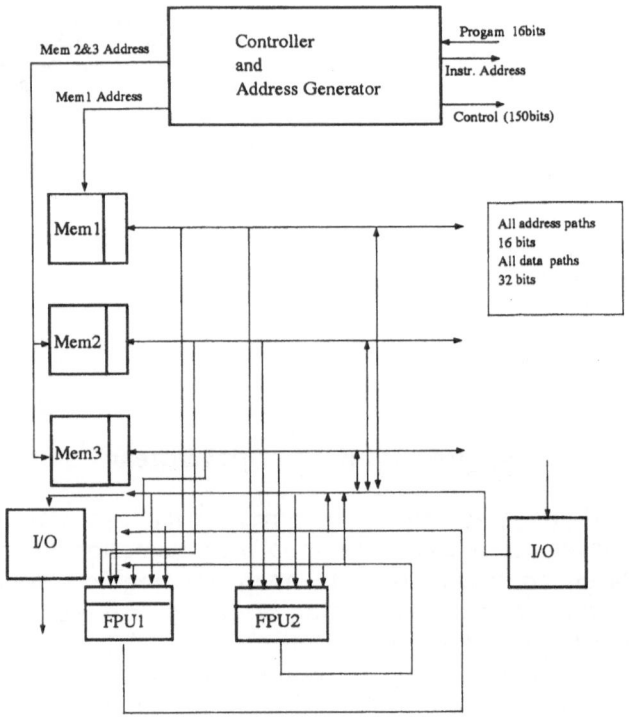

Figure 5   Processing element block diagram

Controller Architecture: As mentioned previously, in order to meet the
reconfigurability requirement, our processor will be
microprogrammable. Fig. 6 depicts the general architecture of the
controller. It consists of the program counter, instruction buffer, address
decoding stage and microprogram ROM and RAM. The ROM is used to
hold general purpose instructions useful for example for register loading
and initialization. The microprogram RAM is used to hold custom
microprograms, written to execute particular algorithms with maximum
efficiency. The controller also includes a status register and has some
external connections for control such as a HALT connection. In the
following sections we will discuss specific implementations of the
controller and its modifications dictated by the technology used.

Instruction Set The design of the processing element is mainly aimed at
the implementation of systolic algorithms. It is therefore natural to
provide highly optimized instructions for the realization of these
algorithms. On the ohter hand, we provide instructions for arithmetic
and memory access, useful for initialization and non pipelinable

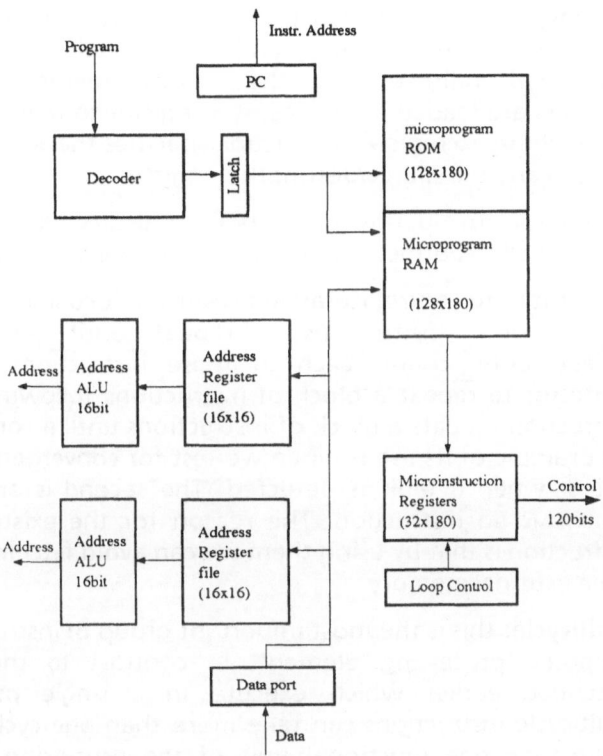

Figure 6 Controller and address generator block diagram

algorithms. These simple instructions can also be used for rapid prototyping and testing of algorithms thus avoiding the need for a detailed microprogram development that would lead to a new multicycle instruction. The machine level instructions of our processor will belong to the following groups:

1. **Register to Register Instructions:** As can be seen from Figure 5, there are 6 groups of registers: the address, multiplier, adder and accumulator groups as well as the I/O group. Each register to register instruction consists of 16 bits.

    The 4 most significant bits identify the instruction group. The rest are divided in two 6 bit fields: the source field and the destination field. The three high order bits identify the register group while the remaining three identify the individual register. The I/O instructions use the source/destination bits to set the I/O direction bits (in 2D arrays) since there is only one input and one output register.

2. **Memory to Register Instructions:** This instruction group is used to access the external memory and the internal ROM. Addressing is performed indirectly, via the address registers. The address registers are loaded by means of a register to register instruction. The memory to register instruction specifies the address register to be used and the source/destination register.

3. **Arithmetic:** to operate on pairs of registers. Our instruction set includes add, subtract and multiply operations.

4. **Repetition:** to control iterative execution. It contains two different type of instructions: repeat_until_condition and repeat_until_count. Each of these instructions prepares the processor to repeat a block of instructions following it. The first instruction repeats a block of instructions until a condition is met. An example of its use is when we test for convergence in a nerual net or when a fault is detected. The second is analogous to a FORTRAN do instruction. The reason for the existence of these instruction is that by using them we can avoid fetching instructions from external memory.

5. **Multicycle:** this is the most important group of instructions for the proposed processing element. In contrast to the instructions described earlier which execute in a single machine cycle, multicycle instructions can take more than one cycle and involve more than one functional unit of the processing element and multiple sources and destinations.

All instructions except for the multicycle instructions will be executed in one machine cycle. The multicycle instructions will be microroutines requiring more than one cycle to complete.

As an example of the multicycle instructions consider the following: The gen_step instruction is a generic instruction capable of implementing the three basic algroithms in neural networks. To understand its design, consider the basic algorithm steps:

MVM_step   Input to treg, Fetch weight from memory
            Multiply/Accumulate
            Output treg, store accumulator (optional)

VMM_step   Input to areg, Fetch weight from memory
            Fetch g from memory, Multiply/Accumulate
            Output Accumulator, Store accumulator to memory

OPU_step   Input to treg, Fetch ΔW from memory to accumulator
            Fetch g form memory, Multiply/Accumulate
            Output treg, Store accumulator to memory

From the above steps we can define the generic step instruction:

Gen_step   Input to Raddress, Fetch form Maddress to Raddress
Fetch from Maddress to Raddress, MAC
Output from Raddress, Store Raddress to Maddress

In the above, Raddress is a register address and Maddress is a memory address.

Memory Architecture: As can be seen in Fig. 5 the processing element has three local memory banks. In order to minimize the number of address bits, we adopted a *segmented memory space* approach. In this, the processing unit generates the offsets for words contained in a 64K segment while the ring controller sets the 16bit segment number which is common to all processing elements. In this way, we can provide an address space of 4Gwords.

As can be seen from the gen_steps instruction, in some operations we may need to perform two memory reads and one memory write in the same cycle (e.g. outer product computation in recursive ANNs). In this case, either one of the two memory banks should be dual ported to support one read and one write simultaneously or we should have an additional memeory bank. Since it is possible to share the address of a source and a destination in different memory banks, the local address space is divided in two: memory bank A and memory banks B & C.

Communication architecture.: To avoid clock skew problems and to be able to combine processing elements of different speeds, we adopted an *asynchronous communications protocol*. The architecture of the I/O interface is shown in Fig. 7. Each processor that needs to communicate with its neighbor issues an OREQ signal and puts its data on the bus and at the same time loads it to the output FIFO. If the corresponding processor is ready to accept the data, it issues an IACK signal to clock its output FIFO. If no IACK signal is received and the processor needs to send additional data, it loads it into its output FIFO and continues execution unless the FIFO is full in which case it waits until a slot is available. On the other hand, the receiving processor, when it receives the OACK signal, checks its input FIFO. If the FIFO is empty and it is executing an input instruction, it puts the data directly on the appropriate bus to its eventual destination. Otherwise, if it is not ready to process the data and its FIFO is not full, it accepts the data by issuing an IACK signal and puts it into the FIFO. In case the FIFO is full it simply ignores the OACK signal and continues processing.

Floating Point Units: Detailed scheduling of PE operations depends on the FPU architecture. For example, if the FPU has only two input ports (as is the case with most FPUs) the multiplier operands and the adder operand must be loaded in different cycles. At the same time, depending on the depth of the execution pipeline, the adder operand will be

Systolic I/O Organization

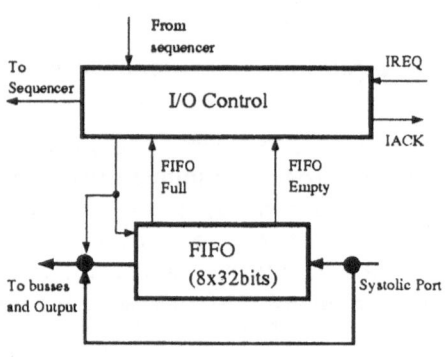

Input:
 If IREQ=true
  If FIFO = empty Then
                   { Read port; Bypass FIFO; IACK=true }
                Else
                   { If FIFO= full Then { IACK = false }
                                Else
                                   { Read Port; Write FIFO; IACK= true}
                }

Output:
 Set OREQ = true
 Write to Bus
 If OACK= true Then { continue }
                Else
                   { If FIFO = NOT full Then {Write FIFO; OREQ=false}
                                   Else {Halt}
                }

Figure 7   I/O control

needed a number of cycles later than the multiplier operands thus requiring the use of FIFO buffers.

Catalog Parts vs. Custom Parts.   In addition to performance requirements, the choice of a particular technology for processor implementation, imposes new constraints. For example, the Transputer design example shows that a single chip, commercially available PE solution is not capable of achieving the performance requirements but on the other hand gives us a lot of flexibility in mapping algorithms and partitioning large problems. In addition, commercially available chips are designed to address a much broader class of computational needs therefore they are results of very different design tradeoffs that result in underutilizing parts of the chip (e.g. memory management units) and overutilizing others that may not have adequate performance (e.g. FPU).

Multichip implementations based on all catalog microprogarammable parts (e.g. AMD 293xx, ADSP 32xx, ADSP 14xx) can meet the performance requirements for a single PE but the resulting processing element is large, usually occupying a whole PC board. In such a case, signals have to travel further, therefore delays are increased and power requirements are much higher since long wires have to be charged thus making *system* design harder although PE design is facilitated. For this reason, designs using all catalog parts are excellent for rapid prototyping and building small demonstration systems that achieve high performance levels.

On the other hand, a fully custom designed chip would contain all the architectural elements needed for both good PE design and good system design. Such a project however is both time consuming and difficult to achieve the performance levels attained by special purpose commercially available FPUs because of the necessarily small area devoted to each functional unit. Especially in the area of FPU design, it would be very difficult to match the performance of the Weitek, Analog Devices and AMD floating point units.

A compromise approach is to use custom designed chips for *parts* of the PE so that we can maximize performance and flexibility and use commercially available high performance chips for the FPU. This approach can utilize hardware very efficiently, is more economical and requires much less effort than a fully custom approach. The main drawback of this approach is the increased bandwith requirements since the custom designed part of the PE will have to provide a large number of I/O pins to control the FPU and memories. It is however the best approach for moderately large systems that have to provide very high performance.

In the following sections we will describe two designs implementing the architecture described above that exemplify the tradeoffs discussed in this section.

A Catalog Parts Design. We designed a board level processing element based on microprogrammable, commercially available building blocks in order to provide a platform where hardware and software features of our array can be tested. To achieve this goal the design emphasizes flexibility and performance but in doing so, its cost, size and power consumption is increased making the construction of a large system very difficult.

The block diagram of the processing element is shown in Fig. 8 and the functional blocks are briefly discussed below.

Microsequencer and Microcode Memory: The microsequencer is responsible for generating microcode addresses and handling interrupts occuring when we initially load the microprogram RAM and during I/O, when the I/O buffers are full. The microsequencer operates at a speed of

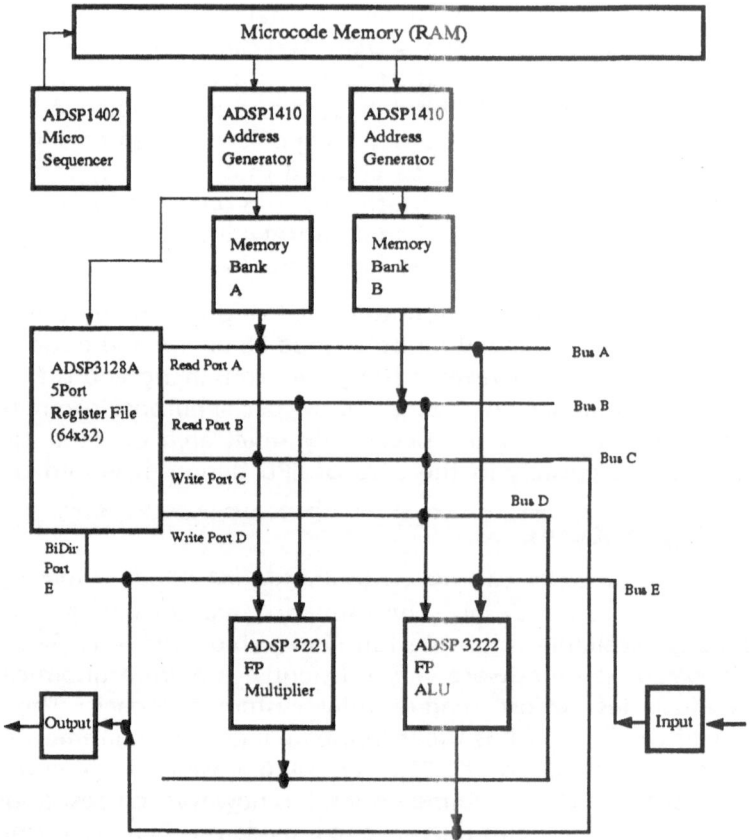

Figure 8   Board level prototype

50 nsec. The microcode memory consists of 512×180 bits and contains *horizontal* microcode to reduce delays and simplify hardware construction. Its access time is 30 nsec.

Address Generator and Data Memory: The data memory of our system is treated as a large circular buffer. That is, when addressing reaches the end of buffer, it wraps back to the beginning. The operation implementing this function can be described by:

$$Y \leftarrow R_{nj} \text{ IF } (R_n \geq C_i) \text{ THEN } R_n \leftarrow I_{kj} \text{ ELSE } R_n \leftarrow R_n + B_m$$

where Y is the output address bus, R_n is an address register, C_i is a limit register containing the limit address of the circular buffer, I_k is an initialization register and B_m is an increment register. The architecture of the address generator chip used is specially tailored for the execution of this operation.

It can generate addresses in 30 nsec and the multiplicity of address, limit and increment registers facilitate the implementation of partitioning algorithms that map large problems on a small physical array.

The data memory consists of three banks of fast SRAM (30 nsec). One bank is addressed independently while the remaining two, share the address bus. In ANN simulations, the third memory bank is used to store the change of the weights ΔW_{ij} during learning. Since we read weight Wij in order to calculate ΔW_{ij}, it is not necessary to recompute the address. In addition, the data memory also belongs to the ring controller address space so that data can be loaded directly during the initialization phase.

Execution Unit: The main difference between the board level PE and the architecture presented previously is the fact that we are using two floating point units instead of a single multiplier/accumulator. The main reason for this difference is the fact that most commercially available FPUs have at most two input ports. Since in some operations (e.g. OPU updating) we need to load three numbers to the execution unit, if the FPU had only two ports, we would need two cycles to complete an operation while with the present design, we can complete an operation in one cycle. At the same time, in non- ANN applications the same PE can process two independent data streams by executing two floating point operations at the same time thus doubling the throughput.

A Partial Custom Design. In order to reduce the size and power consumption of the processing element and at the same time increase the speed and capability of incorporating a variety of arithmetic units, the possibility of an implementation based on both commercially available and custom designed chips is explored. As mentioned previously, because of market demand, floating point units are highly optimized and produced with a variety of capabilities. A custom floating point unit would be very difficult to compete with commercially available ones both in terms of performance and in terms of cost. We decided therefore to use an off the shelf FPU and concentrate our design efforts to the rest of the processing element.

Let us examine now the constraints imposed by an integrated implementation of our processing element and the way we chose to satisfy them:

Flexibility constraints: Price and performance tradeoffs dictate the choice of a commercially available FPU. It would be unwise to tie our design to a particular FPU since technology evolves rapidly and new candidate FPUs may appear. In addition, it would be unwise to tie our processor array to one or two specific applications. It is therefore necessary to prefer a microprogrammable controller as opposed to one based on random logic in order to be able to change the control of the FPU and also change the functions implemented by our processing element.

Size constraints: Design time, yield and power consumption are directly influenced by the size of a chip. To reduce the cost we should use dynamic CMOS logic in our chip since it provides both small size and low power consumption. In addition, the complexity of internal busses should be kept at a minimum since they require a large percentage of the chip area and dissipate a lot of power. One solution to the bus complexity problem is to implement them outside of the chip and simply provide the means to control them.

Speed constraints: High performance is obviously the most important objective of our design. It is however conflicting with the size constraints discussed above. Dynamic logic is in general slower than static logic but requires much less area. Especially in memories such as the microprogram memories in our chip, dynamic logic can provide us with one transistor per bit storage while static logic requires six transistors per cell.

The solution to the conflict between area and speed requirements is based on the fact that the algorithms we need to implement are iterative. It is therefore possible to fetch an instruction, decode it and latch it on a register and then repeatedly issue the same control signals to the rest of the system. Since the multicycle instructions are microroutines and not single instructions, instead of a single control register we can provide a microoperation register bank with some control logic that will implement the repetitive operations. The rest of the chip and especially the memories can be made as slow as required by the size constraints and still maintain the performance level required.

Bandwith constraints: The number of I/O pins on a chip largely dictates its cost since packaging can be more expensive that the actual chip. In addition, driving output pins consumes a large percentage of the total power required by the chip. It is therefore important to minimize the number of pins in our chip. On the other hand, in order to satisfy speed requirements, we need a large number of I/O ports. There are two 32bit data busses for the memory, three (possibly four) 32bit busses for the FPU, two 32bit ports for the systolic communication, two 16bit memory address ports and two 16bit ports for program address and instructions. All of these ports would require 284 pins. In addition, we need a large number of control pins for the FPU (typically 50) and memories. We see that the proposed chip should have in excess of 300 pins.

One possible solution is to multiplex some busses so that they deliver their data in more than one cycle. A solution like this would be acceptable if we could design a chip that would run at least twice the speed of the FPU but since FPU speeds range from 10 nsec to 80 nsec, the speed required for our chip becomes prohibitive. The solution we prefer is to directly interconnect FPU, memory and I/O subsystems without going through the custom chip. In this way we can define a controller chip such as the one depicted in Fig. 6. This chip will be able to control the hardware since it procides 120 control pins and at the same time generate two independent addresses and communicate with the program memory.

Packaging considerations: We defined above a flexible and fast controller for our processing element but we still have not reduced the overall area by much as compared to the board level design. To drastically reduce processing element size we propose the use of Multichip Modules (MCMs). Such modules have been in use in various forms for more than one decade but recent advances make them even more attractive.

In MCMs, bare chips are mounted directly on substrates that have imprinted on them an interconnection pattern and the associated decoupling capacitors and terminating resistors. In essence the MCMs are like miniature printed circuit boards but have a number of distinct advantages shown in the following table:

Parameter	MCM Silicon-on-Silicon	PCB
Line density (cm per cm^2)	400	30
Line width (microns)	10	750
Max. Size (cm)	10	66
Pinout	800-3200	1600-3200
Turnaround time	1-10 days	2-13 weeks
Cost ($ per cm^2)	10	0.03

Although there are alternative technologies such as cofired ceramic substrates, we prefer using silicon-on-silicon technology because of its field programmability by means of fuses and also its inherent fault tolerance because of the possiblity of having redundant paths.

Bonding of the chips can be achieved in a variety of ways. The most common are wire bonding and tape automated bonding. A relatively new technique, flip-chip bonding, has become available (although used at IBM for more than 10 years). In this, bonding pads are at the same side of the chip as circuitry and the chips are bonded face down on matching bonding pads on the substrate. Bonding is achieved using anisotropically

224

conductive adhesives. This method minimizes lead inductance and signal delay and saves on substrate space.

A System Example In this section, the operation of a design using the custom controller described above and the Weitek 3332 FPU will be presented [4].

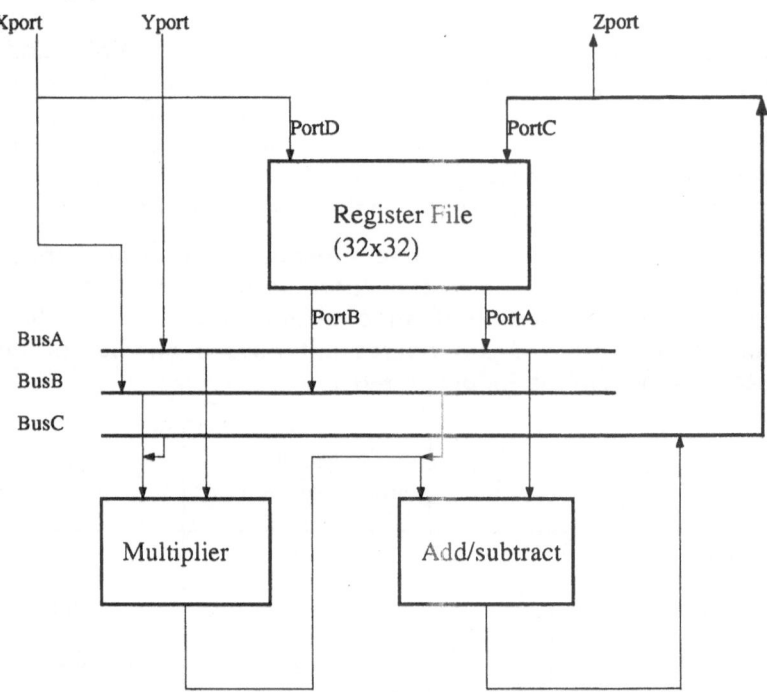

Figure 9   Simplified block diagram of the Weitek 3332 FPU

The Weitek 3332 [3] is a 32bit floating point unit containing a three stage pipeline multiplier/accumulator. In addition, it has two 32bit input ports, a 32bit output port and a four ported register file of 32 32bit words. An important feature that makes it suitable to implement our generic instruction for neural nets is its *double pump* input mode where data can be fed through its input ports at twice the basic speed by first loading a pair of data on the rising edge of the system clock and then loading the remaining pair at the falling edge of the clock. Using this mode we can feed the three inputs of our generic instruction during the same cycle. In the first half cycle, the two multiplier operands are loaded while at the second half cycle, the adder input is loaded. Since the adder input will be needed two cycles later, it is loaded into the register file which in this case is used as a FIFO. In every cycle therefore, the FPU is executing a multiply/add instruction with two operands coming from the input ports and one coming from one of three registers (selectable by microcode) in the register file. In the computation of the retrieving phase, the generic instruction can be greatly simplified by noting that the adder input is the output of the adder during the previous cycle. It is therefore unnecessary to use the double pump mode and the register file. We can instead use one of the temporary registers to hold the accumulator value.

FUTURE WORK

A software environment The architecture described above can be implemented on a small number of boards that will be connected to a workstation. Since the processing elements are microprogrammable, a user has the possibility of defining custom microcode implementing specific algorithms provided that they can be expressed in terms of inner and outer products. Microprogramming however is very tedious and it is worthwhile in real time applications. To make our system useful as a tool for neural network research, we defined a software environment where the user is able to specify an algorithm as a sequence of predefined operations in a high level language and execute it without being aware of the underlying hardware. In such cases, performance is expected to decline but programming effort is reduced significantly. The software environment will consist of:

1. *A Graphical User Interface:* to facilitate display of input and output values and network performance.
2. *A High Level Language:* with principal data types (matrices and vectors) to facilitate programming and at the same time facilitate translation to systolic algorithms. Primitive operations include matrix arithmetic and non linear function implementation.
3. *Algorithm Mapper:* to map the cascade of basic operations into a virtual systolic array. This tool is only possible for the limited class of basic matrix arithmetic operation.
4. *Array Partioning:* to map the virtual systolic array (with unlimited number of processors) to the typically smaller physical array in a way that maximizes performance. Again, this tool is possible for the basic matrix arithmetic operations.

Since the last tool is the most crucial one for the performance of a neural network simulator, we will give an example of the methods used for the case of matrix vector multiplication. Assume that the network contain 12 units and we want to use an array of only 4 processing elements.

Design for the Retrieving Phase The consecutive MVM operations can be partitioned based on the *locally parallel globally sequential* (LPGD) scheme [6]:

1. At (I + 1)-th iteration, the tasks of the 12 units in the network are uniformly distributed at the 4 PEs, and the associated weights are appropriately arranged as before. The twelve activation values of

previous iteration $\{a_j(l)\}$, instead of cycling once in the ring array, are cycling three times (each time with four values) to generate the first four net-inputs $\{u_i(l+1),\ i=1,\ ...,\ 4\}$ ready for nonlinear activation (see Figure 10), i.e.,

$$u_i(l+1)=\sum_{j=1}^{4} w_{ij}a_j(l)+\sum_{j=5}^{8} w_{ij}a_j(l)+\sum_{j=9}^{12} w_{ij}a_j(l)$$

2. The same procedure is again repeated twice to generate the rest two batches of activation values for the $(l+1)$-th iteration, i.e., $\{a_i(l+1),\ i=5,\ ...,\ 8\}$ and $\{a_i(l+1),\ i=9,\ ...,\ 12\}$.

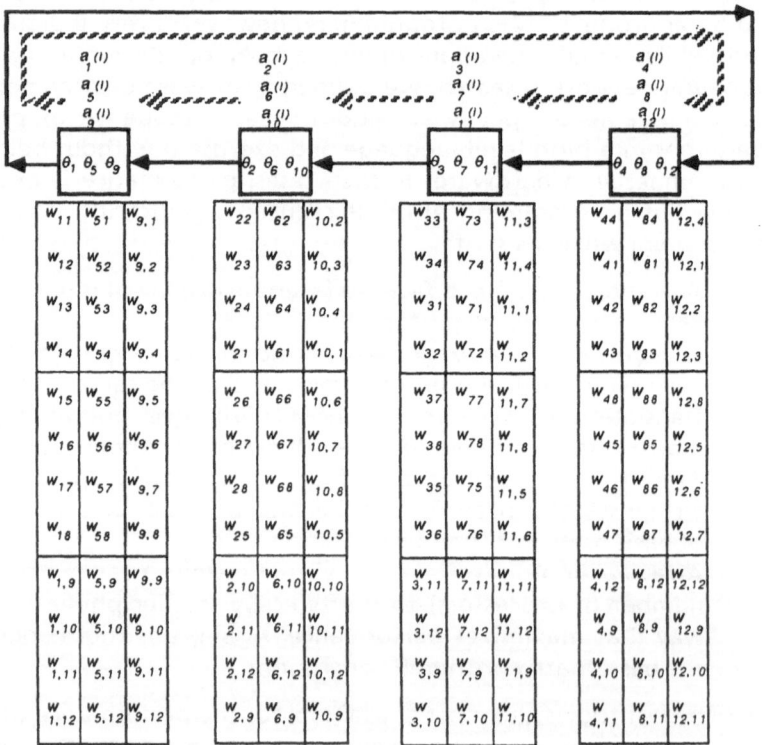

Figure 10   The ring architecture with 4 PEs can be used to implement the consecutive MVM in retrieving phase of an L-layer RBP (or HMM) network with 12 units

CONCLUSION

Although digital VLSI implementations of neural networks suffer from the drawbacks of larger silicon area, relatively slower speed and the greater cost of interconnecting processing units, these difficulties are in no way insurmountable. A possible approach to get around them is discussed in the paper. We believe that digital techniques, when fully developed for neural netwoks, can offer very attractive advantages which can not be matched by their analog counterparts.

REFERENCES

[1]   DARPA Neural Network Study. AFCEA International Press, 1988.

[2]   C. A. R. Hoare. Communicating sequenctial processes. Communications of the ACM, 21:666-677, 1978.

[3]   Weitek Inc. Wtl 3332 32-bit floating point data path. 1988.

[4]   S.Y. Kung, J. A. Vlontzos and J. N. Hwang. A VLSI array processor for neural network simulation. Journal of Neural Network Computing.

[5]   J. A. Vlontzos and S. Y. Kung. Hidden markov models for character recognition. In Proc. IEEE Int'l conf. on Acoustics, Speech and Signal Processing, May 1989.

[6]   S.Y. Kung. VLSI Array Processors. Prentice Hall Inc., N. J., 1988.

[7]   S.Y. Kung and J. N. Hwang. Parallel architectures for artificial neural nets. In IEEE, Int'l Conf. on Neural Networks, ICNN'88, San Diego, pages Vol.2: 165-172, July 1988.

[8]   S.Y. Kung and J. N. Hwang. A unified systolic architecture for artificial neural networks. Journal of Parallel and Distributed Computing, Special Issue on Neural Networks, 6(2), April 1989.

[9]   INMOS Ltd. Transputer Reference Manual. 1987.

[10]   W. D. Mao and S. Y. Kung. Implementation and performance issues of back-propagation neural nets. Technical Report, Electrical Eng. Dept., Princeton University, January 1989.

[11]   J. N. Hwang, J. A. Vlontzos and S. Y. Kung. A systolic neural network architecture for hidden markov models. IEEE Trans. ASSP, Dec.1989

TOROIDAL NEURAL NETWORK: ARCHITECTURE AND PROCESSOR GRANULARITY ISSUES

S. JONES, K. SAMMUT, Ch. NIELSEN and J. STAUNSTRUP

INTRODUCTION

The parallel and distributed nature of large neural network computations means that they execute very slowly on sequential processor architectures. The fine-grained and richly-connected structure of neural networks means they map poorly onto the coarse-grained restricted IO bandwidth found in many MIMD architectures such as the transputer. This has stimulated a wide range of researchers to develop fine-grain parallel processing architectures capable of supporting rich inter-connectivity for the support of neural networks.

Many researchers have addressed the recall aspect of neural networks. However, learning is at least an order of magnitude more complex a processing task than recall. This suggests that any hardware processor for neural networks should support both learning and recall.

Our work aims to develop a programmable neural network engine capable of high-speed support of both learning and recall. Such a machine has applications both as a general purpose neural network engine suitable for real-time applications and also, as an accelerator card which can be attached to a workstation that can speed up neural network design and training.

Key to our motivation is the development of an architecture which based on standard digital technology, can utilise VLSI technology to implement a large number of neurons at a modest cost.

This chapter reports on an architecture under development at Nottingham University, the toroidal neural network processor. The ring interconnection strategy is overviewed and it is argued that this structure satisfies both the rich interconnect and fine grain parallel processing of neural networks, together with satisfying the limited physical connectivity imposed by VLSI packaging technology. The architecture, operation and performance of the toroidal processor are described.

Alternative implementations based both on wafer-scale integration (WSI) and self-timed circuits are described together with the architectural modifications necessary to support these different design styles. The self-timed version is being developed jointly with the Technical University of Denmark.

The final section reports the results of an investigation into establishing the inter-relationships between processor granularity, and process and package technology that was undertaken as part of the development of the first toroidal neural network processor chip.

RING ARCHITECTURES

The ring is a much neglected interconnection strategy. Yet for a wide range of tasks it provides an attractive trade-off between applications requirements and technological constraints. It comprises a linearly connected array of elements with the last element connected to the first.

Such a simple structure has some significant advantages, namely

-- Regular and localised interconnection makes them attractive for VLSI implementation
-- The low IO bandwidth (viz. at the two ends of the string) allows large numbers of elements to be integrated on-chip, yet require relatively few pins for IO.
-- Shifting data between elements results in a high total interprocessor bandwidth for local communications
-- Shifting data along 'n' positions in an 'n' element torus provides total connectivity ($O(n^2)$ separate communications) in $O(n)$ time with constant chip IO.

The ring concept can be extended to higher-order structures. A 2-D ring is a torus and provides a suitable interconnection strategy for neural networks. Higher-order rings are possible. The difficulty with higher-order rings is that they are difficult to implement in planar technologies. However, this does not exclude optical implementation of higher-order rings.

TOROIDAL NEURAL NETWORK

Structure

Figure 1 shows the structure of the toroidal neural processor. The basis of operation is that the states of each neuron revolve around the horizontal axis. The weights associated with each state evolve around the vertical axis. After 'n' cycles (where 'n' is the number of neurons in the network), all (weight, state) pairs have met up at a processing element and either summed (in recall) or modified (in learning).

The Torus architecture will also support Multi Layer Nets (MLNs). This is not a new idea, other researchers working with ring architectures have also implemented MLNs. However, the conventional approach has been

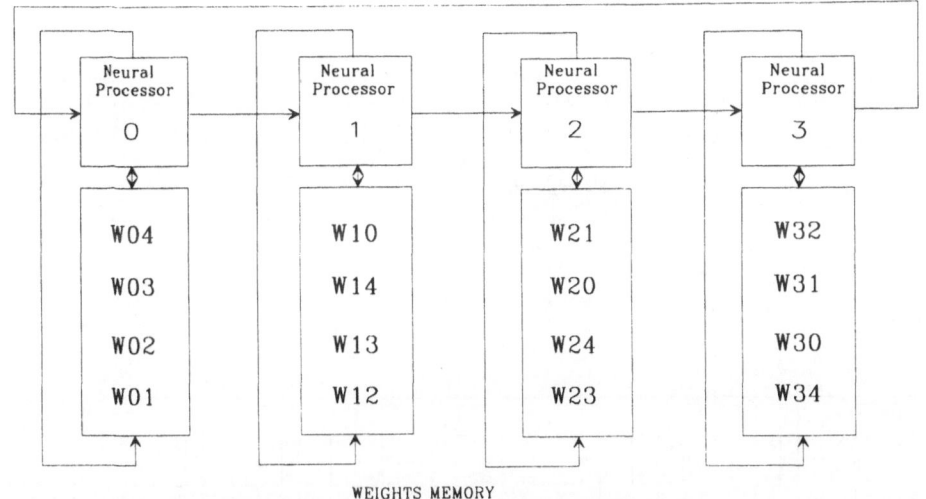

WEIGHTS MEMORY

Figure 1   Toroidal neural processor

to use multiple rings (one per layer) [1] or to use each PE in the ring to emulate a column of nodes [2,3], whereas our approach is to place each layer sequentially in the ring along side its neighbours.

Since the size of each PE's weight memory is equal to the number of PEs in the ring, then each neuron can have weighted connections with every other neuron. If no connection is to be made between two neurons, then the appropriate weight value can be set to zero. Figure 2a shows a 2-4-1 MLN with connections between neurons of adjacent layers only. Figure 2b illustrates how this structure is implemented in the Torus architecture. From this figure it can be seen that the matrix for less than fully connected networks can be sparse in places. It may well be appropriate to use coding techniques to avoid such inefficiencies.

Although this approach has the drawback that it will require one entire Torus cycle (one cycle through the entire ring of PEs) for each layer, it allows the entire net to be implemented in a single ring. In this manner the number of layers which can be implemented will be unrestricted. Furthermore, the allocation of different numbers of neurons to different layers can be easily achieved without the need for re-configuration at the hardware level. The single ring also simplifies WSI and self-timed implementation. This will be discussed in more detail later in the Alternative Implementations section.

All data is pipelined around the array. Globally broadcast signals to each processing element control the operation and sequencing of the ring.

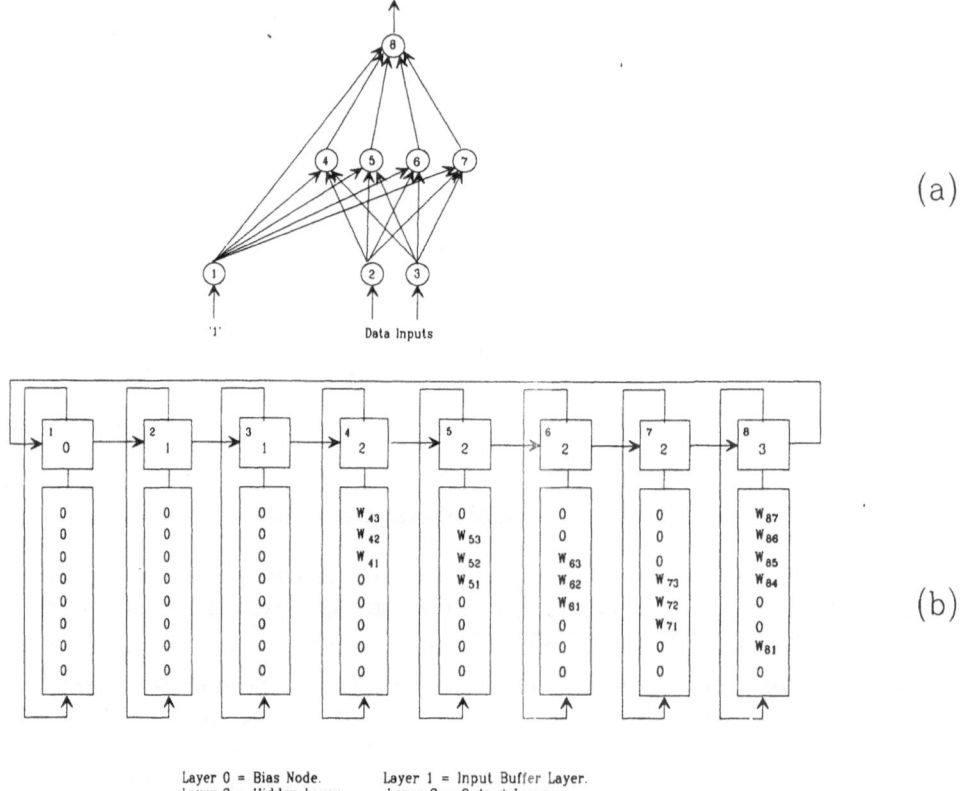

Outputs

Data Inputs

Layer 0 = Bias Node.   Layer 1 = Input Buffer Layer.
Layer 2 = Hidden Layer.   Layer 3 = Output Layer.

Figure 2   Implementation of multiple layer network in torus

Figure 3 shows the structure of the processing element. The parallel multiplier and adder form the arithmetic unit. Three types of registers are provided to allow the neural algorithms to be implemented.

-- Parameter Registers (ie. IPOP, Highlim, Lowlim, Gain, Lrnconst1 and Lrnconst2). These registers contain parameters used by the arithmetic unit and cannot themselves be changed by the arithmetic operations. IPOP is used to store either the input or desired output elements of the pattern vector depending on the layer the PE belongs to. The Highlimit, Lowlimit and Gain registers specify the slope of the activation function while Lrnconst1 and Lrnconst2 determine the learning rate (or any other relevant coefficient)

-- Working registers (ie. Acc, Act, Error). These registers hold the result of arithmetic operations. The Acc (accumulator) contains the intermediate values of a

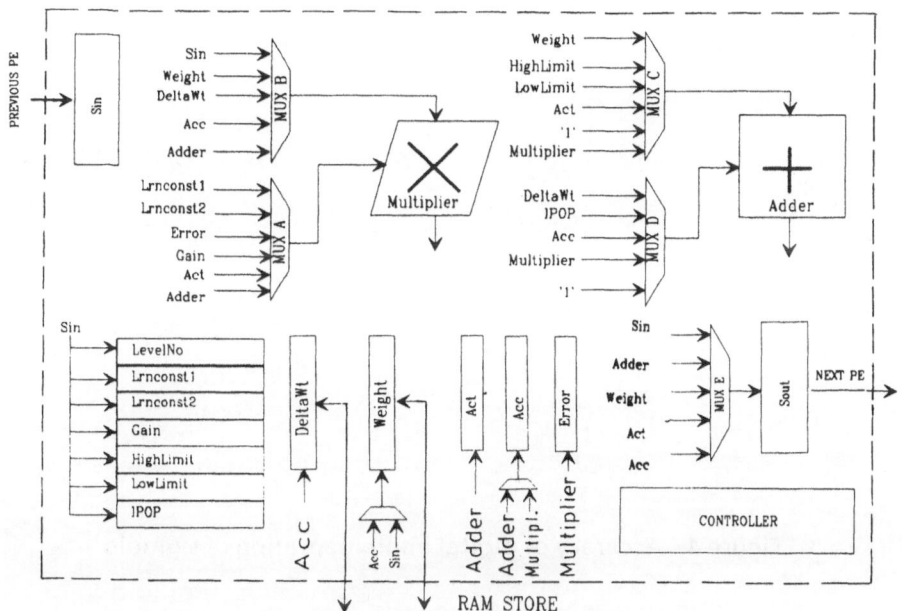

Figure 3   Processing Element Schematic

calculation, while the Act and Error registers are provided to hold the results of the Activation function and Error determination operations respectively.

-- Tag Register. A LevelNo register is also provided to identify which layer the PE has been assigned to. All instructions are tagged so that only the layer which matches the instruction's tag number may execute it.

The activation function is generated using the same digital hardware as used for the Learn and Recall modes of the neural operation. This approach allows a range of activation functions such as step, threshold and non-linear functions to be produced without using any specialised hardware. Using 12 bit precision arithmetic and implementing a squares function, it has been observed that a very close approximation to the sigmoid and hyperbolic non-linearities may be obtained. Figure 4 demonstrates the similarity of the 12 bit square approximation to the sigmoid function. Equation 1 shows the approximation function for the sigmoid, while Equation 2 shows the square function approximating the sigmoid. In (2) λ represents the gain of the slope and X_{SAT} the saturation point.

234

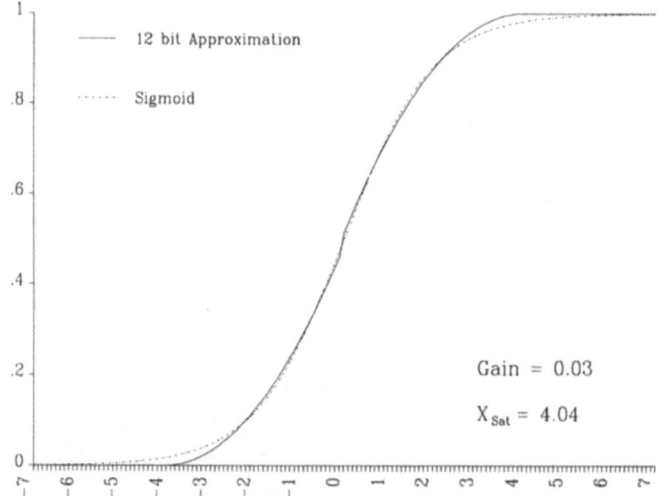

Figure 4   Accuracy of digital implementation of sigmoid

$$f(a) = \frac{1}{1+e^{-a}} \tag{1}$$

$$
\begin{array}{ll}
x = 1 & x \geq X_{sat} \\
x = 1-\lambda(X_{sat}-X)^2 & X_{sat} > x \geq 0 \\
x = \lambda(X_{sat}-X)^2 & -X_{sat} < x < 0 \\
x = 0 & x \leq -X_{sat}
\end{array} \tag{2}
$$

Processor complexity

The current processor is estimated to require a data accuracy of around 18-bits in order to learn successfully over a range of applications. Assuming this, the processor size breaks down as follows (Table 1).

These are figures for a bit-parallel implementation. Pipelining the calculation in the time-domain as well, will allow simpler arithmetic units (eg. bit-serial operation). As will be shown later the precise choice of processor granularity is a complex issue to determine.

Logic Unit	Number of Gates
Multiplier (18 × 18 bits, 18-bit truncated output Baugh-Wooley)	2667
Adder/subtracter (Carry Look-Ahead, 18-bits)	1200
Registers (13 × 18-bit, with clear)	3042
Control logic (10%)	691
Total	7600

Table 1   Processor gate count

Instruction set

The following is a summary of the instruction set currently developed to implement the Back Propagation algorithm (cf. Learning) together with the number of clock cycles which they require.

The instructions are categorised as follows in Table 2. Each instruction is in turn broken down by the Torus array controller into a set of low-level hard wired instructions. Different operations required to support alternative networks can be realised by different combinations of the hard-wired instructions.

The array controller is also responsible for transferring weight data between PE and local RAM store.

Operation

The processor operates in 2 modes, namely recall and learning.

Recall. The equation for recall in an L layer net can be written as

$$X_i(l+1) = f(\sum_{j=1}^{N_1} W_{ij}(l+1) \times x_i(1) + \theta_i(l+1)) \tag{3}$$

where W_{ij} represents the synaptic weight between neuron j of layer l and neuron i of layer l + 1, x is the input or the activated output of a node, and θ is the bias input to that node.

Recall Implementation. The recall operation is implemented in the Torus using 6 distinct phases. At any one time all cells are in the same state.

CLASS	INSTRUCTION	OPERATION	CYCLES
Resets			
	Hard_Resets	Resets all Rregisters	1
	Soft_Resets	Clear	1
	Clear	Clears Scan path (Sin → Sout)	1
Load/Unload			
	Load_LevelNo	Loads contents of Sin into LevelNo	1
	Load_Mem	Loads contents of Sin into Memory	1
	Load_IPOP	Loads Sin into IPOP	1
	Load_Lrnconst1	Loads Sin into Lrnconst1	1
	Load_Lrnconst2	Loads Sin into Lrnconst2	1
	Load_Gain	Loads Sin into Gain	1
	Load_Highlim	Loads Sin into Highlim	1
	Load_Lowlim	Loads Sin into Lowlim	1
	Shift_Out	Shifts out data from pipe	1
	Unload	Place IPOP into Sout	1
	Activate	Place Act into Sout	1
Forward propagate			
	process	Multiply Sin with Weight and add to Acc	1
	Sigmoid	Place Activation result in Act	2
Error propagate			
	Output_Error	Determines BackProp output layer error and places result in Error.	3
	Hidden_Error	Determines BackProp hidden layer error and places result in Error.	2
	Sum_Error	Sums Weighted error and places result in Error	1
Weight Modification			
	Weight_Update	Updates weight and places results in Weight and Delta Weights	5
	DeltaWt_Update	DeltaWt for Cumulative BackProp	3
	MainWt_Update	Updates Weight from DeltaWt	2

Table 2   Torus Instruction set

1. Reset: all registers are cleared

2. Initialise: In this phase the weight matrix of each neuron together with the activation function parameters are loaded in via Sin - Sout (State in port, State Out register) which are linked together in a scan-path.

3. Load: The input vector is loaded into the IPOP registers of the input buffer layer.

4. Process: In process phase each processor sends its current state (in the Sout) register to the downstream processors. Each processor receives the neighbours data from the Sin port. This is multiplied by the incoming weight on the Weight register. The product is formed and the sum of products accumulated in the Acc register. After N cycles all (weight, state) pairs will have been summed.

5. Activation: The activation function is applied to the contents of the Acc using the approximation method. This will yield the new state of the neuron, which is transferred into Sout.

6. Output: Here, the neural states and weights are output along the scan-path.

If multiple layers are implemented in the torus, then the process and activation phase must be repeated for all hidden and output layers in turn before the result can be extracted.

Learning. Learning is a considerably more complex process than recall. Also, unlike recall, there are a variety of learning algorithms which the Torus must be able to implement if it is to provide true emulation capabilities. The learning algorithms must be broken down into a sequence of simple operations. These operations can then be implemented using a carefully selected set of instructions which make use of the parameter registers provided. Hence, the Torus will be able to implement a range of these learning algorithms using a common and versatile set of instructions.

The standard process used for MLP learning is known as the Error Back Propagation (BP) algorithm [5]. This is an iterative gradient descent method designed to minimise the mean squared error between the desired output and the actual output from the net. In a classification task, the output is a vector that represents a particular class of input vectors. The learning operation is implemented in the Torus using the following method

1. Reset: Reset all registers.
2. Initialise: Initialise all weight vectors with a desired connectivity of small random weights. Load all parameter registers.

3. Load: Load the input pattern vector into the IPOP registers of the input buffer layers and the associated output pattern vector into the IPOP registers of the Output layer.

4. Forward Propagate: Forward Propagate the inputs to the output layer as in the recall mode.

5. Output Error: Determine the error of the result and place error in Error register.

6. Error Back Propagate: The value of the Error registers in the upper layer neurons are multiplied by the weight value. Their product is then added with the value of Sin and the result is placed in Sout. After one Torus cycle, the Sout registers in the hidden layer will contain their respective summed weighted errors.

7. Hidden Error: The hidden error is then obtained from the back propagated value by taking the value from Sin and multiplying it by the differential. The result is placed in the hidden layer's Error registers.

8. Weight Modification: The weight value is updated by calculating the product of the Error value, the learning rate (contained in Lrnconst1), and the lower layer's activation values (together with the bias value) which are circulated around the scan path. The DeltaWt value which contains the value of the previous weight change, is multiplied by the momentum term (contained in Lrnconst2) and added to the previous product. After one Torus cycle all the weights will have been updated.

Alternatively if Cumulative Back Propagation is desired, the DeltaWt value is obtained from the learning rate, error and input product. At the end of a presentation of P patterns the DeltaWt is added to the main Weight. Although this method requires one extra Torus cycle to update the main weight, the overall number of clock cycles required will be less than the conventional method's requirement.

ALTERNATIVE IMPLEMENTATION

Wafer-scale integration (WSI)

As will be shown later, using large (greater than 1 million transistor) chips, the main bottleneck with the Torus approach is the IO pin count arising from the use of weights memory on a separate chip from the processor. Each PE must have access to its own local memory. Unless the memory can be integrated on the same chip as the PE then it will be impossible to integrate a large number of PEs on a single chip due to pinout limitations. Conversely however, if the RAMs were integrated with the PE on the same chip then while the pinout constraint would be simplified, the RAM would take up a considerable portion of the chip. This would again limit the PE count.

These problems can be circumvented by significantly increasing the die area available by implementing the Torus as a WSI device. The only pinout required from the wafer would then be limited to the control signals and the data paths. The number of PEs (together with local RAM) that could be integrated on the same silicon slice would then be dramatically increased. However it must be remembered that since the total size of the RAM memory is a square of the number of PEs in the Torus, then the memory will always be the limiting factor to the integration count.

Apart from the increased integration count, WSI allows the natural system bandwidth to be captured because none of the data paths need to be restricted by pinout limitations. Moreover interprocessor communications and processor/memory bandwidth are significantly faster because all connections are local and do not have to pass through IO pin drivers.

Although WSI holds many attractive features, the large silicon area will cause a rather low yield of working PEs unless appropriate fault tolerance is implemented.

Fortunately the Torus architecture is well suited to a fault tolerant WSI implementation because of its one dimensional structure. Detailed research in linear array and ring implementation [6,7] has shown that using a linear reconfiguration algorithm ('The Catt Spiral') a high proportion of working PEs can be configured into a fault-free ring . Furthermore, the natural fault tolerance of large neural nets permits correct operation even if some of the neurons fail after learning.

All the PEs simultaneously execute the same instruction and data is passed to a specific PE by placing it in the correct position in the data pipe input sequence. The shift register implementation of the RAM elements allows easy mapping of the input stream into the memory and (vice-versa). Both these features eliminate the need for specific PE and memory addressing hardware.

The long distances which global signal lines have to travel across the wafer causes clock skewing and synchronisation problems. These difficulties can be avoided by using slow global and fast local clocking strategies. In addition local instruction stores may be implemented and instruction clusters (i.e learn or recall) can then be selected using a mode select signal line. These features are currently under investigation and should ensure that high throughput can still be achieved.

Self-timed implementation

Self-timed circuits are asynchronous (i.e. not controlled by a global clock). In general, this means that their operation speed is determined by their average processing times rather than their worst case as in synchronous circuits. Furthermore, large synchronous processor arrays in CMOS VLSI pose formidable problems in terms of clock loading and current spikes on the power distribution network. Using Self-timed circuits avoids the clock loading and distribution issue and eases power spikes by 'smearing' the current requirement over a larger proportion of the computation cycle (cf. synchronous CMOS where almost all power is consumed during the clock transition).

In addition to these general advantages of self-timing, there are specific advantages in using self-timed circuits that arise from both the nature of the neural calculation, namely

-- The neural calculation is based around vector-matrix calculations. However, where the network is less than fully connected many of the elements in the matrix will be zero. Of course multiplication by 0 is faster than multiplication by any other numbers. This means that the average speed of calculation can be significantly less than the worst-case speed. Thus, self-timing can achieve large speed improvements.

-- In a neural network there is no separation of data and control, the current state implicitly controls the computation. This fits nicely with the approach we use for designing self-timed circuits where data and control are mixed to avoid delay dependencies between control logic and the data path.

The following section provides background information on the building blocks used in the construction of the self-timed ring- based neural network.

Self-timed neural networks

Self-timed circuits are data driven, so a computation is performed when input data is ready. Outputs initiate other computations which in turn can trigger further computations. So whereas a synchronous circuit computes on the arrival of a clock signal, a self-timed circuit proceeds as soon as the input data is ready. It is important that inputs change otherwise it is not possible to detect their arrival. A commonly-used protocol is where a stream of inputs alternate between a distinct value E (empty or reset) and proper values, for example

$$v1 \ E \ v2 \ E \ ... \ vk$$

Using such an alternating protocol sometimes called return to zero coding, the arrival of a new proper value is detected by the input signal being different from E. There are many encodings which can be used to implement such an alternating protocol, an overview is given in [8].

The fundamental circuit structure for self-timed circuits is a queue of elements maintaining the alternating protocol. In a stable configuration, the nodes at the tail of the queue all hold the same value and the nodes at the head of the queue alternate between empty and proper values.

Input of a proper value can only occur when the two first queue elements are empty. Similarly, input of an empty value can occur only when the first two nodes hold a proper value. When a new proper value is added to the queue, it is propagated from the tail of the queue towards the head until it reaches an element whose successor's output is empty. During this process each queue element alternates between holding a proper value ($\neq$ E) and having the empty value (E). This alternation makes it possible to detect changes, which underpins the correct operation of a self-timed circuit. The queue is made out of building blocks of the form

$$(in = E)(\neq succ = E) -> out: = in$$

When $(in = E) \neq (succ = E)$, either $((in = E)$ AND $(succ \neq E))$ or $((in \neq E)$ AND $(succ = E))$, then in both cases it is safe to propagate the value of in to out. There is a very simple realisation of such a queue element using Muller-C elements.

The building blocks presented below are simple variants of the queue element, but together they form a powerful toolset for the construction of self-timed rings.

The pipeline element. The queue element can be extended with a combinational function, which computes a new value to be output, instead of just copying the input as it is done in the queue

$$(in = E) \neq (succ = E) -> out: = F(in) .$$

The function F can be arbitrarily complicated and time consuming. However, there are some restrictions on the realisation of it;

-- F must be E-preserving, so F(E) = E and no other value must yield E

-- F must be monotonicity preserving.

A monotone signal changes from E to a proper value without passing through other proper values, or conversely it changes from a proper value to E without passing through other proper values. A function is monotonicity preserving if it gives monotone outputs assuming it is provided only with monotone inputs. So F can take an arbitrarily long time to change provided it changes the output monotonically.

Join and split elements. Two sequences following the alternation protocol can be merged with a join element

$$(in1 = E) \neq (succ = E) \text{ AND } (in2 = E) \neq (succ = E) -> out: = F(in1,in2)$$

Similarly a sequence can be distributed to two different destinations with a split element

$$(in = E) \neq (succ1 = E) \text{ AND } (in = E) \neq (succ2 = E) \rightarrow$$

$$out1, out2: = F1(in), F2(in)$$

Self-timed ring structures

By connecting the first and last elements, a sequence of three or more elements can be made into a ring. Indeed, such a self- timed ring with 3 elements has been used to construct a high- performance floating-point division chip [9].

In a large ring many elements execute in parallel. How many depends on the variation in processing times. One pipeline element can become a bottleneck if it is consistently slower than all other elements. However, if on average the elements take the same time then a significant speed up is possible, whereas in a synchronous version all processors would have to perform at the rate of the slowest element.

A pure ring is insufficient for practical computation because it has no capabilities for IO. However, the join and split structures provides the capability for supporting IO. Moreover, they can be used to form higher-order rings (Figure 5).

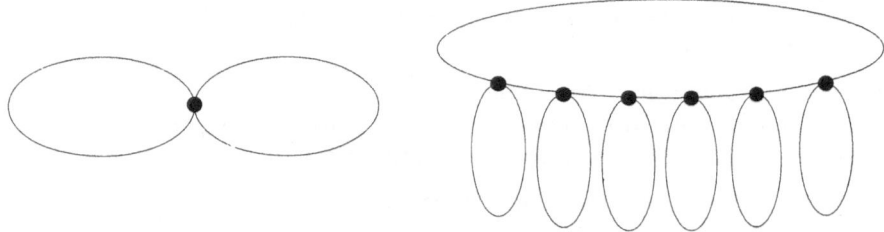

Figure 5   Use of Join and Split Elements for forming higher-order rings

Self-timed toroidal neural network

The toroidal architecture described in the previous sections is very well suited for a self-timed realisation. While it is simple to build self-timed rings, the control signals used in the synchronous version must be avoided, because they represent a centralised delay dependence. It is for example, essential that all synchronous processors at any one instance are in the same mode. In a self-timed implementation it is axiomatic that one cannot assume control signals (eg. soft-reset) arrive simultaneously at all processors. On the contrary, the circuit must work properly regardless of the delay on the wires connecting processors and the difference in the speed of the processors. There are two main ways in which this requirement can be satisfied;

-- Merger of the control and data paths. This can be done by tagging each data element with a special control token. Each piece of data is followed by a control signal, which determines the operations to be performed on that specific piece of data.

-- Distribution of the control to each of the processors in the ring. This then limits dependencies to those of passing data between adjacent processors.

The control of a larger system may utilise both strategies depending on the origin of the control.

A self-timed neural network has been designed using this approach. The overall architecture is similar to the synchronous version except that learning is not supported (Figure 6).

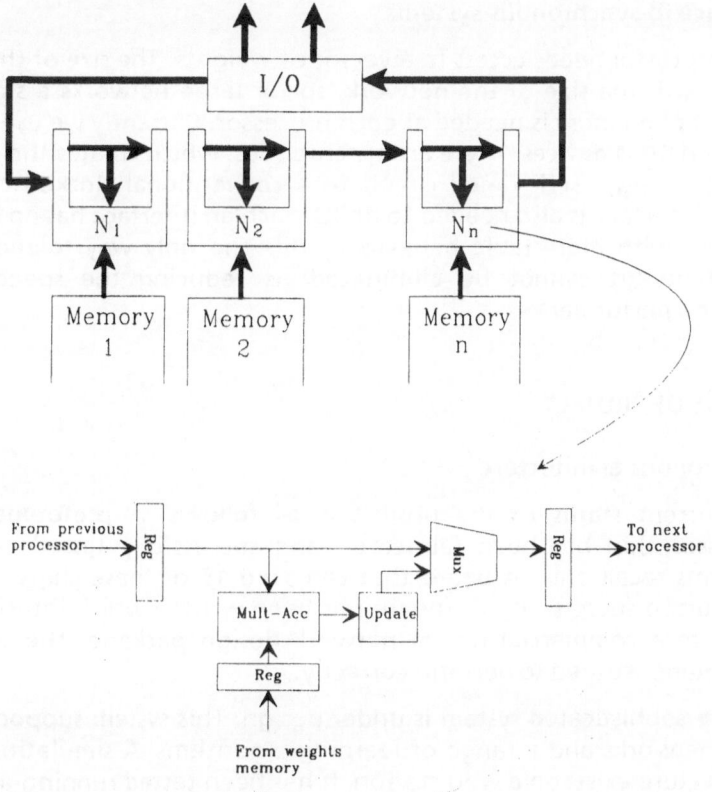

Figure 6   Self-timed processor architecture

Each processor has access to its own memory containing weights, elements to control the operation of the update function and signals to control the operations performed on the data. By programming the contents of the local memory it is possible to dynamically reconfigure the operation of each processor to emulate multiple neurons. Furthermore, it is possible to reconfigure even though other processors are still computing and passing on state elements. Inputs to the reconfiguring processor will just queue up until it is ready to resume normal operation.

The input and output of the neural states in the ring is controlled by a front-end processor attached to the IO unit. This unit tags the circulating state elements with a token telling the processor the status of the particular state element (eg. whether it is a new element in the ring). These tokens together with the control signals from the local memory enable the processors to determine the correct behaviour according to the current input data.

Interface to synchronous systems

Each processor needs access to a vector of weights. The size of this vector grows with the size of the network, so for large networks a significant amount of memory is needed at each processor. Currently we use off-chip standard RAM devices. These are synchronous, which creates the problem of interfacing a self- timed circuit to a conventional clocked circuit. A similar interface is also needed to do IO. Such an interface has an inherent problem with metastable behaviour [10]. The only way to reduce the probability (it cannot be eliminated) is reducing the speed, again lowering performance.

STATUS OF PROJECT

Synchronous architecture

The current status of the project is as follows. A prototype neural processor chip has been fabricated and successfully tested. The chip performs recall only. A board that can hold 16 of these chips has been constructed successfully. Software has been written which interfaces the board to a commercial neural network design package. The torus has been demonstrated to perform correctly.

A more sophisticated system is under design. This system supports multi-layer networks and a range of learning algorithms. A simulation of this architecture exists on a workstation. It has been tested running a number of demonstrator algorithms and performs identically to commercial neural network CAD packages.

Funding has been received to permit the construction of a hardware prototype of the above system. It will be complete by the third quarter of 1991.

Current research investigations are concerned with the following aspects;

-- Identifying suitable instruction, architectural and circuit strategies for the support of the widest possible range of learning algorithms,

-- Assessing strategies for efficiently supporting large networks on smaller processor arrays.

Wafer-Scale

In addition to the above research topics, there are a number of specific research issues needed to be addressed in the development of a WSI version;

-- Partitioning of instructions to permit a slow trans- wafer instruction stream and a fast local instruction execution,

-- Techniques for optimising the use of the on-wafer memory (eg. coding or varying accuracy data representations).

Self-timed

The self-timed research is focusing on the development of a self- timed version of the multi-layer learning machine so as to increase our expertise and understanding of the design issues involved in self-timed systems implementation.

PROCESSOR GRANULARITY ISSUES

This section reports the results of some recent experimentation on a recall-only version of the processor.

Objective

The objective of this experimentation was to assess hardware performance trade-offs in the processor. Analysis revealed that performance is determined by the following factors.

-- The number of processors that can be integrated in the system. This is determined by the number of processors that can be integrated on a chip. This is in turn limited by 2 possible factors, namely the number of transistors that can be integrated while still achieving adequate yields and the maximum pinout of the chip which restricts the number of processors as each one requires pins for access to its portion of the off-chip weights memory.

-- The speed at which each processor operates. The processor spends the majority of its time in process mode, consequently process cycle time dominates overall performance. There are 2 possible limiting factors, namely the delay in computing the multiply- accumulate operation of Acc : = Acc + (Sin.Weight) and the delay in transferring data from the weights memory to the Weight register on the processor.

-- Since the multiply-accumulate operation and the fetch operation for the next cycle can be overlapped, the maximum processor speed is the greater of these 2 values.

A number of design factors are pertinent to determining processors/chip and processor speed, namely data precision of the state and weight variables; speed and complexity of the multiplier-accumulator, and update logic; access speed of the weights memory; data bandwidth between the processor and weights memory.

The data precision is determined by the application data and convergence criteria of the neural system. The access speeds of the memory assuming DRAM technology is somewhere around 100ns.

However, the data bandwidth factor has a number of important implications for system performance, including; having a wide datapath (eg. 16-bit) maximises the bandwidth, but requires the most pins and the largest processor. This will tend to reduce the number of processors per chip but maximise performance per processor. Having a narrower datapath (eg. 2-bits) and taking say, 8 cycles to calculate the weight-state product, requires less pins and reduces the processor size. This will tend to increase the number of processors per chip but reduce performance per processor. Figure 7 illustrates the options possible.

Consequently, the performance of a particular partitioning strategy is dictated by 4 main factors: the integration limit, the package pinout, the processor to weights-memory bandwidth and the processor speed. Key to the successful application of such an architecture is exploring the interaction between these factors to identify appropriate processor designs.

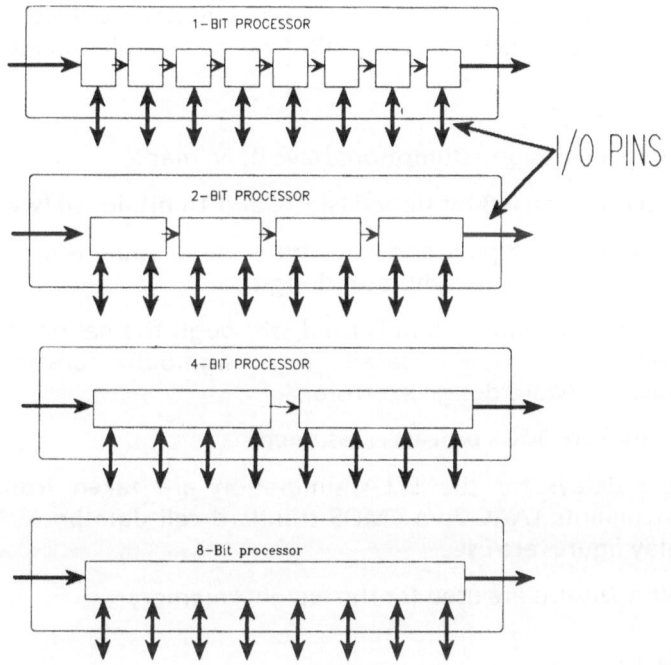

Figure 7   Processor granularity design options

Methodology

In total 5 different designs were considered each with the same state and weights precision (8 and 16 bits respectively), but differing in the width of the datapath between the processors and the weights memory. There were 5 main variants defined, namely 1, 2, 4, 8 and 16-bit datapath.

Each variant was designed and simulated using ELLA, starting with behaviourial models, decomposing the designs to functional level and then gate level.

The behaviourial models were simulated for simple neural network applications using a neural network simulation package for verification. The functional and behaviourial models were cross- checked for consistency prior to the gate level designs being undertaken. The gate level models were similarly cross-checked against to verify design correctness.

The design proceeded in 3-stages, namely;

-- The design of a single neural processor as detailed above,
-- The specification of a neural chip based on contemporary ASIC technology incorporating as many processors as constrained by pinout and yield,

-- The estimation of performance of a single-board neural computer based around such ASICs and DRAM-based weights memory.

Processor design

The following design assumptions have been made.

-- Data precision is 8-bit signed (state) and 16-bit signed (weights).

-- A parallel multiplier and carry look-ahead adder is used as the basis of the multiplier-accumulator design.

-- A 2-state update system is used, although the design style used can allow more sophisticated (eg. sigmoid) functions with a straightforward design extension.

-- A 2-micron CMOS process is assumed.

-- Gate delays for the ELLA simulation are taken from the Texas Instruments TAAC 2μm CMOS standard cell data book. Typical gate delay figures are used.

-- 100nS DRAMs are used for the weights memory.

Results

Figure 8 summarises the cycle time (time in which 1 multiply- accumulate takes place) time/connection and transistor requirements for the 5 processor variants.

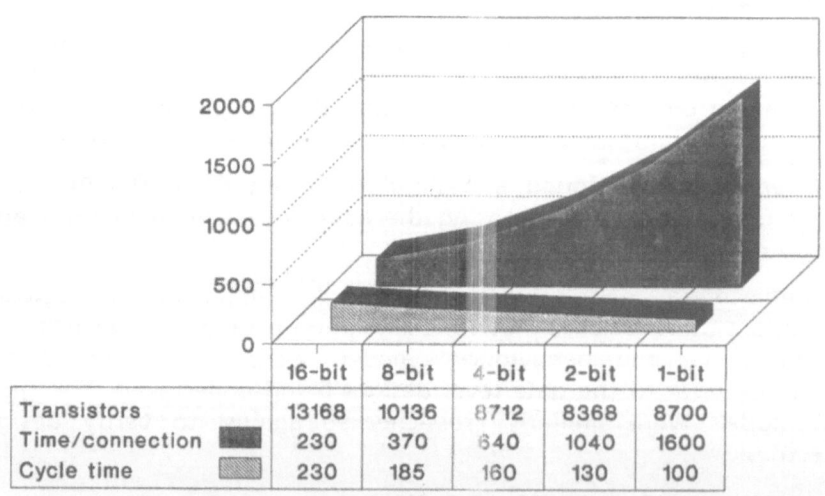

	16-bit	8-bit	4-bit	2-bit	1-bit
Transistors	13168	10136	8712	8368	8700
Time/connection	230	370	640	1040	1600
Cycle time	230	185	160	130	100

Figure 8   Processor cycle time (speed and logic)

As might be expected as the processor-weights memory datapath decreases, the time per connection increases as extra cycles are needed. However, as Figure 10 shows, this scaling is not linear. This is because as the datawidth decreases so does the number of stages of logic in the multiplier-accumulator, allowing faster operation. Indeed, the simplest version '1-bit' substitutes AND gates for the multiplier, this version has a multiply-accumulate cycle time almost exactly that of the memory access.

With all variants, the operation of the processor is slower than the memory access. Consequently, the limit on cycle time is always imposed by the processor. We have attempted to build a reasonably fast processor (eg. a parallel multiplier) and we have attempted to minimise our critical path delays.

In general the narrower datawidth processors require less logic, but the 1-bit processor is slightly larger than the 2-bit processor. This is because the relatively small saving in multiplier logic does not offset the extra logic required to deal with a 16-stage multiply-accumulate cycle.

Chip design

A 128 pin package is assumed. As is common with most large- packages, around 10% of the pins are required for power and ground. The pins on the package are allocated thus; 10 power pins (8% of total pins); 2 clock pins; 3 control pins; 8 pins for Sin; 8 pins for Sout. This leaves 97 pins for the weights-processor IO.

Current ASIC technology imposes an integration limit of somewhere around 200,000 transistors. Assuming this limit it is interesting to see the characteristics of the different processors. Figure 9 gives these details.

This data represents approximately the number of processors that could be integrated on a single medium-cost ASIC using 1990 technology.

The 16-bit and 8-bit variants have the same characteristics since they are pin-limited. The 2-bit and 1-bit variants have a significantly reduced number of processors due to the yield constraint of current technology. The 4-bit version is very slightly yield limited losing just one neuron.

In general if a better process technology is available, then the 1 and 2-bit options are the appropriate designs. This is because the finer-grain processors are relatively faster (eg. the 1-bit cycle time is quicker than the 16-bit cycle time - Figure 10). It is only worthwhile moving to a finer granularity design if sufficient processors can be integrated to compensate for the increase in connection time.

Increasing the pinout on the chip alters the position (but not the shape) of the curves and makes the wider bitwidth processors more suitable at higher integration densities.

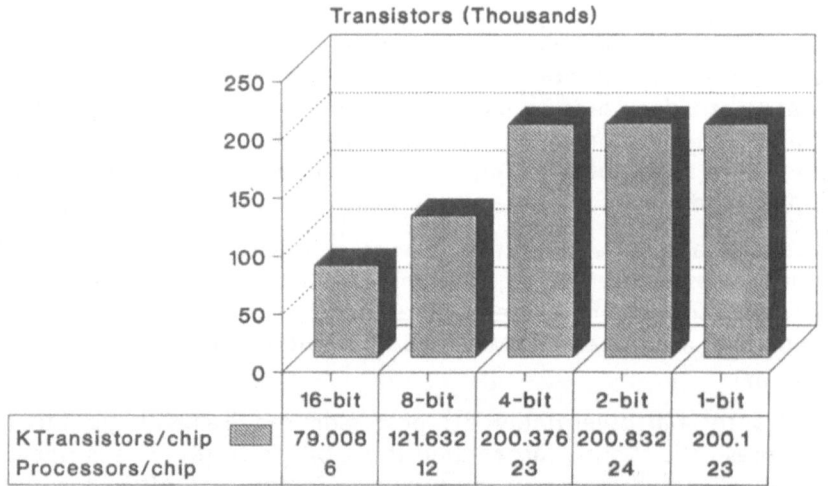

	16-bit	8-bit	4-bit	2-bit	1-bit
K Transistors/chip	79.008	121.632	200.376	200.832	200.1
Processors/chip	6	12	23	24	23

Figure 9   Chip details (128 pins 200K transistors)

Board design

The following assumptions have been made for estimating the performance of the system.

32 processor chips and associated DRAM devices can fit on a single board. Preliminary board-planning indicates that this is achievable using SIMM modules for the RAM.

System clock speed is equal to the maximum chip speed (ie. the performance is not system clock limited). Since the processor speed is around 10MHz this is realistic.

The board controller comprises a single chip microcontroller and memory address decode logic.

System performance

Figure 10 gives details of the maximum system performance (measured in connections/second in process mode) using the 5 processor variants. From this figure it can be seen that the peak system performance is achieved with the 4-bit option. Other strategies are less efficient.

The 16-bit and 8-bit variants cannot integrate enough processors per chip as they are pinout limited (resulting in few neurons per board) and also have a cycle time (resulting in relatively slower operation).

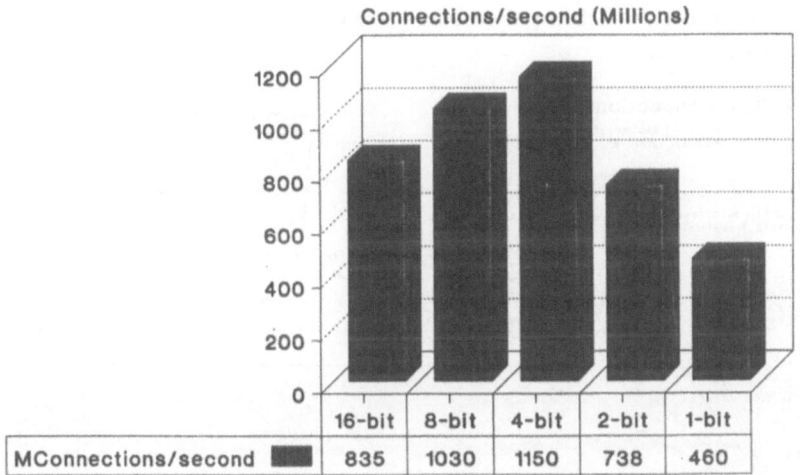

Figure 10   System performance (32 processor-chip system)

The 2-bit and 1-bit variants cannot integrate enough processors per chip because they are yield limited. Furthermore, the yield limit with current technology is sufficiently strong to offset their reduced cycle time.

With current technology the 4-bit variant seems to offer the right tradeoff between IO complexity and logic requirements.

Analysis

The work has necessarily been restricted and a number of assumptions as to packaging and logic count has had to be made. Nonetheless, a number of relevant points can be seen.

The proposed linear systolic architecture does indeed appear to provide a high-performance low-cost neural network computer. Indeed, its peak performance (4-bit 1150 Million Connections/second) is over 2 orders of magnitude faster than a typical neural network engine based on sequential signal processing chips.

The optimum processor granularity varies as a function of process and package. A better process tends to favour finer-grain processors. A better package (in terms of number of pins) in turn favours the more bit-parallel processors. This suggests that as technology improves, the fine-grain processors should be adopted. However, as Figure 11 shows, the move to a finer-grain processor should only be done when a certain packing density has been reached. This point is when the increase in the number of processors offsets the decrease in individual processor speed caused by multi-cycle processing. For the processor version described above, the move from 4-bit to 2-bit processors is when the process can support 320K + transistors.

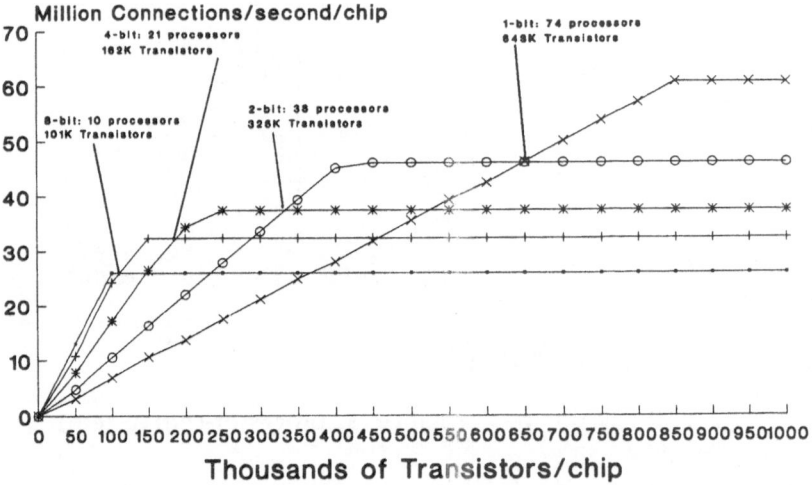

Figure 11   Comparative performance vs transistors/chip

Using current ASIC technology it is interesting that only around 750 processors can be realised on a single board system. With most neural applications requiring several thousands of neurons, this means that faster systems need to develop fault-tolerant (large-area IC) processors or high-speed multi-board systems techniques or alternatively, to develop techniques to efficiently overlay a large richly-connected neural net onto a smaller physical array. Nonetheless for modest size networks extremely fast performance can be achieved.

In this study, we have neglected the overhead of loading and unloading the array. When the number of neurons is less than or equal to the size of the array, this is low. However, when the number of neurons is greater than the number of PEs in the array, then depending on the size of the array and the connections between them, this overhead may be quite large.

CONCLUSIONS

Details of architectural and implementation issues concerning ring-based processor arrays for neural network calculation have been presented.

It has been argued that the ring is a powerful structure for the implementation of parallel computer systems. Furthermore, it provides the rich connectivity needed by neural networks allied to suitability for the exploitation of VLSI technology.

ACKNOWLEDGEMENTS

The authors would like to thank all 20 members of the Parallel and Novel Architectures Group at Nottingham University for their support of this work. This work has been funded in part by a NATO award for co-operation between the University of Nottingham and the Technical University of Denmark, the UK Science and Engineering Research Council, the Danish Technical Research Council and the Nottingham University Research Fund.

REFERENCES

[1]   S.Y. Kung, J.N. Hwang, 'Parallel Architectures for Artificial Neural Nets', IEEE Int. Conf. on Neural Nets, San Diego, Volume 2, pp 165 - 172, July 1988.

[2]   H. Kato, H. Yoshizawa, H. Iciki, K. Asakawa, 'A Parallel Neurocomputer Architecture towards Billion Connection Updates per Second', Technical Report, Computer-based Systems Laboratory, Fujitsu Laboratories Ltd, Kawasaki, 1989, Japan.

[3]   D.A. Pomerleau, G.L. Gusciora, D.S. Touretzky, H.T. Kung, 'Neural Network Simulation at Warp Speed: How We Got 17 Million Connections Per Second', IEEE Int. Conf. on Neural Networks, San Diego, Volume 2, pp 143 - 150, July 1988.

[4]   'The DARPA Neural Network Study', AFCEA International Press, Nov 1988.

[5]   D.E. Rumelhart, G.E. Hinton, R.J. Williams. "Learning Internal Representations by Error Propagation", Ch. 8. in Parallel Distributed Processing, Volume 1, Eds McClelland & Rummelhart, MIT Press 1986.

[6]   R.C. Aubusson, I. Catt, 'Wafer-Scale Integration: A Fault- Tolerant Procedure', IEEE Journal of Solid-State Circuits, Volume SC-13 No 3, June 1978.

[7]   R.J. Westmore, "A Scaleable Waferscale Architecture For Real Time 3-D Image Generation", IEEE International Conference on Wafer Scale Integration, pp 95 - 100, Feb. 1989.

[8]   T. Verhoeff, 'Delay-insensitive codes - an overview', Distributed Computing 3 (1) 1988.

[9]   M. Horowitz, T.E. Williams, 'A Self-timed Chip for Division', in Proceedings of the Conference on Advanced Research in VLSI, Computer Science Press 1987.

[10]   T.J. Chaney, C.E. Molnar, 'Anomalous Behaviour of Synchroniser and Arbiter Circuits', IEEE transactions on Computers, 22(4):421-422, April 1973.

UNIFIED DESCRIPTION OF NEURAL ALGORITHMS FOR TIME-INDEPENDENT PATTERN RECOGNITION

U. RAMACHER   B. SCHÜRMANN

ABSTRACT

Results obtained by Pineda for supervised learning in arbitrarily structured neural nets (including feed-back) are extended to non-supervised learning. For the first time a unique set of 3 equations is derived which governs the learning dynamics of neural models that make use of objective functions. A general method to construct objective functions is outlined that helps organize the network output according to application-specific constraints. Several well-known learning algorithms are deduced exemplarily within the general frame. The unification turns out to result in economical design of software as well as hardware.

INTRODUCTION

A first glance at the many formulas presented in this chapter may discourage a reader. The reason that we include a paper on the algorithmic analysis of neural algorithms for time-independent pattern recognition is based upon our belief that an understanding of the derivation of neural algorithms is indispensable for the designers of fast and flexible neurocomputers [1]. Generally, a versatile neurocomputer architecture must be built on the elementary compute-intensive algorithmic strings shared by all known neural paradigms. In addition, a set of universal substrings has to be derived from those algorithmic strings which are not common to all the models. Elementary strings and substrings would compose the functional blocks, which have to be implemented in silicon, whereas concatenation of strings or substrings would be initiated by instructions. As the variety of models can be created to a large extent by dedicated sequences of such instructions, the design of a neurocomputer must be preceeded by a careful analysis of the neural network models in use.

In the subsequent sections a unified description of arbitrarily structured neural nets controlled by supervised as well as non-supervised learning is developed. The compute-intensive neural strings are identified. In the subsequent chapter of this book is the architectural mapping of the strings into hardware presented.

PRELIMINARIES

We consider neural nets the recall behaviour of which is described by the following equation:

$$P_i(D_t) y_{ip}(t) = f_i\left(\sum_{j=0}^{N} W_{ij} \cdot y_{jp}(t) + Y_{ip} \right), \quad 1 \le i \le N, \ p \in P . \quad (1)$$

A neuron's individual input is denoted by Y_{ip}, its collective input by y_{ip}, and its discriminator function by f_i. N designates the number of neurons and P the index set for the patterns. P_i is a polynomial in the time-derivative operator D_t.

In this paper it is assumed that the patterns to be recognized contain no information that varies with time. Thus, the time-derivative operator describes only the swing-in behaviour of the network itself, which may be oscillatory or even chaotic. Because of the boundedness of the discriminator function f_i, the asymptotic activity of a neuron is composed of a constant and an oscillatory or chaotic function which may show up for recurrent networks. Depending on the size of the oscillations or fluctuations possibly occurring in the network, its output neurons may contain some small amount of noise (which may be helpful during learning), or the output neurons do oscillate. In this case a redesign of the network has to be considered. As irregularities are easy to detect, we will assume that the recall equation (1) is asymptotically constant.

Further, equation (1) holds for the case that the line delay between neurons is negligible with respect to a neuron's computation time. This restriction has been made for sake of simplicity.

In the sequel we will use a more compact formulation of the asymptotic static version of equation (1) :

$$y_{ip} = f_i\left(\sum_{j=-1}^{N} W_{ij} \cdot y_{jp} \right), \quad 1 \le i \le N , \quad (2)$$

with

$$W_{i\,-1} := 1 \ , \ y_{-1\,p} := Y_{ip} = neuron\ individual\ input \ , \quad (3a)$$

$$W_{i\,0} := neuron\ individual\ threshold\ , \ y_{0\,p} := 1 \ . \quad (3b)$$

Next we consider special cases of equation (2).

Example 1: Perceptron [2]

The weight $W_{i\,-1}$ is set zero, only one input and one output layer exist. Redefine $In_{jp} := y_{jp}$, $Out_{ip} := y_{ip}$. The weights can be expressed analytically as a function of the inputs:

$$f_i^{-1}(Out_{ip}) = \sum_{j=0}^{N} W_{ij} \cdot In_{jp} \qquad (4a)$$

$$\Leftrightarrow \sum_{p \in P} f_i^{-1}(Out_{ip}) \cdot In_{pk}^{-1} = \sum_{p \in P} \sum_{j=0}^{N} W_{ij} \cdot In_{jp} \cdot In_{pk}^{-1} =: \overline{W}_{ik} \quad . \quad (4b)$$

Because the inverse of the pattern matrix is generally of pseudo-type [3] one cannot set the right-hand expression of equation (4b) equal to W_{ik}. Instead, one redefines the weights. Using the definition of a pseudo-inverse, $In \cdot In^{-1} \cdot In = In$, however, shows that the redefined weights fulfil the recall equation (4a). Then, by inserting for Out_{ip} the reference values t_{ip}, one can directly start to compute the weight values.

Example 2: Multi-layer Perceptron [4]

Denote by k_v the neuron index of the v-th layer, by N_v the number of neurons in this layer and by L the number of layers.

$$y_{k_v p} = f_{k_v} \left(\sum_{k_{v-1}=0}^{N_{v-1}} W_{k_v k_{v-1}} \cdot y_{k_{v-1} p} \right) , \quad 1 \le k_v \le N_v , \quad 1 \le v \le L \quad . \quad (5)$$

If one could specify the outputs of each layer in advance, by kind of a partial analysis of the pattern problem, one would make use of the weight formula (4b) of the Perceptron several times and compute all the weights. Unfortunately this knowledge is still lacking.

Example 3: Hopfield [5]

Y_{ip} represents the individual input of a neuron, the collective part of the input of a neuron comes from the outputs y_{ip} of all the other neurons:

$$y_{ip} = f_i \left(\sum_{j=0}^{N} W_{ij} \cdot y_{jp} + Y_{ip} \right) , \quad 1 \le i \le N , \quad p \in P \quad . \quad (6a)$$

Solving for the weights gives now:

$$\sum_{p \in P} \left(f_i^{-1}(y_{ip}) - Y_{ip} \right) \cdot y_{pk}^{-1} = W_{ik} \quad . \quad (6b)$$

Again, the remarks concerning the handling of a pseudo-inverse matrix apply and only in the supervised mode, i.e. for an auto-associative net, is it that this formula may be helpful.

The above examples show that a "non-learning"analytical expression for the weights of arbitrarily structured, but one-layer networks exists in case of supervised learning and that the weights are given as a sum of products which factorize according to the indices of the weight W_{ij}.

Next we turn to general multi-layered networks. To our knowledge Pineda [6] was the first one to extend correctly the backward error propagation algorithm [4] from pure feed-forward multi-layer perceptrons to arbitrarily structured networks that may include feed-back connections. He did, however, consider supervised learning only. In order to provide the basis for an understanding of our generalization to non-supervised types of learning and subsequent unification of both we review the derivation of Pineda´s gradient-descent type of learning rules.

REVIEW OF PINEDA´S WORK

Supervised learning aims at mapping a set of patterns to be learned onto a set of reference patterns which is to appear as the output of the neural network under consideration. The idea is to look for weights that approximate the desired mapping behaviour best. Often one specifies an error function E by

$$E(t) = \sum_{p \in P} \sum_{k \in \Omega} [t_{kp} - y_{kp}(t)]^2 , \quad \Omega := set\ of\ output\ neurons , \quad (7)$$

with t_{kp} the reference states, and proceeds by searching for weights that, firstly, guarantee asymptotically stable solutions and, secondly, minimize the error. Thus, the time derivative of E is to be inspected,

$$\frac{dE(t)}{dt} = \sum_{i,j=1}^{N} \frac{\partial E}{\partial W_{ij}} \cdot \frac{dW_{ij}}{dt} , \quad (8)$$

which can be shown to fulfill both conditions if one sets:

$$\frac{dW_{ij}}{dt} = -a \cdot \frac{\partial E}{\partial W_{ij}} . \quad (9)$$

For, the time-derivative of the error function E is now given by

$$\frac{dE(t)}{dt} = -a \sum_{i,j=1}^{N} \left| \frac{\partial E}{\partial W_{ij}} \right|^2 , \quad (10)$$

which is negative ($a > 0$), and the weights are updated such that E approaches asymptotically a minimum:

$$W_{ij}(t + \Delta t) = W_{ij}(t) - a \cdot \Delta t \cdot \frac{\partial E}{\partial W_{ij}} . \quad (11)$$

From equation (7) and (9) one derives the equation from which Pineda started:

$$\frac{dW_{ij}}{dt} = -a \cdot \frac{\partial E}{\partial W_{ij}} = 2a \sum_{p \in P} \sum_{k=1}^{N} [t_{kp} - y_{kp}] \cdot X_k \cdot \frac{\partial y_{kp}}{\partial W_{ij}} \quad , \quad (12)$$

with X_k the characteristic function of the output neurons.

Next the computation of the derivatives of an activity y with respect to the weights is to be performed. Using equation (2) one obtains:

$$\frac{\partial y_{kp}}{\partial W_{ij}} = f'_{kp} \cdot \left\{ \delta_{ik} \cdot y_{jp} + \sum_{l=0}^{N} W_{kl} \cdot \frac{\partial y_{lp}}{\partial W_{ij}} \right\} \quad , \quad (13)$$

$$\text{with } f'_{kp} := \frac{\partial y_{kp}}{\partial a_{kp}} \, , \, a_{kp} := \sum_{l=0}^{N} W_{kl} \cdot y_{lp} \, .$$

Note that for a sigmoid function, for instance, its derivative at point y_{kp} is given by $y_{kp} \cdot (1 - y_{kp})$.

Instead of trying to solve equation (13) for all indices k, p, i, j, Pineda was able to derive an equation similar in structure to equation (13), but which contained less indices. This procedure is briefly reviewed below.

Equation (13) can be rearranged such that we can solve for $\partial y / \partial W$:

$$(13) \Leftrightarrow \sum_{l=0}^{N} [\delta_{kl} - f'_{kp} \cdot W_{kl}] \cdot \frac{\partial y_{lp}}{\partial W_{ij}} = f'_{kp} \cdot \delta_{ik} \cdot y_{jp} \quad .$$

Let us denote the term in brackets by L_{kl} and multiply both sides by L^{-1}_{mk} (L^{-1} is the pseudo-inverse matrix of L, and the same remarks apply for its handling as above). Summing over k gives:

$$\sum_{k=0}^{N} \sum_{l=0}^{N} L^{-1}_{mk} \cdot L_{kl} \cdot \frac{\partial y_{lp}}{\partial W_{ij}} = \sum_{k=0}^{N} L^{-1}_{mk} f'_{kp} \cdot \delta_{ik} \cdot y_{jp} \quad .$$

Finally, this way one obtains:

$$\frac{\partial y_{mp}}{\partial W_{ij}} = L^{-1}_{mi} \cdot f'_{ip} \cdot y_{jp} \quad . \quad (14)$$

Inserting the right-hand expression into equation (12) then yields:

$$\frac{dW_{ij}}{dt} = 2a \sum_{p \in P} \sum_{k=0}^{N} [t_{kp} - y_{kp}] \cdot X_k \cdot L^{-1}_{ki} \cdot f'_{ip} \cdot y_{jp} \quad . \quad (15)$$

Note that the time-derivative of a weight W_{ij} separates as to its indices, as was observed to happen for one-layer networks. Thus we define the quantity:

$$\Delta_{ip} := 2 \cdot \sum_{k=0}^{N} [t_{kp} - y_{kp}] \cdot x_k \cdot L_{ki}^{-1} \cdot f'_{ip} . \qquad (16)$$

One could proceed now with computing the pseudo-inverse matrix, which is a nasty business. On the other side, the mapping

$$\left(\frac{\partial y_{mp}}{\partial W_{ij}} \right)_{m=1..N} \rightarrow \Delta_{ip}$$

is a contraction, thus one expects an equation for Δ_{ip} which is structurally similar to equation (13). Reversing the operations between the two equations yields:

$$\sum_{i=0}^{N} L_{in} \cdot \frac{\Delta_{ip}}{f'_{ip}} = 2 \sum_{k,i=0}^{N} [t_{kp} - y_{kp}] \cdot x_k \cdot L_{ki}^{-1} \cdot L_{in} = 2(t_{np} - y_{np}) \cdot x_n .$$

If we now use the definition of the matrix element of L one gets:

$$\sum_{i=0}^{N} [\delta_{in} - f'_{ip} \cdot W_{in}] \cdot \frac{\Delta_{ip}}{f'_{ip}} = \frac{\Delta_{np}}{f'_{np}} - \sum_{i=0}^{N} W_{in} \cdot \Delta_{ip} = 2(t_{np} - y_{np}) \cdot x_n .$$

The equation for Δ_{np} finally reads:

$$\Delta_{np} = f'_{np} \left[2(t_{np} - y_{np}) \cdot x_n + \sum_{i=0}^{N} W_{in} \cdot \Delta_{ip} \right]. \qquad (17)$$

This is Pineda's fixpoint equation for Δ_{np}. Collecting the results we get for the update:

$$W_{ij}(t+1) = W_{ij}(t) + \alpha \cdot \sum_{p \in P} \Delta_{ip} \cdot y_{jp} . \qquad (18)$$

The reason for preferring equation (18) over equation (12) is obviously the reduced set of indices present in equation (17).

In case of a net with feed-back, equation (17) is a genuine fix point equation. If it has solutions, equation (13) will have. Thus the stability of the neural net under consideration hinges on the existence of fixpoints for equation (17).

BACKPROPAGATION REVISITED

Let us consider a pure feed-forward Multi-layer Perceptron [4] . The associated learning rule is obtained by starting the iteration of equation (17) by setting $\Delta_{ip} = 0$. This gives the well-known error correction for the output layer:

$$\Delta_{k_{out}p} = f'_{k_{out}p} \cdot (t_{k_{out}p} - y_{k_{out}p}) \ . \tag{19}$$

This inserted into the right-hand expression of (17) gives for the last but one layer:

$$\Delta_{k_{out-1}p} = f'_{k_{out-1}p} \cdot \sum_{k_{out}=0}^{N_{out}} W_{k_{out}k_{out-1}} \cdot \Delta_{k_{out}p} \ , \tag{20}$$

and, generally, for the i-th layer:

$$\Delta_{k_i p} = f'_{k_i p} \cdot \sum_{k_{i-1}=0}^{N_{i-1}} W_{k_{i-1}k_i} \cdot \Delta_{k_{i-1}p} \ . \tag{21}$$

Since feed-back is not present, the iteration stops at the input layer. The corresponding weight update formula reads according to (18):

$$W_{k_i k_{i-1}}(t+1) = W_{k_i k_{i-1}}(t) + \alpha \cdot \sum_{p \in P} \Delta_{k_i p} \cdot y_{k_{i-1}p} \ . \tag{22}$$

GENERALIZATION TO NON-SUPERVISED LEARNING

Pineda´s derivation of the recursion formula (17) has been obtained under the assumption of supervised learning.

It seems that Pineda was not aware that his derivation is the key to treating supervised as well as non-supervised learning in arbitrarily structured networks on the same footing.

In case of non-supervised learning the task is to prescibe or influence the process of self-organization of a network such that a set of patterns is learned and the network acquires the generalization characteristics the user is longing for. Instead of prescribing the activation values of the output neurons (as is done by supervised learning) one could prescribe a behavioural function for the output neurons or, generally, the topology of the network. By imposing such a relationship between, say, the output vectors belonging to various patterns, one could look for weights which approximate the desired relationship optimally.

For instance, if a set of patterns to be learned is judged to be maximally "non-similar" in human terms, one could request that the output vectors of an intermediate network associated with these patterns should feature minimal cross-correlation:

$$E = \frac{1}{2} \sum_{\substack{p,q \in P \\ p \neq q}} \left(\sum_{k,l} y_{k\,p} \cdot y_{k-l\,q} \right)^2 . \tag{23}$$

Obviously it will depend very much on the skills of the user to find application-specific constraints that direct the process of self-organization of a network and let him find an appropriate objective function. Generally, the relationship we are interested in may be termed an objective function which is composed of several real-valued subfunctions:

$$E = \sum_{k,p} E_y(y_{k\,p}) + \sum_{k,l} E_W(W_{kl}) + \sum_{k,l,m,p} E_{yW}(y_{k\,p}, W_{ml})$$

$$+ \sum_{k,l,m,n} E_{WW}(W_{kl}, W_{mn}) + \sum_{k,l,p,q} E_{yy}(y_{k\,p}, y_{lq}) + higher\ order\ terms . \tag{24}$$

Then this function E is to be optimized, i.e a set of weights to be found that maximizes or minimizes the objective function.

More involved and sophisticated objective functions can be thought of, clearly, but the intent of this paper is to deduce the learning rule of an objective function exemplarily.

Let χ denote the characteristic function of the set Ω of indices belonging to the neurons for which we want to define an objective function E. The set Ω could comprise more than just output neurons. Looking back to equations (8-10), we conclude that stability and maximization (minimization) is assured if we choose:

$$\frac{dW_{ij}}{dt} = (-)a \cdot \frac{\partial E}{\partial W_{ij}} , \quad a > 0 . \tag{25}$$

Since E may depend on the weights implicitly via the activities or explicitly, we get:

$$\frac{\partial E}{\partial W_{ij}} = \left(\frac{\partial E}{\partial W_{ij}} \right)_{impl} + \left(\frac{\partial E}{\partial W_{ij}} \right)_{expl} .$$

Because of the structure of equation (24) the implicit term can be rewritten as

$$\left(\frac{\partial E}{\partial W_{ij}}\right)_{impl} = \sum_{p \in P} \sum_{k \in \Omega} H_{kp} \cdot \frac{\partial y_{kp}}{\partial W_{ij}} , \tag{26}$$

with some function H_{kp} computed directly from equation (24).

The derivative of a neuron´s activity with respect to the weights has been computed already (equation (14)). Consequently, one gets:

$$\frac{\partial E}{\partial W_{ij}} - \left(\frac{\partial E}{\partial W_{ij}}\right)_{expl} = \sum_{p \in P} \Delta_{ip} \cdot y_{jp} , \tag{27}$$

with Δ_{ip} defined by the expression:

$$\Delta_{ip} := \sum_{k=1}^{N} H_{kp} \cdot \chi(k) \cdot L_{ki}^{-1} \cdot f_{ip}'.$$

Copying the procedure of equations (16-17) results in the equation:

$$f_{np}' \cdot \left\{ H_{np} \cdot \chi(n) + \sum_{i=1}^{N} W_{in} \cdot \Delta_{ip} \right\} = \Delta_{np} . \tag{28}$$

This equation is the generalized form of the recursion equation (17) and gives rise to a generalized version of the learning rule (18):

$$W_{ij}(t+1) = W_{ij}(t) - \alpha \cdot \left[\sum_{p \in P} \Delta_{ip} \cdot y_{jp} + \left(\frac{\partial E}{\partial W_{ij}}\right)_{expl} \right] . \tag{29}$$

Since symmetry or anti-symmetry of the weight matrix is inherited by the matrices $\partial W/\partial t$ and $(\partial E/\partial W)_{expl}$, the following corollary can easily be proven:

$$\sum_{p \in P} \Delta_{ip} \cdot y_{jp} = {}^+_- \sum_{p \in P} \Delta_{jp} \cdot y_{ip} \Leftrightarrow \begin{matrix} W \text{ is symmetric} \\ W \text{ is anti-symmetric} \end{matrix} . \tag{30}$$

In other words, the computational load is cut to half.

Example 4: Hopfield network with Kosko´s learning rule [7].

A fully connected one layer network is considered. Let us start from the objective function

$$E_0 = -\frac{1}{2} \sum_{p,q \in P} \sum_{k,l=1}^{N} \left(y_{kp} \cdot W_{kl} \cdot y_{lq} - \frac{\beta}{a} W_{kl}^2 \right).$$

For the recursion equation one gets :

$$f_{np} \cdot \left\{ \frac{1}{2} \sum_{l=1}^{N} \sum_{\substack{q \in P \\ p \neq q}} (W_{nl} + W_{ln}) \cdot y_{lq} + \sum_{i=1}^{N} W_{in} \cdot \Delta_{ip} \right\} = \Delta_{np} ,$$

and for the update rule:

$$\frac{\partial W_{ij}}{\partial t} = -\beta \cdot W_{ij} - a \cdot \left(\sum_{p \in P} \Delta_{ip} \cdot y_{jp} - \sum_{\substack{p,q \in P \\ p \neq q}} y_{ip} \cdot y_{jq} \right).$$

In case that we assume the weight matrix to be anti-symmetric, the fixpoint of the recursion equation is zero. Because the time derivative of W inherits the anti-symmetry of W one gets the equation:

$$\frac{\partial W_{ij}}{\partial t} - \frac{\partial W_{ji}}{\partial t} = -\beta \cdot (W_{ij} - W_{ji}) .$$

From this follows an explicit solution for the weights:

$$W_{ij}(t) = exp(-t \cdot \beta) .$$

Thus, asymptotically all weights are zero in this case.

With symmetric weights one obtains:

$$f_{np} \cdot \left\{ \sum_{l=1}^{N} \sum_{\substack{q \in P \\ p \neq q}} W_{nl} \cdot y_{lq} + \sum_{i=1}^{N} W_{in} \cdot \Delta_{ip} \right\} = \Delta_{np} .$$

Imagine now we would introduce into the last equation a term X_{nq} and requested it to compensate the membrane potential of the n-th neuron:

$$\sum_{l=1}^{N} W_{nl} \cdot y_{lq} + X_{nq} = 0 .$$

Introducing an extra term in the recursion equation means introducing an extra term in the objective function:

$$\frac{\partial E_1}{\partial W_{ij}} := \sum_{\substack{p,q \in P \\ p \neq q}} \sum_{k,l=1}^{N} \delta_{kl} \cdot \delta_{pq} \cdot X_{kq} \cdot \frac{\partial y_{kp}}{\partial W_{ij}} . \tag{31}$$

Now, for a Hopfield net

$$\sum_{l=1}^{N} W_{nl} \cdot y_{lq} = f_{nq}^{-1}(y_{nq}) - Y_{nq}$$

holds. Therefore we set:

$$X_{kq} = Y_{kq} - f_{kq}^{-1}(y_{kq}) .$$

By integrating equation (31) we arrive at Kosko's objective function for a Hopfield net [7]:

$$E = -\frac{1}{2} \sum_{\substack{p,q \in P \\ p \neq q}} \sum_{k,l=1}^{N} \left(y_{kp} \cdot W_{kl} \cdot y_{lq} - \frac{\beta}{2} W_{kl}^2 \right.$$

$$\left. + \delta_{kl} \cdot \delta_{pq} \cdot \int \left\{ Y_{kq} - f_{kq}^{-1}(y_{kq}) \right\} dy_{kq} \right). \tag{32}$$

Similarly, one may obtain the objective function for the "detailed balance" network [8] which is a generalization of Kosko's network.

Example 5: Kohonen network [9]

We consider the objective function

$$E = \frac{1}{2} \sum_{p \in P} \sum_{k=1}^{N} a_{kp} \cdot \sum_{l=1}^{N} \left(W_{kl} - Y_{lp} \right)^2 , \tag{33}$$

where Y_{lp} is the l-th input of the p-th pattern and W_{kl} the weight between the l-th input and the k-th neuron. The function a_{kp} may have the Gaussian form

$$a_{kp} = exp \left[-\left(\frac{\vec{W}_k - \vec{Y}_p}{\sigma_p} \right)^2 \right] , \tag{34}$$

where the arrow means vector notation. This leads to a clustering of the feature maps around the input pattern vectors Y_p. Since the inputs Y_{lp} are independent of the weights, only $(\partial E/\partial W_{ij})_{expl}$ contributes to the learning rule. Hence, we obtain:

$$W_{ij}(t+1)= W_{ij}(t) - \sum_{p \in P} \left[1 - \left(\frac{\vec{W}_k - \vec{Y}_p}{\sigma_p} \right)^2 \right] \cdot a_{ip} \cdot (W_{ij} - y_{jp}) \cdot \quad (35)$$

Note that the function $[...] \cdot a_{ip}$ is of mexican hat type.

UNIFYING SUPERVISED AND NON-SUPERVISED LEARNING

It has become clear by now that the formal frame of equation (24) for defining an objective function applies as well to supervised learning, since the error function defined in equation (7) is of type E_y.

The open question then is to determine what is (not within) the range of non-supervised learning. Included are all models which specify a cost function, a Ljapunov function or any other kind of objective function. The Boltzmann net , the family of associative nets, the Multi-layer Perceptron etc. are all contained. One can as well derive hybrid learning rules by devising hybrid objective functions, containing for example Hopfield-type and error functions [9] .

COMBINING NEURAL INFORMATION PROCESSING WITH CLASSICAL OBJECTIVE FUNCTIONS

Having derived a generalized learning rule and recovered well-known networks as special cases, it is left to find general methods to construct objective functions that help organize the network output according to application-specific needs. An interesting class of objective functions can be obained by the following reasoning.

In image processing and recognition there are plenty of operations M which are defined at pixel level. These low-level operations are usually very time-consuming. Also, these operations are universal in the sense that they can be applied to any image. Contrarily, most applications deal only with a very limited subset of pictures. Thus it is natural to ask for an application-specific modification M_{appl} of M such that it is less universal. This can be achieved by defining the operation M (convolution, Fourier transform, etc.) at the output level of a neural net and let the neural net operate at the pixels and adapt to weights such that the features and properties of the unmodified operation are reproduced optimally.

Let us illustrate the idea outlined above by three more examples.

Example 6: Orthogonality

We assume a set of patterns to be prepared which are maximally distinct in human terms. The objective function is defined by :

$$E = \frac{1}{4} \sum_{\substack{p,q \in P \\ p \neq q}} \left(\sum_{k \in \Omega} y_{kp} \cdot y_{kq} \right)^2 =: \frac{1}{4} \sum_{\substack{p,q \in P \\ p \neq q}} S_{pq}^2 . \tag{36}$$

The scalar product over the outputs is to become minimal. This simple objective function yields:

$$H_{np} = \sum_{\substack{q \in P \\ p \neq q}} y_{nq} \cdot S_{pq} .$$

From this follows the recursion equation:

$$f_{np}' \left\{ \sum_{\substack{q \in P \\ p \neq q}} S_{pq} \cdot y_{nq} \cdot \chi(n) + \sum_{i=1}^{N} W_{in} \cdot \Delta_{ip} \right\} = \Delta_{np} . \tag{37}$$

Since there is no explicit dependence on the weights, the update rule for the weights is formally identical to Pineda's rule (18).

It is left to the reader to derive the terms H_{np} for the objective function expressing orthogonality of the patterns:

$$E = \frac{1}{4} \sum_{k,l} \left(\sum_{p \in P} y_{kp} \cdot y_{lp} \right)^2 .$$

Example 7: Convolution

Let us consider two patterns and define the objective function of cross-correlation:

$$E = \frac{1}{2} \left(\sum_{k,l = -\infty}^{\infty} y_{lp} \cdot y_{k-lq} \right)^2 =: \frac{1}{2} C_{pq}^2 , \tag{38}$$

with, for instance,

$$y_{vr} = 0 \ if \ v \notin \Omega .$$

We are looking for weights which create output patterns with minimal similarity. Computation of H_{np} and H_{nq} yields:

$$H_{np} = \sum_{l=-\infty}^{\infty} y_{l-nq} \cdot C_{pq} , \quad H_{nq} = \sum_{l=-\infty}^{\infty} y_{l+np} \cdot C_{pq} .$$

Inserting the two values into the recursion equation (28) yields the equations:

$$f'_{np} \cdot \left\{ C_{pq} \cdot \sum_{l=-\infty}^{\infty} y_{l-nq} \cdot x(n) + \sum_{i=1}^{N} W_{in} \cdot \Delta_{ip} \right\} = \Delta_{np} \quad . \quad \text{(39a)}$$

$$f'_{nq} \cdot \left\{ C_{pq} \cdot \sum_{l=-\infty}^{\infty} y_{l+np} \cdot x(n) + \sum_{i=1}^{N} W_{in} \cdot \Delta_{iq} \right\} = \Delta_{nq} \quad . \quad \text{(39b)}$$

Example 8: Covariance

This time, we require that the covariance over the outputs be minimal. Let $\mu_{pq}(k)$ be a distribution for the overlap of the p-th and q-th pattern (class) over the output neurons defined by:

$$\mu_{pq}(k) := \frac{y_{kp} \cdot y_{kq}}{\langle yy \rangle_{pq}} \, , \quad \langle yy \rangle_{pq} := \sum_{k} y_{kp} \cdot y_{kq} \, . \quad \text{(40)}$$

Note that $\mu_{pq}(k)$ depends on the set of weights. This (p,q)-symmetric distribution can be measured at the net's output neurons. Now let $\rho_k(p)$ be a reference distribution which may fulfill, for instance, the relations

$$\sum_{k} \rho_p(k) \cdot \rho_q(k) = \delta_{pq} \, , \quad \sum_{p \in P} \rho_p(k) \cdot \rho_p(l) = \delta_{kl} \, . \quad \text{(41)}$$

In other words, we specified a complete set of non-overlapping distributions ρ_k. The corresponding expectation values are defined by

$$\langle y \rangle_p := \sum_{k} \rho_p(k) \cdot y_{kp} \quad .$$

Now, we wish to find a set of weights such that asymptotically

$$\mu_{pq}(k) \rightarrow \rho_p(k) \cdot \rho_q(k)$$

holds. Equivalently, the objective function

$$E = \frac{1}{4} \cdot \sum_{\substack{p,q \\ p \neq q}} \left(\langle yy \rangle_{pq} - \langle y \rangle_p \langle y \rangle_q \right)^2 =: \frac{1}{4} \cdot \sum_{\substack{p,q \\ p \neq q}} Cov_{pq}^2 \quad \text{(42)}$$

is to approach a minimum. From this follows:

$$H_{np} = \sum_{\substack{q \in P \\ p \neq q}} Cov_{pq} \cdot \left(y_{nq} - \sum_{l} \rho_q(l) \cdot y_{lq} \cdot \rho_p(n) \right) \quad .$$

Ideally, the covariance should approach zero in the course of the gradient descent.

We leave it to the reader to consider a covariance over the patterns.

CONCLUSIONS

As the main result of the paper, we have obtained a general update equation for the connection weights of artificial neural networks which, for the first time, comprises supervised as well as unsupervised learning in a natural manner. This update equation originates from taking the gradient of a quite general objective function with respect to the weights, and consists of two terms which are the implicit and explicit derivatives of the objective function with respect to the weights.

The implicit term is formally identical to a result obtained by Pineda for supervised learning, but is more general than the latter because it is based on a more general objective function than supervised learning may specify. In contrast to supervised learning, a general objective function has an additional term, namely the explicit one.

When considering specific objective functions, specific learning rules emerge from the update equation. This way, many well known neural learning algorithms are reproduced by our 3-equation formulation of neural learning dynamics.

Another major result is that our formulation provides a new way for classical signal processing to introduce its means and methods to neural signal processing. By specifying the objective functions known from classical pattern recognition at the outputs of a neural net, a new class of learning rules can be obtained.

Besides, the unified description as presented in this paper results in an economic software and hardware design. It is only three equations, namely formulas (2), (28) and (29), that have to be considered.

$$y_{ip} = f_i \left(\sum_{j=-1}^{N} W_{ij} \cdot y_{jp} \right) \tag{2}$$

$$f'_{np} \left\{ H_{np} \cdot \chi(n) + \sum_{i=1}^{N} W_{in} \cdot \Delta_{ip} \right\} = \Delta_{np} \tag{28}$$

$$W_{ij}(t+1) = W_{ij}(t) - \alpha \cdot \left[\sum_{p \in P} \Delta_{ip} \cdot y_{jp} + \left(\frac{\partial E}{\partial W_{ij}} \right)_{expl} \right] \tag{29}$$

The compute-intensive operations implicitly and explicitly present in these equations are the multiplication of matrices and scalar product operations. Consequently, this set of compute-intensive operations has to be implemented by specially designed VLSI circuits [11].

ACKNOWLEDGEMENT

We wish to thank our colleagues Dr. J. Anlauf and Dr. H.G. Zimmermann for helpful discussions concerning the examples 8 and 5.

REFERENCES

[1] U. Ramacher, "Hardware Concepts for Neural Networks", in: R. Eckmiller (ed.), Advanced Neural Computers, pp.209-218, North-Holland 1990

[2] M.L. Minski, S.A. Papert, "Perceptrons: An Introduction to Computational Geometry", MIT Press, Cambridge, expanded edition 1988

[3] T. Kohonen, "Self-Organization and Associative Memory", Springer, Heidelberg, Berlin, 3rd edition 1989

[4] D.E. Rumelhart, J.L. McClelland, "Parallel Distributed Processing, Explorations in the Microstructure of Cognition", Vol. I, Chap.8, MIT Press, Cambridge, 1986

[5] J.J. Hopfield, "Neurons with graded response have collective computational properties like those of two state neurons", Proc. Nat. Acad. Sci. USA 81, pp. 3088-3092, 1984

[6] F.J. Pineda, "Generalization of Backpropagation to Recurrent and Higher Order Neural Networks", in: D.Z. Anderson (ed.), Neural Information Processing Systems, Am. Inst. Phys., pp. 602-611, New York 1988

[7] B. Kosko, "Adaptive Bidirectional Associative Memories", Applied Optics 26, pp. 4947-4960, 1987

[8] B. Schürmann, "Stability and Adaptation in Artificial Neural Systems", Phys. Rev. A 40, pp. 2681-2688, 1989

[9] T. Kohonen, "The neural 'phonetic' typewriter", Computer 21, pp. 11-22, 1988

[10] B. Schürmann, J. Hollatz, U. Ramacher, "Adaptive Recurrent Neural Networks and Dynamic Stability", in: L. Carrido (ed.), Proc. XI Sitges Conf. on Neural Networks, 1990, Springer, Heidelberg, to appear

[11] U. Ramacher et al., "Design of a 1st Generation Neurocomputer", in: VLSI DESIGN OF NEURAL NETWORKS, edited by U. Ramacher, U. Rückert, Kluwer Academic Publishers, Nov. 1990

DESIGN OF A 1st GENERATION NEUROCOMPUTER

U. RAMACHER, J. BEICHTER, W. RAAB
J. ANLAUF, N. BRÜLS, U. HACHMANN, M. WESSELING

ABSTRACT

The analysis of today´s neural paradigms brings to light a set of elementary compute-intensive algorithmic strings which are shared by all neural models and, thus, make sense to be implemented in hardware. 2-D arrays composed of a specific VLSI Neural Signal Processor MA16 that integrates these elementary strings as hard-wired functional blocks present a favourable solution to the architectural problem of mapping neural parallelity and adaptivity into silicon. The proposed neurocomputer concept is sizeable independently to the applicational domain in terms of processing power, memory size and flexibility, and is designed for throughputs that enable the user to access real-world applications in reasonable time. At the chip site, throughput rates of the order of 500 MC/sec (1 Connection = 16 bit) are achievable with 1µm CMOS technology. 2-dimensional systolic arrays composed of 16x16 MA16 chips will allow for processing of 128 GC/sec.

INTRODUCTION

The response and the characteristics of present models of artificial neural nets are primarily investigated by simulation on vector computers, workstations, special coprocessors or transputer arrays. The fundamental drawback of such simulators is that the spatio-temporal parallelism in the processing of information that is inherent to the neural net is lost entirely or partly and that the computing time of the simulated net especially for large associations of neurons (tailored to application-relevant tasks) grows to such orders of magnitude that a speedy acquisition of "neural" know-how is hindered or made impossible. Figure 1 shows the performance obtainable with commercially available simulators [1,2] in terms of weights and weights per second. This must be contrasted with the applicational needs. It becomes obvious that today´s hardware capabilities are limiting the development of neural network research.

An appreciable reduction in computing time for the simulation of neural nets and thus the handling of largish tasks or those that are to be executed in real-time become possible with specially designed neural hardware. It can be tailored
-- for an application asking for a single neural network fixed in size and topology,
-- for an application which calls for a set of different neural networks which are programmable as to size and concatenation,

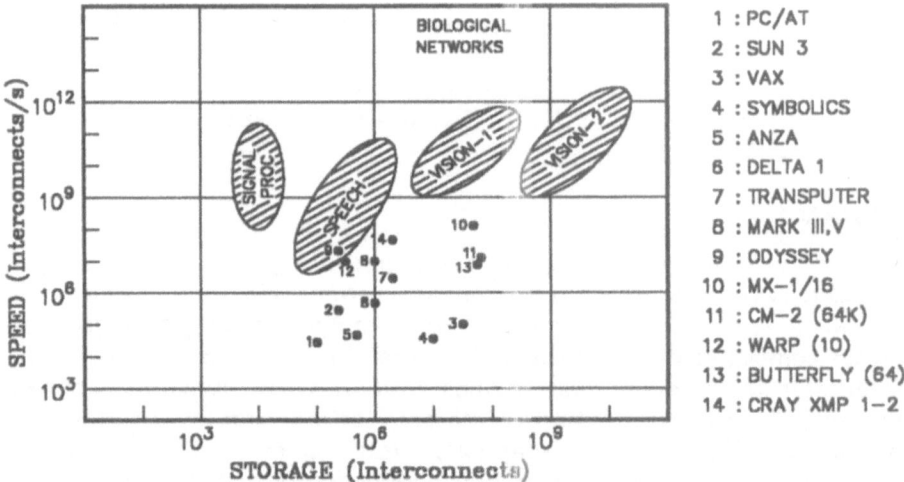

Figure 1   Computational capabilities of neural network simulators
            versus computational requirements of some applications

--   or be designed to serve general-purpose neurocomputing.

Depending on the size and characteristics of the application under
consideration, the weights may be stored on a neural chip. Principally,
storing the weights on the chip results in very fast signal processing.
However, the costs per bit would be higher than if off-chip storage in
DRAMs were used. On the other hand, off-chip storage introduces special
interface costs and timing constraints (see Figure 2).

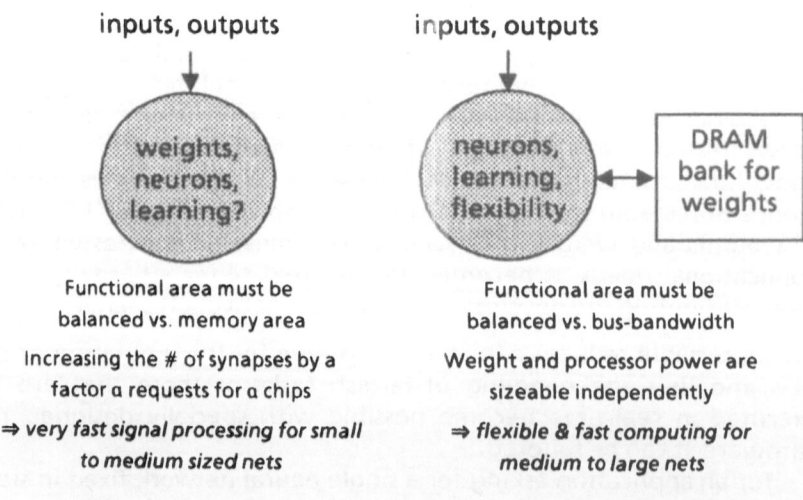

Figure 2   On-chip versus off-chip storage of weights

Consequently, the real-time constraint and the size of the application under consideration will have to be examined and contrasted with the implementation potential of CMOS technology.

Today´s CMOS technology provides sufficient speed of about 10^9 to 10^{11} weights/second and about 200kbit weight memory per neural chip [3] . With Wafer Scale Integration technology 160 Mbit-DRAMs can be fabricated [4]. Since DRAM technology is not satisfactory for fast logic, SRAM technology is preferable for which a WSI potential of about 32 to 40 Mbit can be estimated. If half the wafer were reserved for weight storage, the other half for the emulation of (various) neural algorithms, we could expect about 20 Mbit for storage of weights. For vision, this would equal about 2 million weights, which is not enough (see Figure 1, vision1: static pictures). In such a case, one has the choice either

-- to load repeatedly new sets of weights into the chip´s memory,
-- to distribute the weights over a sufficiently large number of chips,
-- or to distribute the weights off-chip in DRAMs and use the chip area for neural processing exclusively.

Loading new weights from off-chip memory into the chip´s memory would be paradoxical, since the data throughput would equal that of a chip which features the same amount of functionality, but has no on-chip memory for weights.

On the other hand, if 10^8 weights were to be stored on neural chips, the question arises if such a multi-chip system is not too costly and voluminous compared to a system which has the weights stored off-chip in cheap DRAMs and the chip area reserved for signal processing only. For vision, the latter system would consist of 400 4Mbit-DRAMs (or 10 160Mbit-DRAMs), whereas about $1.6/2 \times 10^{9-5} = 8000$ neural chips were needed. Since most of the chip area is needed for weight storage and little area left for computation [3,5] the system which has memory and processing power separately implemented would also need much less than 8000 chips for neural computation.

In summary, the first and second applicational species of neural hardware mentioned above are candidates for on-chip integration of weights provided the application is not too large (in terms of weight memory); applications like speech or vision overtax in any case the integration potential of present technology as well as that of future 0.3-μm technology by whole orders of magnitude. Instead, storage of weights and computational circuitry have to be implemented separately.

In the latter case, the chip architect will use the saved area for emulation of learning algorithms and a large variety of neural networks. This is urgently needed, since the size of large nets will not permit simulation (especially of the learning phase) within a reasonable period of time and therefore the weights cannot be determined by simulation. In consideration of the resulting lack of neural engineering know-how for applications as vision or speech, it is thus a matter of designing a system and circuit architecture that,

-- firstly, produces a sufficient measure of flexibility and expansion capacity for coping with the known paradigms and those to come,

-- and, secondly, supports in optimal fashion the massive parallel networking present in the search as well as the learning mode.

The ultimate purpose of neurocomputer design is to avail the user of sufficient power and flexibility so that any application-specific study can be undertaken and an application-specific neural system architecture be worked out in terms of dedicated software and hardware.

VLSI architectures designed to meet these goals will obviously look different from those considered for single-chip integration. Systems which are built up from these VLSI chips and additional software and hardware will be called henceforth (general-purpose) neurocomputers. The hardware part is named neuroemulator, the software part is called neural programming environment. This chapter of the book focusses mainly on the hardware part.

ALGORITHMIC ANALYSIS OF NEURAL PARADIGMS

Generally, a versatile neurocomputer architecture must absorb in silicon the elementary algorithmic strings shared by all known neural paradigms. In addition, a set of universal substrings has to be derived from those algorithmic strings which are not common to all the models. Elementary strings and substrings would compose the functional blocks, which have to be implemented in silicon, whereas concatenation of strings or substrings would be initiated by instructions.

Thus, a careful analysis of existing neural models has to be conducted at first. Considering the applicational range of time-independent pattern recognition, for instance, it can be shown that three equations govern the dynamics of neural networks [6]:

$$\text{Search:} \qquad y_{ip} = f_i \left(\left[\sum_{j=-1}^{N} W_{ij} \cdot y_{jp} \right] \cdot \lambda_i \right) \qquad (1)$$

$$\text{Learning:} \qquad \Delta_{np} = f'_{np} \cdot \left\{ \frac{\partial E}{\partial y_{np}} \cdot \chi(n) + \sum_{i=1}^{N} W_{in} \cdot \Delta_{ip} \right\} \qquad (2)$$

$$\text{Update:} \qquad W_{ij}(t+1) = W_{ij}(t) - a \cdot \left\{ \sum_{p \in P} \Delta_{ip} \cdot y_{jp} + \left(\frac{\partial E}{\partial W_{ij}} \right)_{explicit} \right\} \qquad (3)$$

Here the notation

$$W_{i-1} := 1 \ , \ y_{-1p} := Y_{ip} = neuron \ individual \ input \ ,$$
$$W_{i0} := neuron \ individual \ threshold \ , \ y_{0p} := -1 \ ,$$

is used. A neuron´s collective input is denoted by y_{ip}, and its discriminator function by f_i. N designates the number of neurons and P the index set for the patterns. χ is the characteristic function of a subset Ω of the set of neurons.

The function E is a general objective function which may be composed of several real-valued subfunctions:

$$E = \sum_{k,p} E_y(y_{k\,p}) + \sum_{k,l} E_W(W_{kl}) + \sum_{k,l,m,p} E_{yW}(y_{k\,p}, W_{ml})$$

$$+ \sum_{k,l,m,n} E_{WW}(W_{kl}, W_{mn}) + \sum_{k,l,p,q} E_{yy}(y_{k\,p}, y_{lq}) + higher \ order \ terms \ . \quad (4)$$

During learning this function E is to be optimized, i.e. a set of weights to be found that maximizes or minimizes the objective function. The derivative of the objective function is given by the expression

$$\frac{\partial E}{\partial W_{ij}} = \sum_{p \in P} \sum_{k \in \Omega} \frac{\partial E}{\partial y_{kp}} \cdot \frac{\partial y_{kp}}{\partial W_{ij}} + \left(\frac{\partial E}{\partial W_{ij}}\right)_{explicit} \ , \quad (5)$$

which is related to the equations (2-3).

When considering specific objective functions, specific learning rules emerge from the update equation (3). This way, almost all neural learning algorithms are reproduced by our 3-equation formulation of neural learning dynamics. For instance, supervised learning is described by

$$E(t) = \frac{1}{2} \sum_{p \in P} \sum_{k \in \Omega} [t_{kp} - y_{kp}(t)]^2 \ , \quad \Omega := set \ of \ output \ neurons \ , \quad (6)$$

with t_{kp} the reference states.

The compute-intensive operation explicitly and implicitly present in the Equations 1-3 is the multiplication of matrices. The computation of $\partial E/\partial y$ and $(\partial E/\partial W)_{explicit}$, respectively, can be formulated as special cases of multiplication of matrices. All other consecutive computations are uncritical in terms of computational throughput and can be realized by hardware off-the-shelf. Consequently, the operation of matrix-matrix multiplication has to be implemented by specially designed VLSI circuits.

As the use of analog concepts is detracted from, besides others, by the low computing accuracy of only a few percent, which is inadequate for the learning algorithms with its many iterations, and application-oriented problem analysis and modelling of neural network characteristics can not be made independently of circuit design [7], digital realization is believed to be best suited for neurocomputers, i.e. for machines with which to cross into an unknown region and study in breadth and depth the potential of neural networks.

ARCHITECTURAL SOLUTION OF THE INTERCONNECTION PROBLEM

Flexibility and general-purpose action have been shown to be manageable. It is left to design a powerful data flow architecture which supports multiplication of matrices and provides easy connection of these with other operations from Equations 1-3. There exist various ways to implement matrix multiplication. Generally, which one will be chosen depends on the applicational needs, technological constraints and the system environment. In the special case of neural algorithms the challenge comes from massive parallelity, i.e. handling large matrices, and the interconnect problem.

Since the weights will be stored in cheap DRAMs, it is of outmost importance to provide a sufficient communication bandwidth between the memory bank and the processor chip(s). The two bottlenecks encountered are the number of pins for the transfer of weights to the VLSI processor(s), and the cycle time of the DRAMs. With two-fold interleaving of 4Mbit-DRAMs a worst case clock frequency of 25 MHz can be achieved (see below). Functional blocks like multipliers and accumulators, on the other hand, can be designed for much higher clock frequencies. This seems to indicate an imbalance in troughput of the processor chip(s) and the weight bank.

Figure 3 illustrates the interconnection problem. It shows the architectural scheme of 4 scalar product chains [8]. Each column represents a functional block for scalar product operation, multiple use is made of the inputs, but not of the weights. The i-th column is served the weight sequence $(W_{ij})_{j=1..N}$ and the sequence $(y_j, y_{j+1}, y_{j+2}, y_{j+3})_{j=1, 5, 9,..N-3}$ of parallel pixels, say. The advantage of this architecture is a very prompt recognition response. However, the number of pins required (e.g. 256 pins for 16 16-bit weights only) is already close to the limits of standard chip packaging. To make things worse, future 0.3 μm technology will allow for larger multiplier fields to be implemented in silicon. The only solution here is to introduce weight multiplexing. As clock frequencies well exceeding 100 MHz are to be expected with 0.3 μm chip design, the weight multiplexing would require memory banks which have to be much faster than can be realized with interleaved-memory techniques for

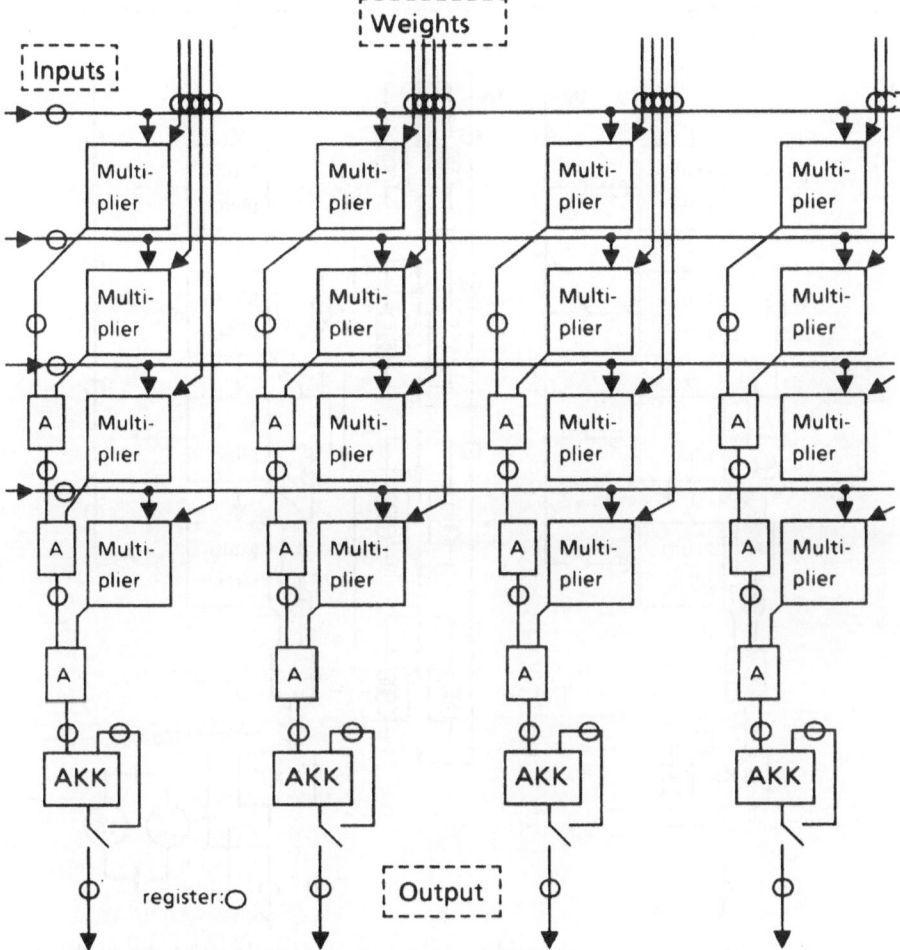

Figure 3   Architectural scheme of 4 scalar product chains

RAMs. Therefore the architecture presented in Figure 3 cannot be recommended, unless new techniques for optical memories with fast read <u>and</u> write cycles will be developed [9].

If single pattern response is not mandatory a more sophisticated architectural mapping of the weighting algorithm can be devised which circumvents the bottleneck mentioned above, but does not lower the chip's computational throughput. To explain the data flow in the revised scalar product chain let us consider first Figure 4a. Since the number of weight ports has been reduced to one, the computational throughput of the chip can be kept constant only if each weight is used 4 times. Let us designate by N_4 the set $\{v: v = 0,4,8, ..., N/4 - 1\}$. During the search phase

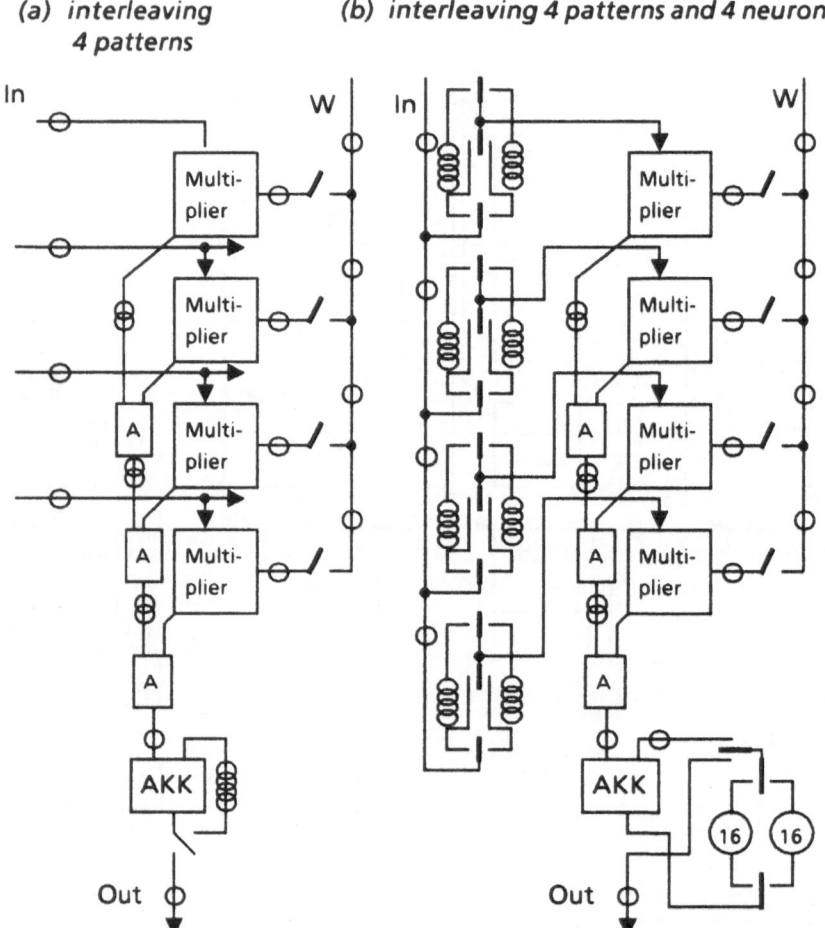

(a) *interleaving 4 patterns*

(b) *interleaving 4 patterns and 4 neurons*

Figure 4   Reducing the number of pads without lowering the throughput

the weights $\{W_{ij}\}_{j=0..N}$ are received serially, whereas the inputs $\{y_{v+k\ p} : k=0..3\}_{v \in N_4}$ arrive v-serial and k-parallel; as mentioned above, y_{jp} denotes the j-th pixel of the p-th pattern. The k-th multiplier picks up W_{ij}, $j = v + k$, and holds it for 4 cycles to multiply y_{jp}, $p = 1..4$, then proceeds by picking up W_{ij+4} and multiplying y_{j+4p}, $p = 1..4$, and so on. The products are summed along the chain for equal p and stored separately in the accumulator's shift register. After completion of the weighting, these registers are read out and the procedure is repeated with the weights of some other neuron, and so on.

Obviously, the trick can be applied to the inputs, too. Each y_{jp} must be stored now and made use of locally 4 times in order to keep the computational throughput constant. Thus the neuron index i, which has been fixed for a chain in Figure 3 and 4a, is to be varied. Since the pattern index was assumed running first, a y_{jp} is read out of the input shift register every 4 clock cycles to multiply the weights $W_{i'j}$, $i' = i, i + 1, i + 2, i + 3$. Thus the single-neuron accumulation of Figure 4a is extended to a 4-neuron interleaved accumulation. Consequently the accumulator's shift registers are increased to 16. When the accumulation is finished, the activation values for 4 patterns of a block of 4 neurons will be stored there intermediately. In summary, the weights and data for the chain of Figure 4b are served according to the temporal order displayed in Figure5. Note that the ith column of a W-matrix and the ith row of the y-matrix contain the data for the ith multiplier.

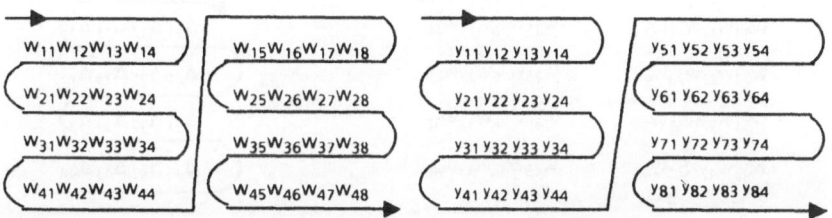

Figure 5   Weight and data schedule for Figure 4b (Equation1)

It needs still to be explained how the chain can be used for learning and update. To calculate a weight update (Equation3) the following identity is to be observed (t means transposition):

$$\sum_{p \varepsilon P} \Delta_{ip} \cdot y_{jp} = \sum_{p \varepsilon P} y_{jp} \cdot \Delta_{pi}^{t} = \sum_{p \varepsilon P} \Delta_{ip} \cdot y_{pj}^{t} . \qquad (7)$$

As the right and middle expressions of (7) are structurally similar to the scalar product formula of the search mode, the systolic chain shown in Figure 4b can be used for weight updating as well. The final computation for each gradient descent step can be performed by preloading 16 weights W_{ij}(old) into the chain's accumulator, as will be described below.

With respect to Equation 2 one obtains:

$$\sum_{i=1}^{N} W_{ij} \cdot \Delta_{ip} = \sum_{i=1}^{N} W_{ji}^{t} \cdot \Delta_{ip} = \sum_{i=1}^{N} \Delta_{pi}^{t} \cdot W_{ij} . \qquad (8)$$

The right expression of (8) recommends interchanging the ports of weights and data as well as changing from row-wise to column-wise read-off of the weights. Since DRAMs are assumed to be used for weight storage, the throughput would become considerably reduced because of the lack of the page mode. Considering the middle expression of (8) global transposition of the weight matrix is not recommendable because control and bus circuitry become costly. Instead, it is desirable to use a fixed schedule for feeding the weights into the chain and have some extra circuitry added to the chip that supports the operation of transposing a matrix.

Since the matrix multiplication, as described above, is worked out in blocks of 16 weights, the transposition of the weights can be performed block-wise on the chip. Contrary to the search mode the 4-stage input shift register of the k-th multiplier is pre-loaded with Δ_{ip}, p = 1..4, k = i modulo 4, which is repeatedly used for multiplying W_{ij}, j = 1..N . Note that the ith row of a W-matrix and the ith row of the Δ-matrix contain now the data for the ith multiplier (see Figure 6).

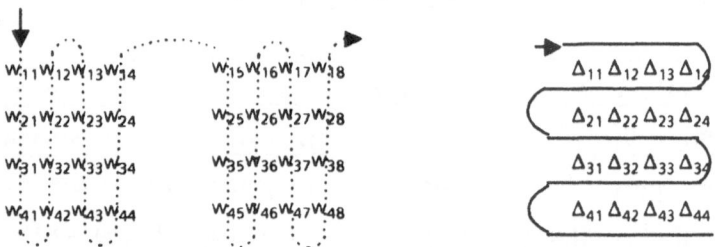

Figure 6   Weight and data schedule for Figure 4b (Equation 2)

Note also that there is a summation rippling along the chain for every pair of indices (j,p). Since the accumulation over index i is interrupted by the index j running from 1 to N, the result must be stored in a FIFO the size of which is proportional to the number of patterns times the number of neurons. Therefore it must be placed off-chip. In the next period, i.e. when running through the next 4 rows of the weight matrix, the content of the FIFO is read out and fed into the chain's accumulator in order to continue the accumulation over the index i.

Figure 7 shows the elementary functional block of a chain which is re-configurable for the three modes corresponding to the Equations (1-3) .

The buffers have dual functions: one half is always allocated to the data transfer with the data channels outside the chain, whereas the other half of the memory provides the chain with data, or is receiving data from it. This data transfer always takes place in blocks of 16 data words which each contain a 4 x 4 submatrix. Buffers B1 to B3 produce a memory capacity of 2x16 data words accordingly.

The nucleus of the elementary chain comprises 4 multipliers and 3 addition units, which carry out the actual matrix-matrix operations. A fifth multiplier does the scaling with λ_i in the search phase and α_i in the update phase. The data flowing from here reaches the accumulator which is configurable for local accumulation (run in the search mode) or distributed accumulation (run in the learning mode via all elementary chains, Equation 2; see below).

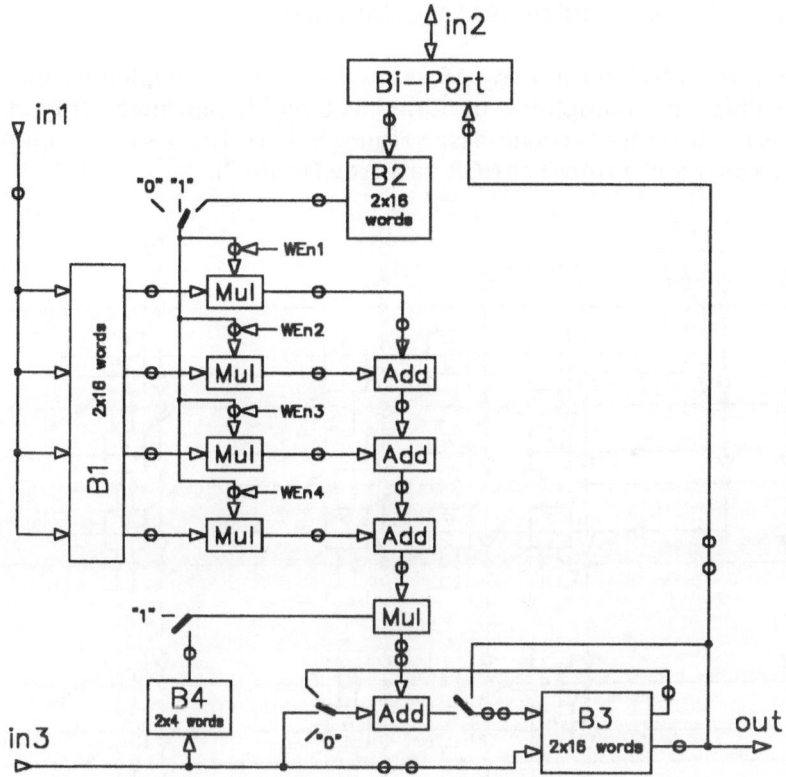

Figure 7   Elementary chain reconfigurable for search and learning

A multiplexer at the output of B2 enables a 4x4 unit matrix to be fed into the multiplier chain. A switch external to the chain (see Figure 10) has the same function in the input path in1. These switches enable the direct transfer of data from B1 or B2 into buffer B3 and on to the ports out and in2. This way many other operations called for by Equations 1-3 can be realized.

It is noticed that the length of the chain corresponds to the degree of interleaving. It will depend on the application under consideration to what extent patterns may become interleaved. The higher the degree of interleaving is set, the longer takes the evaluation of the first pattern. Subsequent patterns are responded to, however, without additional delay.

In summary, albeit a computational throughput, which is equal to the fully parallel architecture of Figure 3, x-fold interleaving allows to reduce the number of weight and input pins by a factor of x. The idea of applying the technique of interleaving to neural network implementation has been first formulated in [10].

ARCHITECTURE OF THE NEURAL SIGNAL PROCESSOR MA16

With 1μm CMOS technology 16 multipliers can be implemented on a single chip. The multiplier is designed as a highly pipelined array of full adders in carry-save technique (see Figure 8) [11]. The basic cell comprises a 1bit adder and a NAND or NOR gate (see Figure 9).

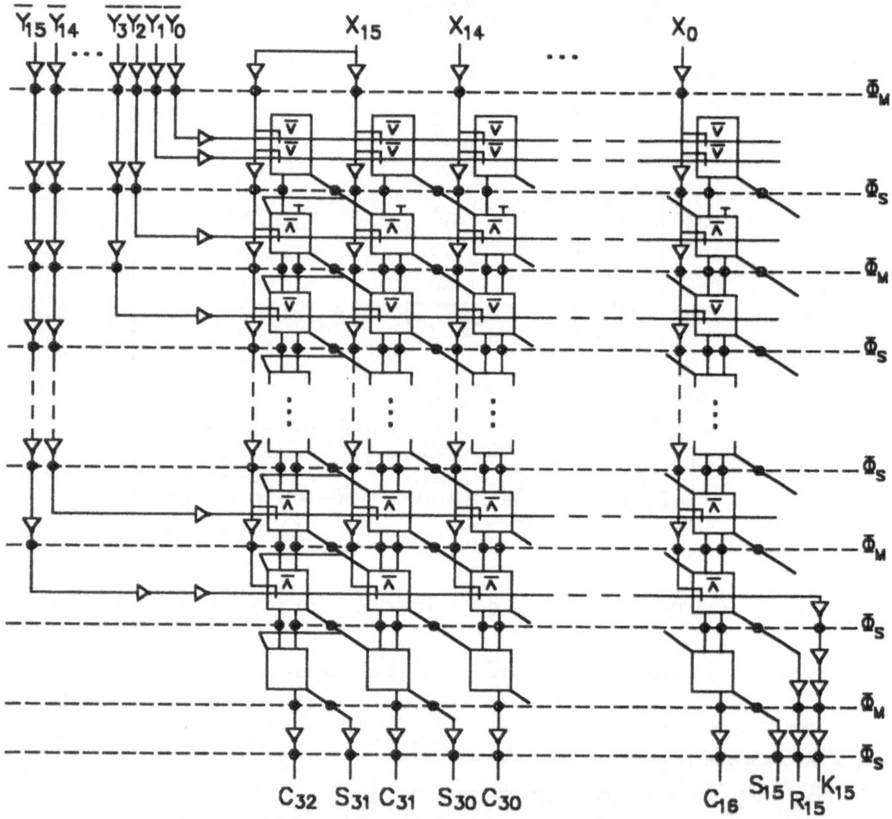

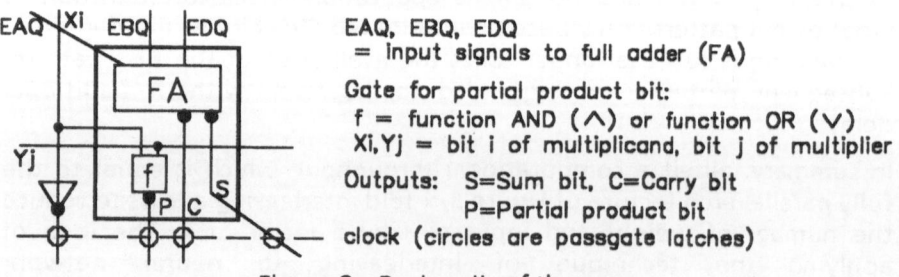

Figure 8   Full adder array of a 16x16 bit multiplier

EAQ, EBQ, EDQ
= input signals to full adder (FA)

Gate for partial product bit:
f = function AND (∧) or function OR (∨)
Xi,Yj = bit i of multiplicand, bit j of multiplier

Outputs: S=Sum bit, C=Carry bit
            P=Partial product bit
clock (circles are passgate latches)

Figure 9   Basic cells of the array

The above multiplier architecture yields a processing speed of 1120 MC/sec. This value is obtained for a worst case clock rate of 70 MHz. However, because of the memory environment the chip will operate at a moderate frequency of 25 to 40 MHz. This clock rate is realizable with today´s 4 Mbit DRAMs with 2-fold interleaving (see below).

The core area of the MA16 chip measures about 98mm^2, comprising 488k transistors. At 25 MHz the power consumption is estimated to be less than 4 Watts (not regarding pad drivers). More details on the circuit design of the chip are to be found in [12].

If 16-fold interleaving is not appropriate, the elementary chain´s length must be reduced which can be taken advantage of by implementing more than 1 chain. Figure 10 shows an arrangement in which the accumulation corresponding to Equations 1 and 3 is performed locally, whereas the accumulation corresponding to Equation 2 is distributed, i.e. rippling along 4 chains.

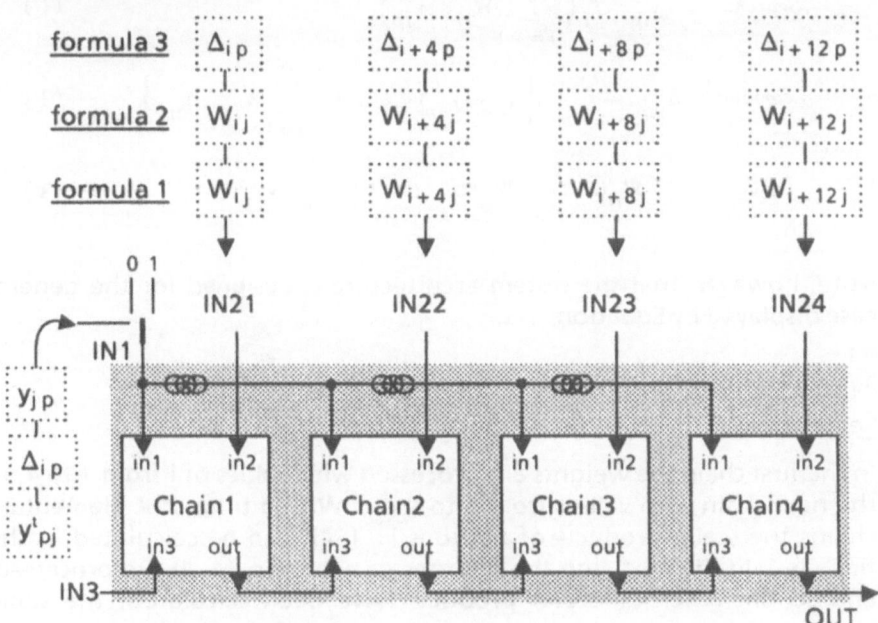

Figure 10   Schematic of the chip MA16 containing 4 chains

The I/O with the environment of the elementary chain takes place via two input ports, one output port and a two-way port. The in1 port is fed by a data channel which provides all chains on the chip with the same data by systolic propagation. The two-way port in2 is the individual interface to the weight memory, which contains the weights W_{ij} and correction values Δ_{ip} of the neurons allocated to the elementary chain. Note that in the update mode the port in2 serves to preload the weights of the last gradient descent step into the chain accumulator as well as to output their updated values. The in3 and out ports form the interface to the neighbouring elementary chains, through which preload data for B3, end results from B3, or intermediate results from the distributed accumulations are passed on.

In the case of distributed accumulation (Equation 2), in each elementary chain there is a 3 clock cycle delay in transferring data from in3 to out due to three pipeline registers along this path. Because of this the input has to be delayed by 3 clock cycles between adjacent chains, too. In addition, two parallel data blocks which run into the in2 ports of two neighbouring chains must themselves produce a delay of 3 clock cycles.

Next, we describe the systolic data and control flow in the 4 chains on a chip according to the operating mode. For lack of space we will specialize to supervised learning which is defined by

$$\text{Search:} \qquad y_{ip} = f_i \left(\sum_{j=-1}^{N} W_{ij} \cdot y_{jp} \right) \tag{1'}$$

$$\text{Learning:} \quad \Delta_{np} = f'_{np} \cdot \left\{ (t_{np} - y_{np}) \cdot \chi(n) + \sum_{i=1}^{N} W_{in} \cdot \Delta_{ip} \right\} \tag{2'}$$

$$\text{Update:} \qquad W_{ij}(t+1) = W_{ij}(t) - a \sum_{p \in P} \Delta_{ip} \cdot y_{jp} \tag{3'}$$

Note, however, that the system architecture is designed for the general case displayed by Equations 1-3.

Systolic Data Flow

Search phase

In the first chain the weights are processed with values of i from 1 to 4, in the next chain with values from 5 to 8 etc. With a total of K elementary chains the scalar products of neurons $i = 1..4K$ can be calculated in the first step. In the next step the neurons $4K + 1..8K(p = 1..4)$ are processed, and so on. When all scalar products have been calculated, the same process can be repeated for the next set of patterns ($p = 5..8$).

At the beginning of the search phase, the λ_i coefficients are preloaded into buffer B4, and buffer B3 is preloaded with initial values. In the case of the Hopfield net, the initial values for B3 may be the individual input values Y_{ip}, or in other cases may contain a value of zero. For preloading the λ_i coefficients, these values are passed in the order $\lambda_1, \lambda_2, \ldots$ to the IN3 port. The first 4 values are taken from buffer B4 of the first elementary chain, the next 4 values pass to buffer B3 and are taken over by buffer B4 of the next chain etc. The same procedure is used when preloading the Y_{ip} values.

The preloading procedure requires a time period of $(64 + 16) \cdot K$ clock cycles. At a later time this operation can take place parallel to running accumulation so that no additional clock cycles are lost. Whilst the last 16 word data block of the preload phase is being fed into port IN3 of the first elementary chain, the first weight block may reach port IN2, and the first input block may reach port IN1. After 16 clock cycles this data is passed from the two buffers of the first chain to its multipliers. Buffer B2 releases the matrix block $\{W_{ij}\}$ in the same way as it was read in. This data is then taken over by the transfer registers allocated to the multipliers in the way described above.

Whilst the first partial sum was being calculated, B2 has read in the next weight block and B1 has read in the next input block. In the same way, in the next set of 16 steps the new partial scalar product is calculated for $j = 5..8$, added up in the accumulator loop and so on.

During these operations, new initial values can already be preloaded into buffers B3 and B4, which will be used for the calculation of the next part matrix. This is the case if the accumulation reaches $j = N$. When this happens a set of 16 end results S_{ip} is present in buffer B3. In the next 16 clock cycles these values are read out of port out, whilst simultaneously the new Y_{ip} inital values are preloaded into in3. Buffer B3 of the second chain lets the 16-value data block $\{S_{ip}\}$ of the first elementary chain pass before it can read out its own data $\{S_{i \cdot p}\}$, and so on.

Learning phase

Firstly the correction values Δ_{ip} are sent to buffer B1 through port In1. Buffer B1 of chain k receives data block $\{\Delta_{4(k-1) + i,p}|\ i = 0..3,\ p = 1..4\}$. This preloading can also be done in the background during active elementary chain operation. In the learning phase Buffer B2 sends the weights out in transposed format, so that now the ith multiplier of the first elementary chain receives weight W_{ij}. At the other input multiplier i receives the correction values $\{\Delta_{i,p}|\ p = 1..4\}$, as in search mode. At the end of the addition chain we have the accumulation of the product $W_{ij} \cdot \Delta_{ip}$ for an index of i from 1 to 4. This partial sum is added to initial values $(t_{np}-y_{np})$ from port in3 and is passed on to the next elementary chain without being stored in the accumulator. This elementary chain is where the intermediate result is added to the partial sum with $i = 5..8$ etc. At the output of the last elementary chain we receive the sum with $i = 1..4K$ (K is the number of elementary chains), which is buffered in an external FIFO memory. When these partial sums are calculated for all $j = 1..N$ and $p = 1..4$, they are fed into the front of the accumulation chain in the next step, to be added to further products $W_{ij} \cdot \Delta_{ip}$ with $i = 4K + 1...8K$ and so on.

After a complete calculation of new correction values (including the external multiplication with f'_{np} , see equation 2') these are to be transferred into the weight memory, so that they can be fed into the processor in place of the weight values during update. To reduce the amount of external hardware needed, this data transfer is implemented in the MA16 chip.

To achieve this, the fully calculated correction values are re-read into the B1 buffer, which also takes place parallel to the active accumulation operation of the chain. At this point the accumulation is suspended for 32 clock cycles. In the first 16 clock cycles a unit matrix is blended in from the B2 side of the multiplier chain, whilst the Δ_{ip} values are read out of B1 in the usual order. After a determined number of clock cycles the data block $\{\Delta_{ip}\}$ runs through the fifth multiplier (which multiplies by 1), the adder (which adds 0) and the B3 buffer (which lets the data pass through) to port in2, which acts as an output. The total delay time of this path is measured such that the data is available at port in2 for precisely the second 16 clock cycles of the data transfer time.

Update phase

Apart from the calculation of the sum (Equation 7), the addition with the old weight W_{ij}(old) (see Equation 3') and rewriting the new weights W_{ij}(new) have to be carried out in the update phase.

The calculation of the scalar product is done locally, as in the search phase. In place of the W_{ij} values, the Δ_{ip} values are now led to port in2, and the y_{ip} values appear at port in1 in transposed order.

In contrast to the search phase, no preloading of data from port in3 into buffer B3 is forseen in this mode. The first partial sum block arrives directly into B3, by having a zero in the second input to the addition loop. In the update phase the fifth multiplier carries out the multiplication with the learning factor a_i, which was previously loaded into buffer B4. The addition with the old weights W_{ij}(old) can be carried out at any point in time (at the beginning, middle or end of the accumulation), by merging in the respective weight submatrix $\{W_{ij}\}$ at port in2, and a unity matrix at data path in1. In this way the required addition is integrated smoothly into the rest of the accumulation. After completing the accumulation for a W_{ij} data block, a new accumulation can be started immediately, whereby the other part of the B3 buffer is used. The updated weights W_{ij}(new) from the previous accumulation can be read from port in2 at any time, whereby, however, the current accumulation would have to be suspended for 16 clock cycles.

Systolic Control Flow

The data buffers B1-4 contain most of the control elements of the elementary chain. The way in which the data are read into, or read out of the buffers, is determined by the mode of operation of the elementary chain. Mode and trigger instructions are to be provided at the buffer inputs and outputs to enable the flow of data to be controlled.

Let us consider the buffer B1 and describe how the systolic data flow structure is inherited by the control flow.

<u>Buffer B1</u>.   The basic cells of the data buffers are formed by slightly modified static standard CMOS memory cells (Figure 11). The nucleus comprises two interconnected inverters which can be set by the left transfer transistor on the signal of the read connection . The state of the memory cell can be transmitted to the write connection using the right transfer transistor.

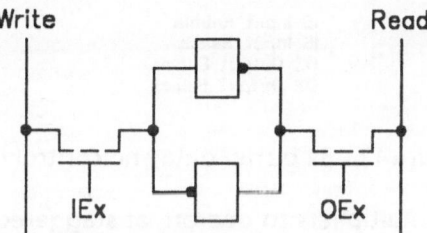

Write   Read

IEx   OEx

Figure 11   Basic cell of the buffer

In the search and update phase, buffer B1 takes over the network input data y_{jp} from port in1 of the elementary chain, and the Δ_{ip} correction values in the learning phase. On the output side these values are passed on to the 4 multipliers.

The memory comprises an arrangement of 16 blocks of two memory words (Figure 12). The division of the memory blocks into two halves makes it possible to write into one half of the memory whilst simultaneously reading out of the other half. These blocks are laid out in a 4x4 matrix whereby four blocks at a time have access to a communal data output. The input data is led to the buffer via a communal input. In normal mode the data is to be distributed in columns, and in transposed mode in rows.

To effectuate this, in the control line for the write transfer (IE for normal or TIE for transposed input) a single input is fed in at the beginning of a load cycle, which propagates itself in columns (IE) or rows (TIE) via the registers of the memory field. This impulse is repeated every 16 clock cycles. To control the four-channel output of data to the multipliers, a clock impulse is given to each OE control chain which is repeated every 4 clock cycles.

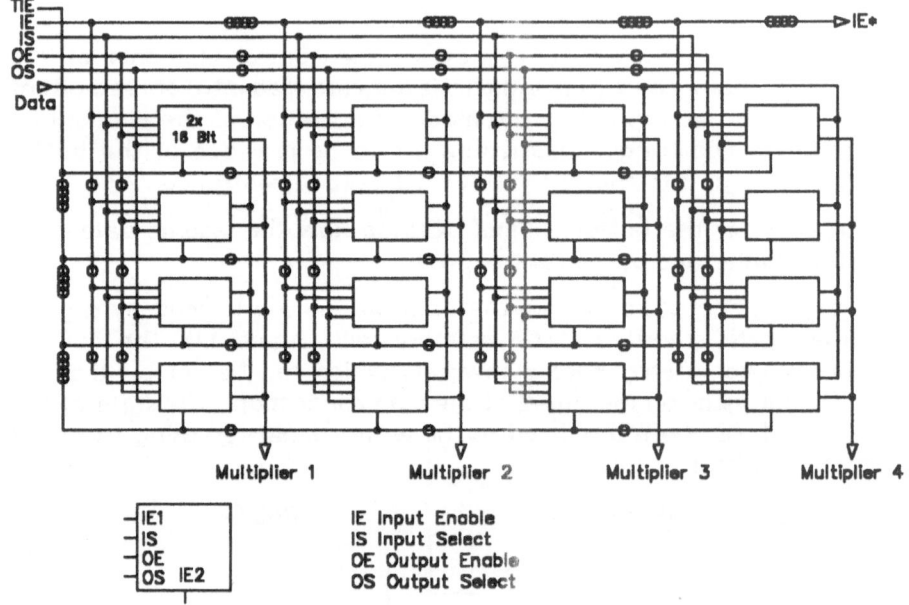

Figure 12   B1 buffer data and control flow

To allow for the 4 multipliers to operate at staggered intervals, the read operations from output to output must each be staggered by one clock cycle. As there is no time stagger when writing to the memory chains, we are not always writing into one half of the memory and reading from the other half simultaneously, but may at some point be reading and writing within the same half of the memory, but never within the same memory cell.

The internal implementation of the memory blocks is shown in Figure 13. The allocation of which of the two memory cells is to be used for writing to and which cell is to be read from is determined by the control lines IS (Input Select) and OS (Output Select). Releasing for writing or reading is done by the control lines IE1 and IE2 (Input Enable, IE = IE1 v IE2) or OE (Output Enable). From this control information the signals for the transfer transistors are produced using simple logic. The control signals for the input transistors are linked to a clock impulse to guarantee synchronised setting.

More involved, but following the same principles is the control of the remaining buffers B2-4.

Halting accumulation.   For the external storage of the weights, dynamic RAM chips are used. These can be read from quickly, as long as a row (page) of the memory matrix is being read. Additional time is needed if a

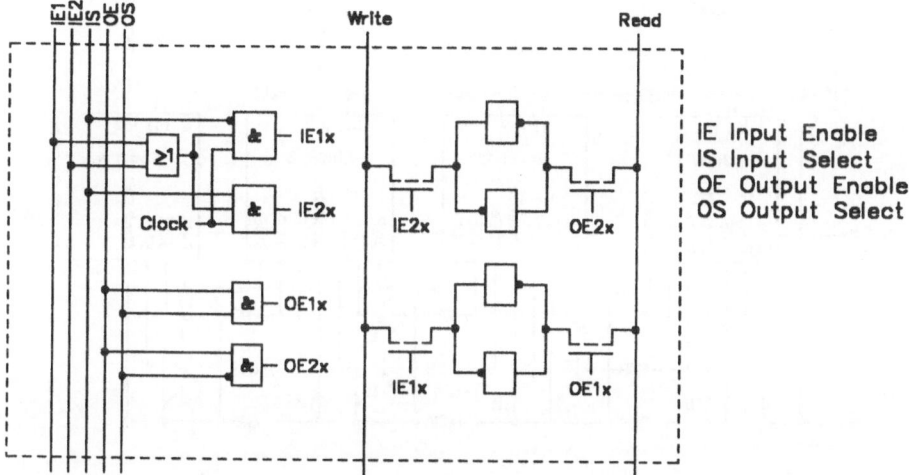

IE Input Enable
IS Input Select
OE Output Enable
OS Output Select

Figure 13   Internal memory block for buffer B1

page change takes place, meaning that the flow of data is interrupted. For this reason it has to be made possible to put the elementary chain into a wait state, i.e. the data buffers of the elementary chain may not deliver or accept any data for a given number of clock cycles. In this mode the data buffers stop automatically if they are provided with no trigger impulses. If a buffer has received one of the required trigger impulses, it is active for 16 clock cycles and cannot be stopped. Subsequently the buffer will cease to operate until a new impulse is given.

Distribution of control information on the chip.   Controlling is made easier by the fact that all elementary chains on the chip carry out basically the same operations, meaning that the control instructions for the chains are similar and only have to be generated once on the chip. However, this can only be taken advantage of if the control signals systolically propagate themselves in the same way as the data do. This can be achieved simply by providing three pipeline registers in all control lines connecting two chains, as done in the data lines (Figure 14).

MA16 Chip Controller

The chip-internal control word is unsuitable for an external driver, as apart from the large word width of 25 bits, a complicated external controller would be needed. It would be desirable to have a more strongly coded and therefore narrower control word, which can also handle several clock cycles with a single instruction, so that an ordinary commercial microprocessor would do. This requires an intelligent chip controller which will supply the internal control lines with sequences depending on the external driver.

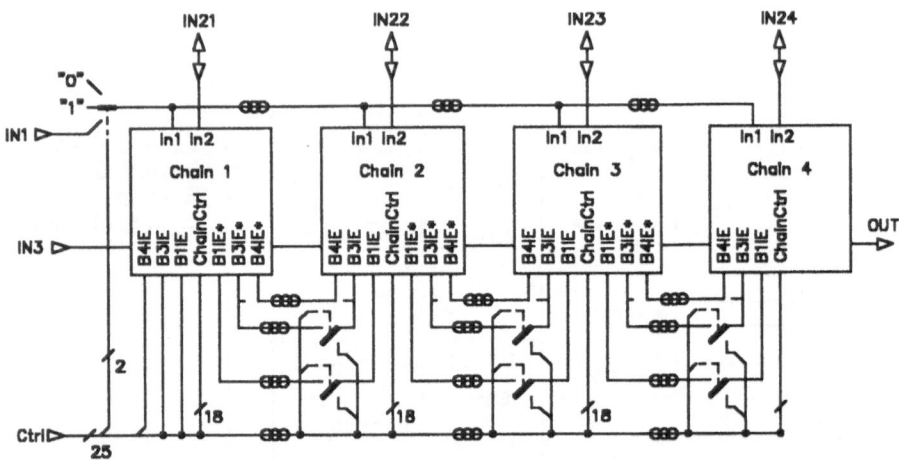

Figure 14   Elementary chain setup on the chip with systolic propagation of control signals

It is not really useful to code all possible combinations of operation into a single instruction, as various operations can run simultaneously and independently of each other. It is therefore better to provide the operations running in parallel with their own instruction sets, each representing a group of control lines. The state of the control word can thus be described by the following set of instructions (Table 1) .

Control line	Description	External Instructions	Bits
Mode	Chip Mode of Operation	SRC (Search) REC (Learning) UPD (Update)	2
AcC	Accumulation control	RAC (Run Accumulation) HAC (Halt Accumulation)	1
LoC	Load control	RLS (Run Load/Store) HLS (Halt Load/Store)	1
LoOp	Load operation	LOD (Load Data) LDC (Load Coefficient) STO (Store Data) NLS (No Load or Store)	2
RegSel	Register select	UR1 (Use Register Set #1) UR2 (Use Register Set #2)	1

Table 1   MA16 chip instruction set

The first field of the control word contains the mode of operation. The accumulation control fulfils two objectives: if the instruction on this line is changed from HAC to RAC, accumulation is started and the buffers start to read and write using the given time offset. The HAC instruction suppresses the required trigger impulses for the buffers, so that the chip falls into an accumulation wait state.

The same applies to RLS and HLS in the loading operation.

The loading operation itself is selected by LoOp. The appropriate instructions invoke operations which are also dependent on the mode of operation of the chip:

Search mode

LOD — Preload of start values Y_{ip} from IN3 in B3; store the previously calculated endresults S_{ip} to OUT (Execution time : 4x16 clock cycles for the whole chip).

LDC — Preload of scaling coefficients λ_i in B4 (4x4 = 16 clock cycles).

STO — Storing of accumulation results, contents of B3 are not overwritten (Execution time : 4x16 = 64 clock cycles).

NLS — No load/store operation, data is sent from IN3 to OUT with the usual time delay.

Learning mode

LOD — Preload of Δ_{ip} correction values via IN1 into B1 (4x16 = 64 clock cycles).

LDC — No meaning in this mode (= NLS).

STO — Data transfer of Δ_{ip} correction values from buffer B1 through IN2 to the weight memory (16 + 16 = 32 clock cycles, additional HAC instruction needed).

NLS — No load/store operation.

Update mode

LOD — Old weight values W_{ij}(old) inserted into update accumulation (multiplication with unity matrix, 16 clock cycles).

LDC — Preload of learning factors a_i in B4 (4x4 = 16 clock cycles).

STO — Store new weights W_{ij}(new) from B3 through IN2 to the weight memory (16 clock cycles).

NLS — No load/store operation. Data is sent from IN3 to OUT with the usual time delay.

During the time needed to execute the load operations, because of the systolic propagation of the load instructions it is not possible to suspend them with HLS. If one of the load instructions (SRC,LOD), (SRC,STO), (SRC,LDC), (REC,LOD), or (UPD,LDC) reaches the control inputs at the same time as HLS the operation is not carried out until LoC changes to RLS. In the other load instructions the LoC input is meaningless, as these operations only have an execution time of 16 or 32 clock cycles, independent of the number of elementary chains in the neuro emulator. The execution times of the operations listed above are 16xK or 4xK clock cycles, where K is the number of elementary chains.

The last control bit RegSel selects one half of the buffer for accumulation. The corresponding instructions UR1 and UR2 only affect the same buffers used in the load operations in the individual modes of operation. In the search and update phase these are B3 and B4, in learning mode B1. In update mode, a change of instruction on this line also causes accumulation to be initialised. The two halves of those buffers that are directly coupled to the accumulation (B1 and B2 during search and update, B2 during learning) are changed automatically every 16 cycles when a RAC instruction is received.

The function of the chip controller is to convert the instructions described above into sequences of signals for the control lines. As the operations of the elementary chain can be roughly divided up into the two groups of accumulation and load operations it also makes sense to divide up the sequence production for each of the two groups (Figure 15). Depending

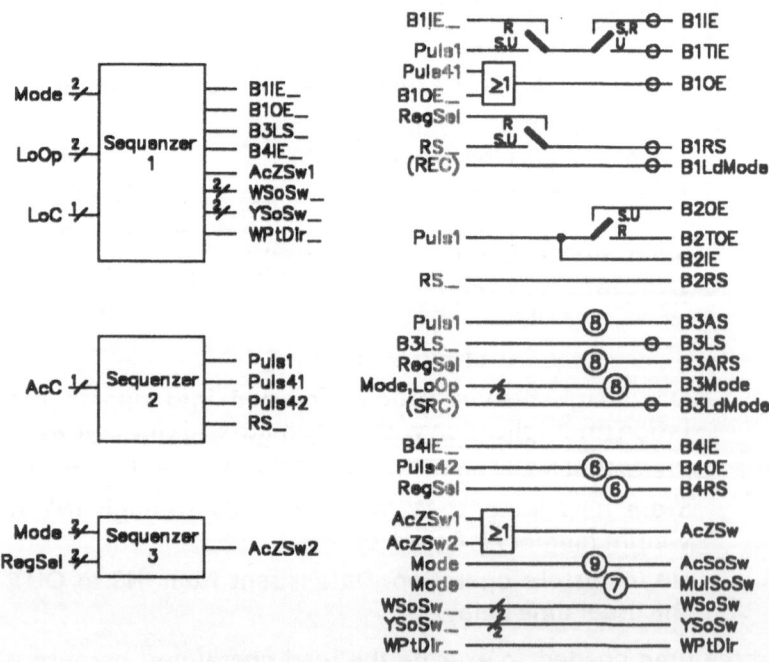

Figure 15   MA 16 chip sequencers

on the mode of operation, some of the control lines may serve both groups of operations. In these cases the signal sources for the lines will be switched over according to the mode.

The signals for the load operations are created by sequencer 1 and the signals for the accumulation operations produced by sequencer 2. Sequencer 3 has the single function of controlling the initialisation of the accumulators in update mode.

To achieve the timing precision needed to drive the function blocks in the elementary chain it is necessary in some cases to delay the control signals by several clock cycles in relation to the chip driver. This can be done by inserting additional delay registers in the control lines or by appropriate programming of the sequencers. The control signals which can be obtained from the chip instructions by using direct logic and the trigger signals for the accumulation operations are brought with delays through register chains into the appropriate time relation. As the load operations each only have to be executed once, it is simple to generate appropriate control signals in the sequencer at the right point in time.

The sequencers can be implemented as sequential logic circuits, for which simple design concepts are available. For the chip controller sequencer the Mealy method must be chosen, as an input signal can function as an

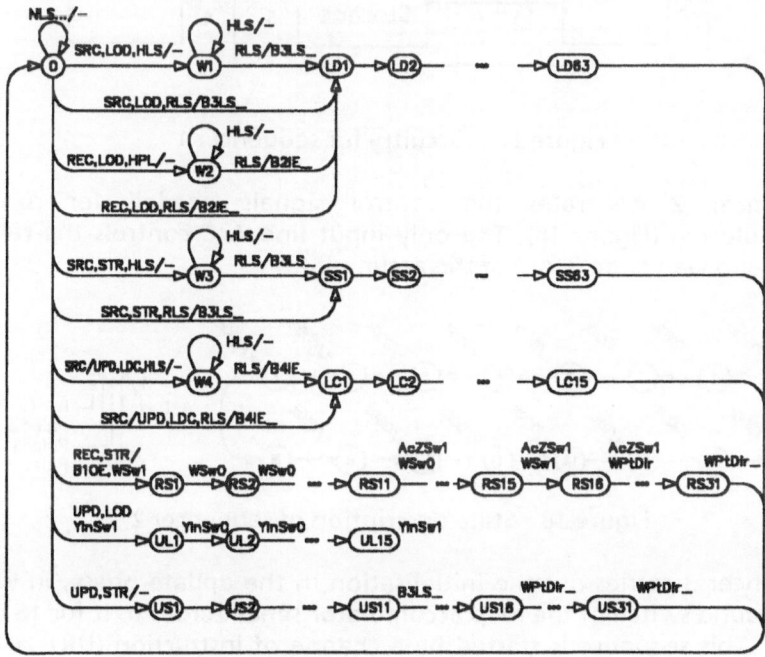

Figure 16   State description of sequencer 1

output signal with the same frequency and the Mealy automat uses less state memory in most cases.

Figure 16 shows the state diagram for sequencer 1. With its 222 states it can manage with a minimal state memory capacity of 8 bits. If several systematic features of the state diagram are exploited, the number of memory bits increases but the amount of input and output logic is drastically reduced (Figure 17). The methodology is in the fact that 6 chains occur in the state diagram, in which there is an unconditional jumping from one state to the next, independent of the input signals. This sequence can be implemented in a simple fashion using a modulo 64 counter, which can be started and reset. The additional state memory stores the input instructions of the Mode and LoOp control words in 4 of its 5 bit cells. The fifth bit cell represents the wait state which delays the execution of some instructions in the case of HLS.

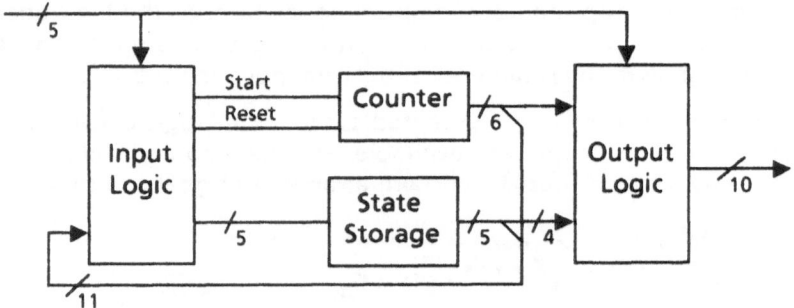

Figure 17   Circuitry for sequencer 1

Sequencer 2 generates the control signals needed for running accumulation (Figure 18). The only input line AcC controls the release (RAC) and wait state (HAC) of the accumulation.

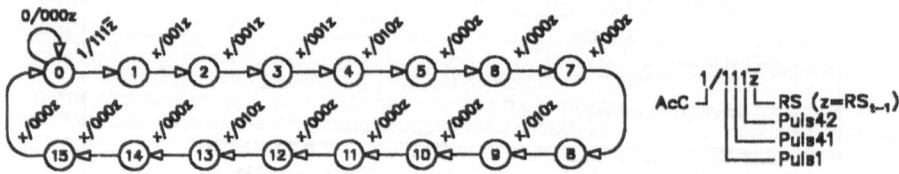

Figure 18   State description of sequencer 2

Sequencer 3 carries out the initialisation in the update phase, in which the ground switch of the loop accumulator sends zeroes to it for 16 clock cycles. This sequence is started by a change of instruction (UR1-> UR2, UR2 -> UR1) on the RegSel control line (Figure 19).

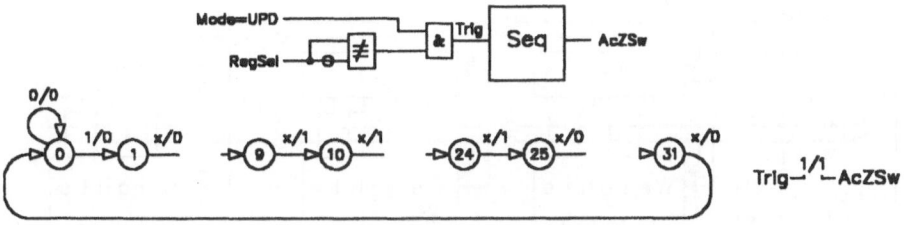

Figure 19   State description and circuitry for sequencer 3

BOARD ARCHITECTURE OF THE NEUROEMULATOR

The systolic architecture implemented at chip level (see Figure 10,14) can be extended to the board level. Figure 20 displays a linear array of chips MA16. The neuroemulator built from this linear configuration consists of DRAM banks for weights and data, of an arithmetic unit which executes the non-compute-intensive operations like table look-up of transfer functions, of interface circuitry for communication with the host or a pre/post-processing unit, and of a central board controller which itself is under the control of the host.

The weight ports IN21 to IN24 are connected to local memory blocks, which have no other links to the outside world. Every data transfer such as the input and output of weight values must therefore take place through the processors. This, however, causes little additional time loss, as load operations involving the weight memory disable the calculation activity of the chips anyway.

In the recursion phase the Δ_{ip}'s are calculated by repeated accumulation runs along all MA16 chips. A FIFO memory is provided to store the partial sums of the Δ_{ip}'s between two consecutive accumulation runs. The calculation of the Δ_{ip}'s may be completed by a multiplication with the derivative of the neuron's discriminator function, and then the Δ_{ip}'s are temporarily stored in a separate memory. Finally the Δ_{ip}'s are transferred into their main memory that gathers all patterns (not only four as the temporary memory) for the update phase. This memory is placed parallel to the weight memory, since in the update phase the Δ_{ip}'s are needed instead of the weights.

Like the temporary memory of the Δ_{ip}'s the memory bank of the y_{jp}'s is arranged in a central place. Compared to a distributed arrangement with a local memory for each elementary chain or chip MA16, this is advantageous because now the memory management as well as the eval-

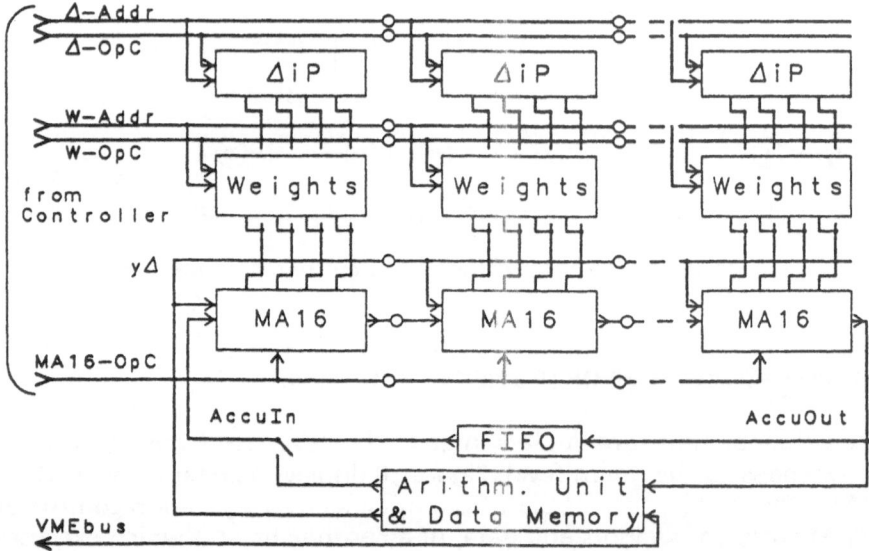

Figure 20   Layout of the linear array of chips MA16, the memory and the external arithmetical unit at board level

uation of the discriminator function have to be performed only once instead of many times. Furthermore an expensive - in terms of components, wiring, and board area - multiplexing of the data paths is avoided because in all three phases (search, learning, and update) the ports IN3 and OUT can be kept connected directly to the neighboured elementary chains.

A 2-Dimensional Array Of MA16 Chips

For a linear array as described above the performance can be increased by two means:

- adding further chips MA16 with the corresponding weight memories augments the computation power as well as the maximum size of a neural net to be emulated;

- in case an increase of the data throughput isn't demanded the size of the weight memory can be enlarged by simply providing additional memory modules for the chips MA16.

But there might also be applications that ask for a higher computation power without needing a larger weight memory. In this case each weight memory module has to supply two or more chips MA16, which operate on different sets of patterns. Figure 21 shows the corresponding architecture of the neuroemulator's 2D-version .

As before each MA16 has got a Δ_{ip}-memory of its own. In order to avoid bus contention when the MA16s access their Δ_{ip}-memory simultaneously, the ports IN2 mustn't be connected directly to each other but must be separated by (bidirectional) three-state registers.

While the rows in such a 2-D array of MA16s operate completely in parallel during the search and learning phase, the situation is more involved in the update phase because then the accumulation over the index p (Equation1) is distributed over all MA16s in a column. Each MA16 accumulates over a different set of patterns, thus calculating a part of the final correction value for the weights. Finally these partial sums are systolically transported to the uppermost MA16 where they are summed up in additional cycles, and then transferred to the weight memory.

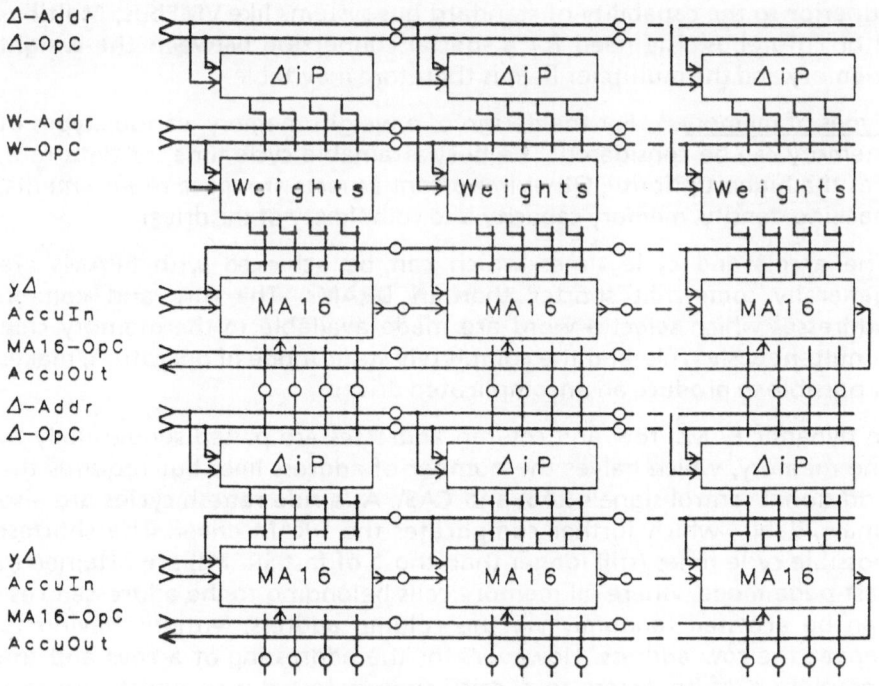

Figure 21   Layout of the 2D array of chips MA16

Memory Architecture

Whereas the three phases of search, learn and update can generally be distinguished from one another, with regard to the weight memory the first two phases are identical. This is because the global transposition of the weight matrix is avoided by changing the configuration of the MA16 chip data flow. During the update phase, both read and write accesses to the weight memory are necessary, whereby the weights are needed in the same order as in the search and learn phases.

The weight values in the weight memory can be organised such that real random accesses are relatively seldom needed in each of the three phases. Most of the accesses to the weight memory are characterized by a linear sequence of addresses.

In cases where the weight values are 16 bit words and with a 25MHz clock, for example, the required data rate of the weight memory which an MA16 supplies is 1600 Mbit/sec. This value is about one or two levels superior to the capability of standard bus systems like VMEbus, Multibus II or Futurebus. The need for a special connection between the weight memory and the multiplier field is therefore inevitable.

Types of memory.   For the design of a weight memory, various types of memory can be considered. The best attainable cycle time, or data rate, has the highest priority. Other important aspects are space requirements, packing density, memory capacity and complexity of the driver.

The access and cycle times which can be achieved with SRAMs are generally somewhat shorter than in DRAMs. The row and column addresses which select a word are made available to the memory chip simultaneously. This, and the completely static mode of operation, makes it possible to produce an uncomplicated driver.

In dynamic RAMs, row and column addresses are passed sequentially to the memory, which halves the number of address lines but requires the additional control signals RAS\ and CAS\. As a rule refresh cycles are also unavoidable, which further complicates the DRAM driver. The shortest possible cycle times (still longer than those of fast SRAMs) are attained in fast page mode, where all memory cells belonging to the addressed row can be accessed randomly via the column address, without having to repeat the row address. However, for the addressing of a row and the completion of an access to a row, time is lost during which no data transfer can take place.

In a video RAM, a whole row of information is taken into static registers which can be accessed by a secondary memory port. This is a serial port which has no random access capability but allows the specifying of a start address for a linear sequence of addresses. The data rates attainable at the serial port are comparable to that of SRAMs. The driver becomes even more complex and external control of the refresh cycles is still necessary.

Finally, a FIFO memory, on the basis of the 1Mbit DRAM, for example, has only serial read and write ports which are more or less independent of each other. The memory no longer has any address inputs at all, only reset and clock inputs for internal address counters. The refresh cycles are generated automatically. The access times are similar to those of video RAM serial ports. The number of pins is less than in a standard DRAM, and the package is smaller. However, because of the dynamic mode of operation of the FIFO RAM, there are several considerable restrictions to be considered.

Selecting a memory type for the weight memory.   As far as speed, memory capacity and space requirements are concerned, video RAMs are comparable to fast SRAMs, but the driver is considerably more complex in the aspects of refreshing, addressing and the controlling of two ports. From a technical point of view video RAMs can therefore not be regarded as an alternative to fast SRAMs.

Because of the high packing density which can be attained, a FIFO RAM appears at first to be ideal for the weight memory, as long as random accesses are kept to a minimum. What disqualifies this particular chip from being used for the intended application is the fact that after breaks of greater than 1ms between accesses, the internal address counters have to be reset. Preloading to an arbitrary startup value is not possible. This makes single-step operation and the debugging of the neuro-emulator virtually impossible.

This leaves only SRAMs and standard DRAMs as alternatives, with which the following solutions can be considered :

α) Fast SRAMs with cycle times of less than 30ns, so that clock rates of around 25MHz can be achieved directly. The disadvantage is that the fast SRAMs available on the market have a memory capacity of 4 to 16 times less than that of DRAMs.

β) Standard SRAMs with cycle times between 50 and 100 ns (a quarter of the memory capacity of DRAMs) require interleaving, for which two-way intermediate storage is needed in the data bus.

γ) Standard DRAMs of the 1Mbit and 4Mbit generations have cycle times of between 45 and 60ns in fast page mode. By interleaving two components at a time, worst case clock rates of 25MHz seem to be realizable. High packing density and good availability are the advantages of standard DRAMs.

DRAM cycle times.   With DRAMs in fast page mode the situation is as shown in Figure 22.The data output is controlled by CAS\, the resulting memory cycle being $t_{CM} = t_{CAC} + t_{DV} - t_{OFF} + t_{CP} + t_T$. The marginal

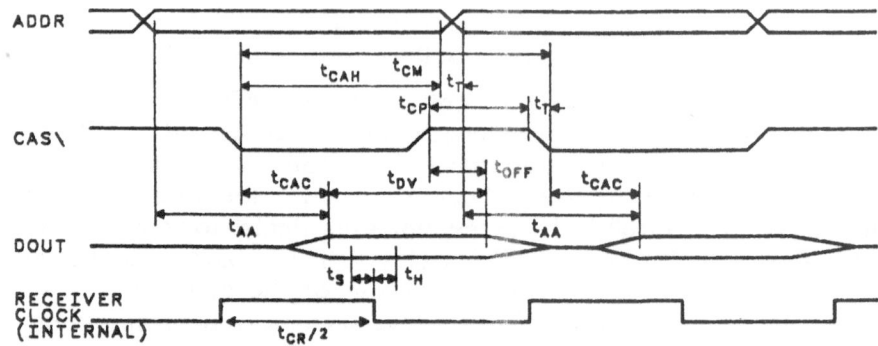

Figure 22   DRAM read access

condition $t_{CAH} + t_T + t_{AA} < t_{CM} + t_{CAC}$ is usually automatically fulfilled. The period of time t_{DV} for which the data must be valid is given by $t_{DV} = t_S + t_H + t_{CR}/2$. Note that for $t_{DV}-t_S-t_H < t_{CR}/2$ a clock edge couldn´t be guaranteed to arrive at the receiver (i.e. the MA16) during t_{DV} at all. If t_S and t_H of the MA16 are defined with respect to the external clock edge the term "$t_{CR}/2$" can be skipped. However, t_{DV} still must take longer than $t_S + t_H$ in order to take into account the remaining timing tolerances.

With a 4Mbit DRAM ($t_{AA} = 40$ns, $t_{CAH} = 15$ns, $t_{CAC} = 20$ns, $t_{OFF} = 0$, $t_T = 5$ns, $t_{CP} = 10$ns) the resulting cycle times are $t_C = 90$ns ($t_C = t_{CM} = t_{CR}$) , or $t_C = 60$ns ($t_C = t_{CM} = 2 \cdot t_{CR}$) .

DRAM interleaving.   Figure 23 shows the timing diagram for 2 interleaved DRAMs. With a cycle time of $t_C = t_{CAC} + t_{DV} + t_D$ the delay time t_D must be considered, in case the deactivation of the data outputs via OE\ takes longer than the activation via CAS\. With $t_D = 5$ns and the other values mentioned above $t_C = 70$ns ($t_{CR} = t_C$) , or $t_C = 47$ns ($t_{CR} = t_C/2$) which is comparable to the values for a fast SRAM without interleaving.

When writing the relationships are similar to when reading. However, somewhat longer setup and hold times (15-20ns as opposed to 10ns) and therefore longer t_{DV} must be taken into consideration for the memory. This is compensated for by the fact that in the MA16 the data output time t_{OUT} (10-15ns) is shorter than the memory read access times. Whereas when reading, t_{AA} and t_{OEA} contain the delay between pin and data output buffer, the data output buffer for the MA16 chip is activated directly by the clock signal during a write operation. The level conversion time from TTL to CMOS for control lines and various other delays don´t exist in this case. In this way, the clock rates for write accesses can be at least as good as those attained when reading.

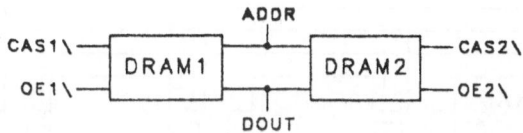

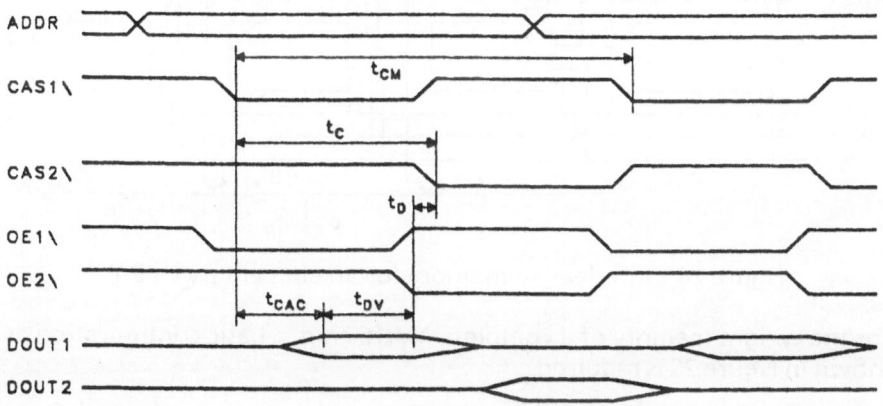

Figure 23   DRAM interleaving

Figure 23 shows us that during a read operation, the data bus is occupied until well into the next address cycle. In a write cycle immediately following this it would therefore come to a bus conflict. For this reason there must be at least one redundant cycle between the two in which neither read nor write takes place. In this phase, the MA16 must also be in a wait state. When switching over from writing to reading, the aforementioned problem does not occur because the data bus is available soon enough to avoid a conflict.

Memory configuration for an MA16 chip.   The weights have a 16 bit word width, meaning that four 1Mbx4 memory chips must be connected up in parallel to be able to serve the weight port of an elementary chain (Figure 24). The capacitive load (without line capacitancies) is about 60pF on the RAS\ and address lines, and half of that on the rest of the control lines.

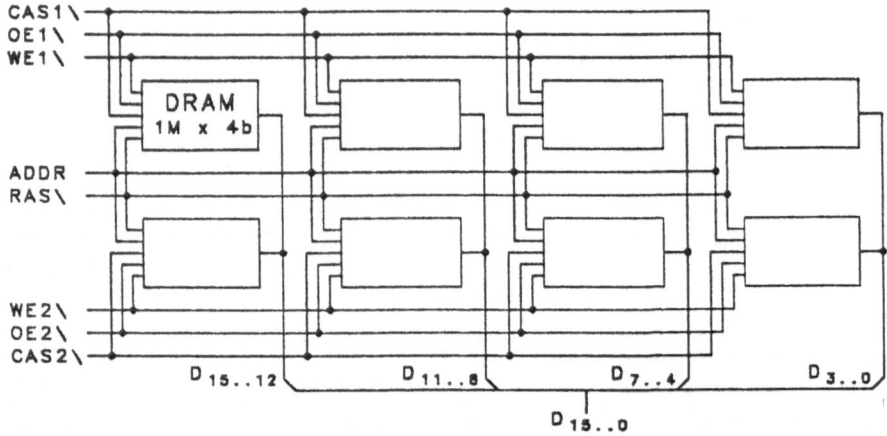

Figure 24   Interleaved memory for an elementary chain

For the weight memory of a complete MA16 chip, a basic configuration as shown in Figure 25 is required.

Because of the systolic data flow, the elementary chains in the MA16 are mutually delayed by several clock cycles, so that the weights from the elementary chains must also be delivered with appropriate delays. One way of achieving this is to delay the memory access to the various weight ports (Figure 25) and therefore transfer the systolic principle to board level. It is not possible to integrate the delay registers between the

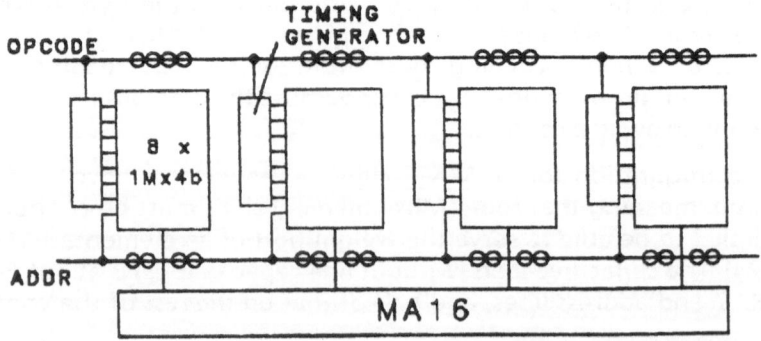

Figure 25   MA16 chip weight memory (Version 1)

memory groups into the MA16 because of the high number of pins (five times 12 address bits and five times 4 OpCode bits), so this would have to be set up on the board.

Another alternative is shown in Figure 26, where all the weight ports of the memory are accessed simultaneously. The required delays in the

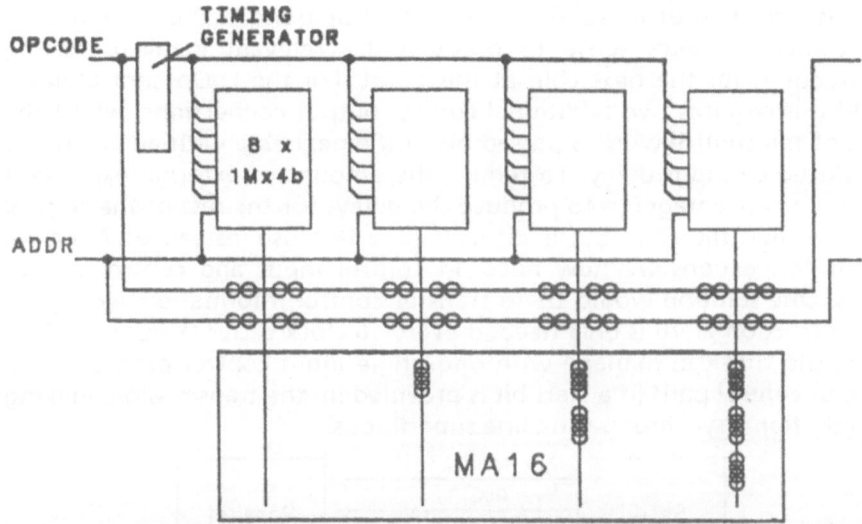

Figure 26   MA16 chip weight memory (Version 2)

elementary chains are integrated into the MA16. For write accesses to the memory the delays must be provided in the opposite order. The write data for the leftmost weight must be delayed by 12 clock cycles while no delay is necessary in the rightmost data path. The delay registers in the address and control lines for the memories can be implemented either in the MA16 (two times 12 and two times 4 pins) or on the board.

Timing generator.   The function of the timing generator is to produce the signals needed to operate the DRAMs. These are derived directly from the system clock, but are occasionally provided with additional delays (Figure 23) needed to guarantee reliable memory operation. The valid combinations of the seven control lines needed for interleaving (RAS\, and two of each of CAS\, OE\ and WE\) are represented in a 4 bit OpCode, which is interpreted by the timing generator to produce the control signals for the memory.

Distribution Of Control Information In Arrays Of MA16 Chips

As the elementary chains of all chips basically carry out the same operations, an external controller only needs to produce control information for the first MA16 chip. These instructions which are encoded by 7 bits can be systolically propagated from chip to chip in the same way as this is done within the chips themselves. It is also necessary at this level to delay some of the load instructions for an additional 64 (LOD, STO) or 16(LDC) clock cycles. Hereby the wait cycles of the load operations must be taken into consideration.

As the decoding of these instructions must be done in the internal chip controller it makes sense to produce the relevant delayed loading instructions for the next chip at this point. For the LoOp control word field this requires two additional control output connections, whilst the rest of the control word is passed on to the next chip with an externally produced constant delay. To reduce the amount of external expense it would be advantageous to produce the delays for the rest of the control word within the chip, but is difficult here because instead of 7 control input connections we now need 14 control input and control output pads. One solution would be to transfer control information serially, as an instruction word is only needed every 16 clock cycles (Figure 27). This way, the chip can manage with one single input control path and one output control path (if a start bit is provided in the transmission, making an additional synchronisation line superfluous).

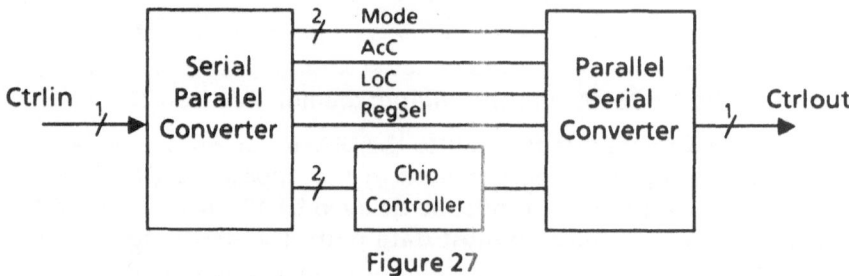

Figure 27

Board Controller and Communication

In the same way as for the MA16 a set of commands can be derived for the weight memory and the Δ_{ip}-memory. These are read and write commands in the fast page mode as well as in the conventional random access mode. Further there is a refresh command and a stop command, the latter switching the memories into the inactive state. The commands are encoded into a 4 bit OpCode that is delivered by the board controller in parallel to the address bits (see Figure 28).

The data unit is controlled by an OpCode that can be viewed as sort of interrupt vector. According to the OpCode, a subroutine is activated on the microprocessor in the data unit that performs the computations as requested by the board controller and generates the addresses for the data memory.

The board controller consists of two parts. A standard CPU with an associated program memory and a sequencer with a related microprogram memory. The sequencer generates the OpCodes for the weight and Δ_{ip}-memory and the array of MA16s as well. It also communicates with the data unit in order to synchronize the data memory within the data unit with the MA16s. This way the sequencer

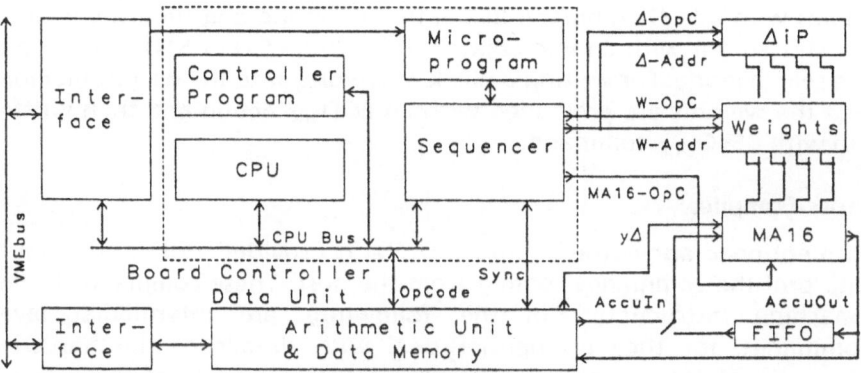

Figure 28   Board Controller

controls the MA16s and all the components that work synchronously with the MA16s. As a consequence the controller CPU is freed from generating OpCodes and memory addresses at a high speed.

The program that is running on the controller CPU has to serve two tasks. Firstly it has to provide the sequencer with addresses (that specify the sequence of OpCodes the sequencer has to produce) and the data unit with OpCodes. To do so the CPU interprets an emulation program written in the machine language of the neuroemulator board (see below). Secondly the controller CPU runs a simple operating system that controls the boot and shutdown procedures of the board and allows for communication with the host. Thereby an emulation program can be downloaded from the host to the neuroemulator and started there.

The data unit collects its input data via a VMEbus interface of its own, e.g. from video/audio interface boards. Besides, the host can download to the data unit's CPU the routines that correspond to the OpCodes delivered by the board controller.

Data stored in the central data memory of the data unit like the y_{jp}'s, the target values t_{ip}, etc, can be accessed easily via the interface by proper memory addressing. The weights (and also the Δ_{ip}'s), however, as they are stored locally with the corresponding elementary chains, can't be transferred so easily. In order to avoid additional components and buses on the board the elementary chains are involved in the data transfer of the weights and the Δ_{ip}'s. This way components and data paths can be utilized that are present anyway. In this case, the ports IN2, IN3, and OUT, and the associate registers of the elementary chains perform a sort of parallel-serial conversion or vice versa. For a weight transfer to the interface, for example, first the weights are loaded into the elementary chains with a LOD command in the update mode and then the weights are shifted out with an immediately following LOD or STO command in the search mode. For writing a block of weights into the weight memory, first the weights are preloaded with an LOD_{SRC} action and then written out with a STO_{UPD} command.

Cross-Compiler

As mentioned above the board controller is provided with a program to interpret the commands coming from the host. These commands (called "machine instructions" in the following) are intermediate-level-commands, for they do neither deal with details of the hardware (memory refresh, page changes in the fast page mode of the memory chips, etc.) nor with a high-level neural net description language.

The following instructions are part of the machine language:

-- Reset the emulator board to a defined state.

-- Receive weights and patterns from the host or send them back to the host. The address range where to store the data (or where to find it) is part of the instructions.

-- Move data between different parts of the memory and between different memories (weight memory, local Δ_{ip} memory, and the memory in the data units). Here again the source address range is part of the instruction as well as the destination address.

-- Select an address where to find the first weight to be used by the MA16 array.

-- Select an address where to find the first pattern in the memory of the data unit to be used by the MA16 array.

-- Select an operational mode of the MA16 array.

-- Transfer data from the host to the data unit (for example, table lookup operations of the transfer function f(..) and its derivative).

-- Download a program from the host for the CPU's in the data units.

-- Select an operational mode of the data unit for the next results coming from the MA16 array.

-- Select an address to store the first result coming from the data unit.

-- Start the array calculation for a given number of data elements. After each calculation all addresses are incremented automatically to get access to the next data elements.

-- Interrupt a running calculation.

-- Control the program flow itself (loop constructs, conditional statements, etc.).

-- Self-test of the board.

The elementary arithmetic operations to be performed by the board (like multiplication of a vector by a scalar or a vector, matrix-matrix multiplications or table look-up etc.) are encoded each by a specific set of machine instructions. The operations of concatenating these arithmetic operations are represented by corresponding sets of machine instructions, too.

To take the burden from the user to parse his algorithms in machine instructions as sketched above, a description language is defined which provides the counterparts of the elementary operations. This language specifies nothing else than the order and multitude of elementary operations present in a specific neural algorithm, their concatenations and the dimensions of the vectors and matrices to be operated on. This way the user is free to implement a large variety of algorithms which comprises the neural net algorithms, of course.

The description language is elementary in the sense that it provides a basic syntactical platform for the creation of "macros" describing a specific neural net algorithm. If the user doesn't like to work out the neural algorithm in terms of elementary operations a graphical user surface may produce this input more easily.

In any case, the task of the compiler is to

-- Allocate memory for the weight matrices of all layers in the network.

-- Allocate memory for the patterns to learn or to recognize in the data units.

-- Allocate memory for intermediate results in the data units and the local Δ_{lp} memory.

-- Calculate a table of the transfer function f(..) and its derivative and to store the tables in the memories of the data unit (possibly different tables for different layers).

-- Create the machine-program by selecting sequences of machine instructions for every network-command and fill in the actual addresses of the weights and patterns needed as well as repeat counts for the number of operations of each kind .

In doing this the compiler is not restricted to formulas (1)-(3). More general code can also be handled. If new network paradigms arise that can only be handled in parts by the neuro computer, the host can do the rest of the calculations. Provisions are made in the compiler to allow for that, too.

The source code of the cross compiler itself is written in C + +. Since there are several C + + compilers available for different machines (including UNIX workstations and PCs) which support this de facto standard, it is possible to provide cross compilers running on all of these computers.

CONCLUSION

A neurocomputer architecture has been described which features the following characteristics:

-- System architecture is based on the compute-bound algorithmic elementary operations shared by all neural networks.

-- Systolic execution of the elementary operations by a specific VLSI Neural Signal Processor MA16, and of the remaining operations by commercially available DSPs or microprocessors.

-- Carrying on the systolic computation to 1- or 2-dimensional arrays of MA16s.

-- Each MA16 is provided with its own off-chip weight memory as well as template memory.

-- Systolic communication and control architecture for the array and the weight memory.

-- VME interface to the host.

-- Neural network description language that specifies the neural algorithms in terms of a sequence of elementary operations and their optional concatenations with DSP or host operations.

-- Cross-compiler that translates the description language into the machine language of the neuroemulator.

The proposed neurocomputer concept is sizeable to the applicational domain in terms of processing power, memory and flexibility. Throughput rates at the chip site of the order of 500 MC/sec (1 Connection = 16 bit) can be realized with 1μm CMOS technology. At the board level a system performance of 128 GC/sec can be achieved with 16x16 of these chips concatenated to a 2-dimensional array, which is sufficient for entering vision 1 (see figure 1). This performance is to be compared with the one offered by Sandy/8 [13] or the RAP [14]. Both realize a ring-systolic architecture similar to our's and Kung's [15,16]. Sandy/8 comprises 256 TMS320C30, RAP uses 4 of these DSPs on a board. In terms of connections per second Sandy/8 is equivalent to 8 MA16, whereas a 8-board RAP with 32 DSPs would equal one MA16.

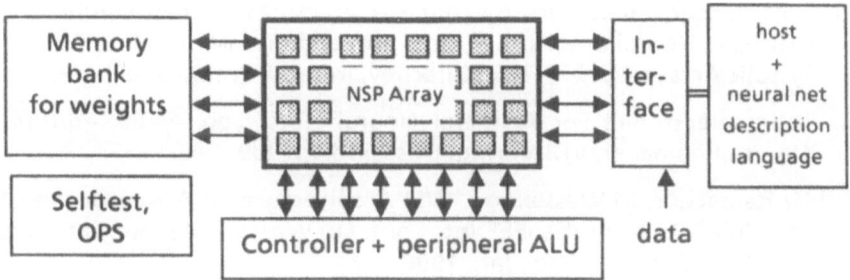

Figure 29   Concept of the SIEMENS Neurocomputer

In terms of network size and support for various neural nets and learning a neurocomputer concept of this kind gets to new dimensions. It will enable the user to access real-world applications in reasonable time and study in breadth and depth the potential of neural networks. SIEMENS' neurocomputer as drafted in Figure 29 can therefore be considered as a research instrument as well as a design platform for working out application-specific neural system architectures in terms of dedicated software and hardware.

REFERENCES

[1] DARPA Neural Network Study, pp.34 (figure 2.14-15), AFCEA International Press, Nov. 1988

[2] ibid.pp.330 (figure 28.5)

[3] H.P.Graf, "A Reconfigurable CMOS Neural Network", Digest of Technical Papers of the Int. Solid State Circuits Conf., vol.33, pp.144, San Francisco, Febr. 1990

[4] L.Curran, "Wafer Scale Integration Arrives In 'Disk' Form", in: Electronic Design, pp. 51-54, Oct 26, 1989

[5] M.Holler et al., "An Electrically Trainable Artificial Neural Network (ETANN) with 10240 Floating Gate Synapses", Proceedings of the IJCNN-89, pp. II-191, Washington DC, June 1989

[6] U.Ramacher, B.Schürmann, "Unified Description Of Neural Algorithms For Time-independent Pattern Recognition", 13th chapter of this book

[7] U.Ramacher, "Hardware Concepts Of Neural Networks", in: ADVANCED NEUROCOMPUTERS, pp.209-218, edited by R. Eckmiller, Elsevier 1990

[8] U.Ramacher, J.Beichter, "Systolic Architectures For Fast Emulation Of Artificial Neural Networks", in: Proceedings of the Int. Conf. on Systolic Arrays, pp. 277-286, Killarney, Ireland, Prentice Hall 1989

[9] "Microelectronics For Artificial Neural Nets", pp. 46-60, editors H. Klar, U. Ramacher, VDI-Verlag, Düsseldorf, 1989

[10] U.Ramacher, M.Wesseling, "WSI Architecture Of A Neurocomputer Module", Proc. of the IEEE Int. Conf. On Wafer Scale Integration, pp. 125-130, San Francisco, Jan. 1990

[11] U.Totzeck, F.Matthiesen, S.Wohlleben, T.G.Noll, "CMOS VLSI Implementation Of The 2D-DCT With Linear Processor Arrays", pp. 937-940, Proc. of IEEE ICASSP-90

[12] J.Beichter, N.Bruls, H.Klar, U.Ramacher, "VLSI Design Of A Neural Signal Processor", to appear in: Proc. IFIP Workshop on Silicon Architectures For Neural Nets, St. Paul de Vence, Nov. 1990

[13] H.Kato et al., "A Parallel Neurocomputer Architecture Towards Billion Connection Updates Per Second", vol.2, p. 47-50, Int. Joint Conf. on Neural Networks, Washington D.C., January 1990

[14] N.Morgan et al., "The RAP: A Ring Array Processor For Layered Network Calculations", to appear in: Proc. Int. Conf. on Application Specific Array Processors, Princeton, Sept. 1990

[15] S. Y. Kung, J. N. Hwang, "Parallel Architectures for Artificial Neural Nets", vol.2, IEEE Int. Conf. on Neural Networks, San Diego, July 1988

[16] U. Ramacher, "Systolic Architectures for Fast Emulation of Artificial Neural Networks", poster paper, presented at IEEE Int. Conf. on Neural Networks, San Diego, July 1988

FROM HARDWARE TO SOFTWARE:
DESIGNING A "NEUROSTATION"

P. BESSIERE, A. CHAMS, A. GUERIN, J. HERAULT,
C.JUTTEN & J.C. LAWSON

INTRODUCTION

Connectionnist algorithms are known to be computationally expensive, for two main reasons. Firstly, Artificial Neural Networks (ANN) are made up of a large number of elements, thus neural network architecture is linked to the concept of massive parallelism. Secondly, due to the iterative algorithms, a large number of processing steps is often necessary in order to ensure convergence and stability in the network. Therefore, hardware supporting these algorithms must be developed together with a well-suited software environment. General-purpose "neuro-emulators" must be used for algorithm explorations and simulations in application-driven implementations. Software packages must allow, in a convivial and interactive way, the description and monitoring (of the evolution) of the simulated neural network. This software layer is fundamental; comparable to the role electronic C.A.D. software plays in microelectronic development.

However, it is clear that neural networks do not constitute the universal solution for complete and complex applications. In fact neural methods must be associated to other, more classical methods such as numerical analysis or Artificial Intelligence, in order to achieve the best results.

In this article, we first describe the requirements for a "neurocomputer". The arguments are based on an overview of typical applications where neural techniques can be used (section "Application Fields"). With some general ideas about the application, it is possible to describe the main features of the "neuro-station" (section "Hardware Specification of a Neurocomputer" and "Environment for Artificial Neural Networks Development").

Secondly, which kind of parallelism, of process control, of communication to use for the "Neurostation" architecture? In section "Preliminary", popular techniques are mentioned. Some choices are made in section "Architecture of a neurocomputer running vector code" leading to the description of SMART (Sparse Matrix Adaptive and Recursive Transforms) as an efficient "Neurostation" core example.

Finally, we describe in section "SOFTWARE ENVIRONMENT" a general programming environment which can be used for different kinds of machines (asynchronous or synchronous parallel machines, sequential workstations...). This is based on the virtual machine concept supported by a high level description language for artificial neural networks.

REQUIREMENTS FOR ARTIFICIAL NEURAL NETWORK DEVELOPMENT

Application Fields

The number of possible Artificial Neural Network (ANN) applications grows with their development. Table 1 summarizes the characteristics of

Applications	Sampling Period (Recall)	Number of		Input Layer Size
		neurons	synapses	
Radar Pulse Identification	50 μs	384	32 k	128
Robot Arm Movement	10 ms < Ts < 1s	312	3600	6
Isolated Words Recognition (1024 words)	10 ms	10 k	280 k	25
Low Level Vision 100 connec./neu.	40 ms	64 k	6400 k	64 k
Risk Evaluation	1 s	6561	≈ 4 M	?

Table 1   Neural networks sizes for typical applications

some applications recently proposed in Darpa study [DAR88]. It groups a panel of low and high level processing applications. Requirements in processing time and in neural network size are very different. For example, for an application in signal processing (radar), time is the most severe constraint, but associated memory size is not critical. At the opposite end of the scale, use of neural techniques in risk evaluation for instance, is based on a very large neural network, but computation time may be of the order of a few seconds. Thus, the constraint is more in memory size than in processing time, if real time processing is not vital.

To sum up, for each specific application a trade off must be made between processing time (computational power), memory size and input-output bandwidth. This compromise concerns general-purpose and dedicated processors. So when considering applications in the hardware environment, these three main features must be balanced. However, the optimal design of dedicated chips benefits from the use of a "Neurostation". Furthermore, whenever the application cannot be well conditionned, a "neurocomputer" very similar to the "Neurostation" is relied upon.

Hardware Specification of a Neurocomputer

Memory size. Neural network architecture and network size determine the number of elements - nodes ("neurons") and links ("synapses"). The set of these elements must be stored in the computer. The memory size is almost directly proportional to the number of "synapses". This number is generally much greater than the number of "neurons". For example, in a fully connected network, the number of "synapses" grows as the square of the number of "neurons".

More precisely, at the data format level, machine representation must be more accurate for the synaptic weights than for the activation states of the "neurons". Indeed, during the learning phase, updating steps of the synaptic weights are relatively small to provide the necessary convergence and stability properties of neural networks. Thus, a floating-point representation of 24 or 32 bits is often necessary; a 16 bit mantissa is required while floating-point is helpful for taking into account proportional effects such as a possible forgetting.

Each synaptic link often requires only two data values; one for the synaptic weight and possibly one for a cumulated incremental value (for example, one synaptic update after p learning iterations). For nodes, different data must be stored: input potential, previous output activation, current output activation, particular threshold (if it is an adaptive one) and error coefficient (in the case of the back-propagation algorithm). So a set of about five variables per "neuron" is often enough at this level. Simulations of neural nets according to a more realistic neural model require a larger set of parameters corresponding to biological data (transfer in the dendritic tree, plasticity of the neural transfer function, more accurate synaptic plasticity, etc).

Communication. Communication speed is necessary to reach the required sampling time. It is a complex problem because in practice, increasing memory size results in reducing bandwith. However, the bandwidth is fixed by the external environment (video rate, etc.) and in the case where memory capacity is not large enough, an additional effort on data transfer is required. Even with enough fast memory, a suitable pipe-line with processing and transfer is often necessary in order to balance transfer and processing rates (see "Pipe-line operation").

Applications	Memory Size	Word bits	In/Out Band-width	Cps Measure (Recall)	MFlops Meas. (Recall)
Radar Pulse Identification	<32 kW	1/neu. 32/sy.	2.5 MBaud	625 MCps	1.2 GFlops
Robot Arm Movement	4.5 kW	32	< 600 W/s	< 350 kCps	< 0.7 MFlops
Isolated Words Recognition (1024 words)	300 kW	32	2.5 kW/s	24 MCps	48 MFlops
Low Level Vision 100 connec./neu.	6.4 MW	1/neu. 8/sy.	1.6 MBaud	156 MCps	312 Mops
Risk Evaluation	4.1 MW	?	< 7 kW/s	4 MCps	8 MFlops

Table 2   Hardware requirements related to implementation of the five typical applications

Computational power. It is very difficult to define the precise relationship between execution time of an algorithm and the theoretical computational bandwidth of the computer supporting the algorithm. The large number of different measurement units is a proof of that [CRU88]. Computational efficiency can be estimated by MFlops or Mips units, but neither of them is really satisfactory. The efficiency of an arithmetic coprocessor (achieving a given number of MFlops) depends on its connection to the host microprocessor. In the same way as the nature of instructions is crucial (CISC and RISC instructions), Mips units are very different according to the diversity of microprocessor architectures. Now, in the field of neural networks, two units have appeared to measure the efficiency of dedicated processors.

The Cps unit (Connections per second) is the most popular and has become a catch phrase for commercial products, but the meaning of Cps varies; it depends on the neural network model (architecture, processing, expected precision...). For example, in the case of a simple direct neural network (without feedback) and without a learning phase, Cps corresponds to the multiplication-accumulation rate! Thus, simulations of recurrent neural networks with a learning phase will require a higher "Cps" to be executed in the same amount of time as the direct network without any learning ability. There is another drawback; Cps does not include system operations supporting inputs and outputs.

Because of the great variety of neural network implementations, there are no realistic benchmarks to classify all the architectural solutions. Nevertheless, in this review we shall use Cps as our measure. In fact, each Cps measure is calculated as a ratio of the number of synaptic weights during the sampling period. So applications with small sampling period or with a large number of synaptic weights are computationally expensive. It is important to note that this Cps measure remains abstract if the Average number of arithmetic Operations Per SYnapses (let this unit be denoted by AOPSY) at each synaptic modification is unknown.

In Table 2, the columns "Cps measure" and "MFlops measure" show the necessary computational power for the recall phase for each application. In this phase, 2 AOPSY (a multiplication and addition) is an optimistic number for a non-recurrent neural network. With the ratio of 2 between Cps and MFlops measurement, we obtain an order of magnitude which is sufficient to analyse, in a critical way, present hardware realizations in this field.

Environment for Artificial Neural Networks Development

Some general ideas. In any application, neural network algorithms are a part of the complete processing chain: the neural network must be connected to its environment (Figure 1). It receives data from one or more sources and drives data outside the network. The nature of input and output data may be different (for example, in a multi-layer architecture). Information types computed in neural networks are only activation states at the neural level and transmission efficiencies at the synaptic level. This restriction implies data transformations before and after the network processing, i.e. at the beginning or at the end of the complete processing. "Real" or symbolic data are processed to suite neural net representation.

Moreover, an important point related to neural algorithm analysis is the way of monitoring the evolution, visualizing multi-dimensional data as "neurons" or "synapses" states [KOR89]. According to behavioural analysis of neural nets, it is fundamental to use such graphic tools and develop new ones. These tools have to help users extract relevant information among the great amount of data derived from neural network states.

Consequently, if the user can see, at a realistic rate, and analyse neural network behaviour, he must be able to modify the behaviour toward a

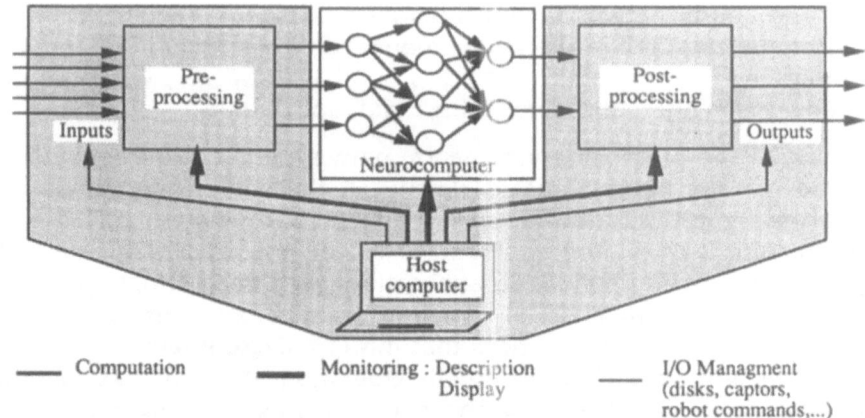

_____ Monitoring : Description
        Display

_____ I/O Managment
       (disks, captors,
       robot commands,...)

Figure 1   Schematic view of the overall environment of a neuro-station

suitable target. This concept of "see and act", is very important, and it is one of the most difficult tasks for a simulation software.

Another important task for simulation software, is to provide users with dedicated objects in order to help in the generation and description of neural networks. Other graphics tools must be developed for this description level [KOR89].

Consequences.   It is clear these two previous tasks (analysis of data & description of networks) act as supplementary software layers between the hardware architecture of the "neurocomputer" and the user, consequently execution time increases and this software environment is an extra load for the "neurocomputer". This effect is particularly important during the learning phase; this phase is very time consuming and generally demands strong external control (user) to converge, stabilize, etc.

Figure 1 sums up these ideas and shows that a lot of extra processing is required in order to turn a "neurocomputer" into a "Neurostation".

A NEUROSTATION CORE

Preliminary

As the first sections have clearly shown the aim of a "Neurostation", as for any workstation, is for the user to move quickly from the idea to visualizing the results. The main steps involved are: programming, compiling, executing and monitoring.

While present day workstations are able to cater for three of the above points, the execution phase remains far to long for most ANN applications. Hence the issue is to speed up programs without jeopardizing the achievements for the three other points. This requires a comprehensive design which is seldom if ever found in commercial products.

There are basically two ways of increasing the power of a computer:

--   operating frequency hikes,
--   multiple engine coupling.

Supercomputer.   The first approach which is tantamount to supercomputing leads to skyrocketting prices both in purchasing and in maintaining the physical environment: as the fastest available technology is used (ECL, AsGa: 100-500 MHz) cooling systems are required. At the other end, in ambiant air, top CMOS technology leads to the best computing power to price ratio (10-50 MHz). But whatever the technology, only pipe-lining allows maximum efficiency. Let us have a quick review of what pipe-lining consists in.

Pipe-line operation.   For instance a processor may require four steps to perform an arithmetic add:

1.   fetching the instruction which codes the operation,
2.   decoding the instruction and fetching the data on which to operate,
3.   adding the data,
4.   writing back the result.

On the one hand, without pipe-lining, the processor will complete all the four steps before going on with the next instruction (Figure 2). A lot of hardware could even be shared among the four stages which are activated sequentially, in order to reduce the overall number of transistors.

On the other hand, using pipe-line operations, each step is performed by a separate stage (Figure 2). Each stage is connected to the following one through a register, and a new instruction can enter the pipe-line as soon as the first step is completed with the current state pushed into the following register. Assuming equal processing time for each stage, instructions will flow smoothly in the pipe-line with an actual throughput four times greater than with the standard architecture.

The second solution, of course, uses more transistors than the first one, but this is no longer an issue. On the contrary, the better structured architecture is easier to implement with a silicon compiler and dedicated stages perfom better.

However, some interlocking mechanism must be provided in order to avoid dependent instructions running wild as the second instruction

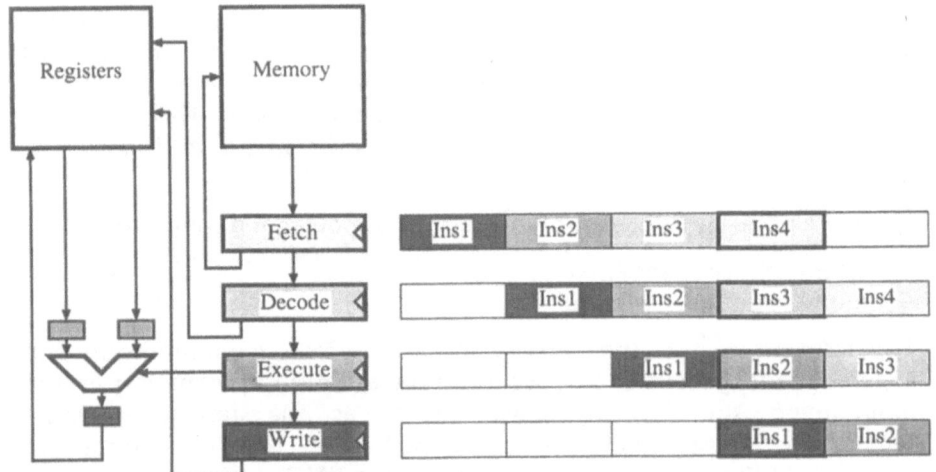

Figure 2   Pipe-line operation

enters the pipe-line before the first one has completed. Though hardwired solutions exist (using register scoreboarding as in 88000® or SPARC® [CYP90]), the most efficient code will only be produced with the compiler ordering the instructions to always keep the pipe-line full of independent instructions (as in mips® [KAN88]). A loop of instructions performing a vector operation is hence easily optimized. A small set of units should even work concurrently, performing many loop steps simultaneously (as in processor arrays [BAR68]).

These techniques are used in supercomputers with vectorizing compilers tracking Fortran code in search of a possible optimization. But programming such monsters becomes tricky as the user must keep a number of constraints in mind, for the compiler not to lose the possible vector operations' trails.

RISC and memory system.   A problem with a very high speed engine is that not only should the pilot have it running smoothly and continously but the engine also should be fed with enough data to crunch. And this has been bothering memory designers for years. A model application given, the tradeoff between speed and cost is usually solved thanks to a good memory hierarchy choice. The hierarchy is commonly made of internal registers, a data cache, a main memory, disk clusters and streamers.

This is clearly illustrated with RISC machines. A RISC processor generally houses many registers (about 32 for instance) on which most instructions operate (Figure 3). With loads and stores being the only instructions to access memory, most instructions execute in one clock cycle. The compiler is charged with properly allocating the registers depending on the probability of reusing the same data. Hence accesses to the cache memory are as limited as possible and once again a mechanism (hardware or software) ensures that the cache is filled with the most likely data to be used within the near future. And so on, up to the streamer. All things considered, everything happens as if all the data were in registers.

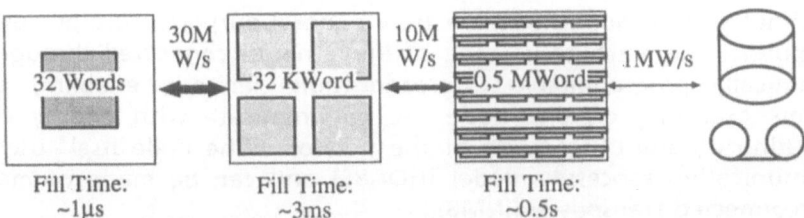

Figure 3   A typical RISC Workstation. Use-Time is supposed to be far greater than Fill-Time

In fact, it is not that simple as the overall efficiency strongly varies depending on the application, the compiling technique, the system memory allocation strategy and the relative size and speed of each memory stage.

Massive parallelism.   Breaking with all previous work, a promising approach appeared some years ago in coupling hundreds or thousands of relatively simple processors running relatively slowly instead of developing a few very high speed engines at all costs (Supernode [MUN89] or CM-1 [HIL85] for instance). This new approach implies programming every existing application from scratch, but that aspect is unimportant as far as neural networks are concerned. On the contrary both are instances of a massively parallel paradigm, hence due to develop in symbiosis.

However, the efficient coupling of a huge number of processors still requires extensive work. Clearly, it is impossible to physically connect each processor to all others [FEN81]. There are basically two ways of bypassing this problem:

using a frame of limited connections regularly reproduced over the network together with a systolic data flow possibly emulating full connection [BLA88],

using dynamically configurable networks with switches or routing schemes (as in telephone commuters) to pass a message from any point to any other location (see [MUN89] and [FAU89] for instance).

Systolic architectures are usually very efficient for tasks such as matrix by vector products, but systolic implementation is limited in the problems it addresses and it should be viewed rather as a building block than as a general-purpose "neurocomputer" on its own (though some flexible architectures are proposed [RAM89][BLA90]). Furthermore, the data sets should be large for systolic speed-up to be effective, but the biggest problem in developing a "Neurostation" is monitoring the execution: the output of internal data with large systolic arrays is usually very complex if not impossible and it often halts processing for a large amount of time.

Machines such as Supernode are truly general-purpose multi-processor computers. The machine is made of many nodes connected through a dynamically programmable interconnection network, enabling any process executing on one node to communicate with one process executing on any other node of the network. The node itself uses a communicating processes model [HOA85] and can be made of many interconnected Transputers [INM89].

The problem is then to map each process upon the network in order to use this last one as efficiently as possible. The problem is further complicated since there is no deterministic way of forecasting when a process will awaken; a slight temporary overload of the network may wreak havoc. Even if mapping is efficient, there must be a lot of overhead for the sake of synchronizing many a priori asynchronous processes (this is a general drawback of message synchronizing schemes). Now the code should have been vectorized in order to execute on an SIMD (synchronous) machine with no overhead. The next part involves tackling this aspect.

Architecture of a Neurocomputer running vector code

What is the best way to efficiently implement neural code in a synchronous machine? Obviously, running a vectorizing compiler on any sequential program is not the answer, but a step in the right direction is made when a vector oriented description is selected. Without the need to break operation into nested loops, much time is gained. The compiler optimizes the code more easily, unmistakably fixing any vector operation. Eventually, the higher level the description, the better.

In this part we focus our attention on how to design a general-purpose "neurocomputer" running vector code. Such a development involves the concurrent design of an architecture, its hardware implementation and the corresponding software layers. However, thereafter the preceding considerations are used to choose first a memory system, then a processor network, next vector operation is explained, a programming model is introduced and finally the first software layers are outlined.

Distributed memory. Clearly running vector code does not mean just another array processor design. These architectures generally offer no suitable solution to the memory system bottleneck. With ANN algorithms there is almost no data reuse while computing the next state of the network: a synaptic weight is used once and only once until the next state evaluation. That is to say, a cache memory must accomodate the whole set of synaptic weights to be effective. The same is true while performing the weight updates. For a network of one million connections, a cache over one million words in size is required -impossibly large. The best solution is to distribute processing along with memory.

The question of the optimal processing level for the Processing Elements then arises (a PE is made of a Processing Block plus a Memory Block and PEs are connected together through a network). It could for instance be a bit level as in CM-1 [HIL85]. As long as logical operations or low precision arithmetic opertations are involved, a low level is appealing since monolithic PEs are feasible (reducing I/O problems [MOR89]). Although, in executing various high precision operations, using a word level is advisable. An handcrafted parallel FPU chip -which is further optimized and easier to use- together with a chip the latest in RAM technology can be chosen. Now 32 bit precision is required, so the second option is chosen.

Network. The mesh connecting the PEs together still has to be chosen. The use of a multiport main memory (on which rely many processor arrays) is rejected due to its complexity and its potential to bottleneck. Message passing is rejected as it leads to asynchronous machines and overhead. Instead systolic architectures are used as a model.

A planar mesh topology is more often used since it matches well with microelectronic constraints. On the contrary, a linear structure is far more flexible since it can easily emulate higher dimension (2D plan, 3D space...) without loss of efficiency. One further point, a problem with systolic architecture often is the cancelation of side-effects which appear at the border, now a line has at most two edges instead of four with a square, eight with a cube... And the decisive point relates to data monitoring, there is no problem accessing any PE of a line from one side, while accessing of an internal PE of a square mesh from one side is much more complex (Figure 4).

The major drawback of a line of PEs when compared with a 2D mesh is time latency. Considering N^2 PEs, it takes N^2 steps travelling from one end of the line to the other, while going from one corner of the square to another (that is to say completing processing) takes between N and 2N steps. Now using a small number of PEs to achieve about 100 MFlops, a processor line (or ring) as in CRASY [GUE88] or WARP [ANN87] should be chosen.

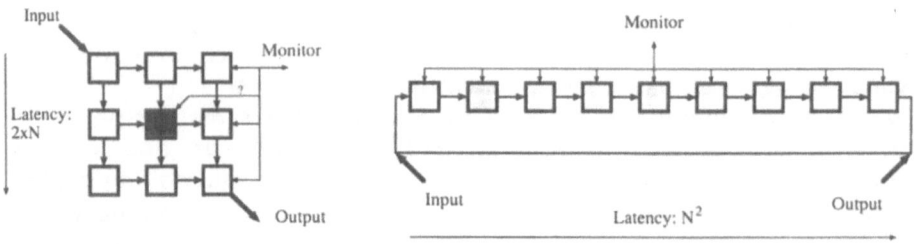

Figure 4   Planar mesh and Ring topologies compared

Vector operation.   Let us now define what we mean by a vector operation. As a matter of fact what we are interested in, as far as vectors are concerned, is nothing but parallelism. This is the reason for our broader definition of vector operation than the one implied by the common mathematical one. We would simply define a vectorial operation as a small set of instructions to repeat many times.

Figure 5 gives two examples. The simpler is a vectorial add, an addition is repeated independently with each pair of components of the two vectors. The second is a scalar product. First the components of each pair are multiplied between themselves, then the results are added amongst themselves. Obviously there won't be any trouble implementing such operations on a line of processors.

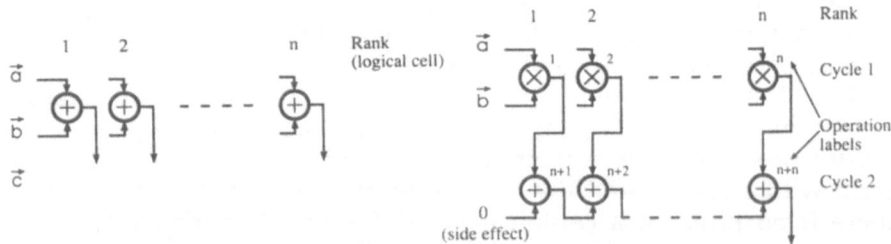

Figure 5   Two examples of vectorial operation

Though, we must state some provisions concerning data dependency: an instruction should use only independent data (as in example 1) or the result of one instruction executed before with the same rank (see instruction labelling of example 2) or even the result of one instruction executing with the preceding rank. To sum up, an instruction labelled $l = k \cdot n + r$ can only use results of the instructions labelled $m = p \cdot n + r$ or $p \cdot n + r-1$, with $m < l$. These provisions allow the implementation of most needed operations using only the direct data paths of the line of processors (no need for skipping one processor to pass a message from its preceding neighbour to its following).

Even parallel conditional operations are valid (Figure 6). First, a vector compare is performed between the terms of each component pair. Then, each logical value is used to select the result between two values. This only requires a few cycles, since arithmetic instructions are assumed to be one-cycle. A MIMD machine would probably not perform better.

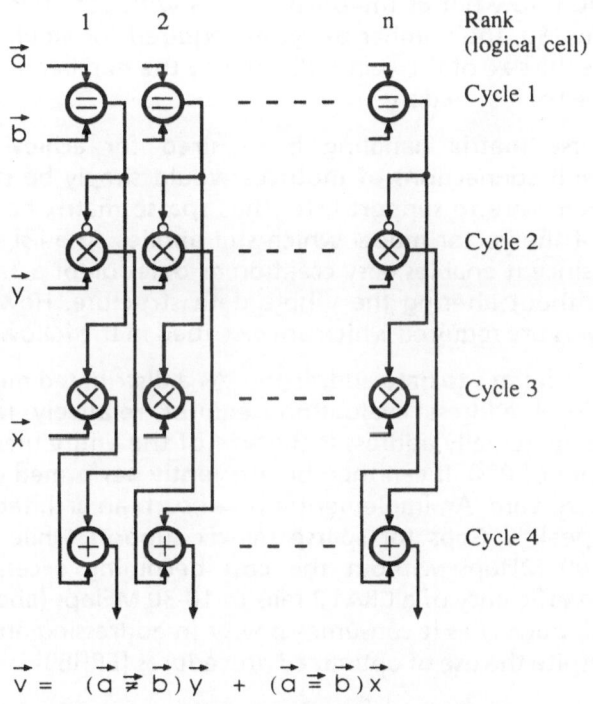

$$\vec{v} = (\vec{a} \neq \vec{b})\,\vec{y} \; + \; (\vec{a} \equiv \vec{b})\,\vec{x}$$

Figure 6 Parallel conditional operation

Programming model. A programming model must then be defined. So far, the architecture described seems very similar to any other systolic architecture. In fact, in a pure systolic model, data travel through the network without instruction flow; each cell is supposed to repeat the same preprogrammed task. This model is generalized with the wavefront model [KUN82] where a wavefront instruction flows across the network, together with the data on which to process (if the mesh relates to an Euclidean topology, the wavefront is an hyperplane of the corresponding space). Hence the architecture is much more flexible. However, if the dimension of the cell array is reduced to a line, it becomes in fact a simple pipe-line.

Obviously, the best way of programming a pipe-line architecture is to use a RISC model. Adhering to that, a way to implement the vector operation must be chosen. The goal is to execute one instruction on each cell, but when the vector size exceeds the cell number, the whole calculus will require many steps (with physical cells emulating many logical ones). Such an operation could be performed with a microprogram, leading to multi-cycle instructions and poor flexibility. It is highly preferable to use a loop (or a sequence) of single-cycle instructions, which will of course increase the instruction flow but as the bottleneck is with data this should cause few problems. So, the number of cycles required for emulating a vector instruction is the size of the vector divided by the number of cells: it is the actual number of instructions.

In fact, sparse matrix handling is required for achieving the best efficiency. Null connections of matrices would simply be skipped using dedicated hardware to support lists. Thus sparse matrix handling shows the power of the vector model which still applies. The list usage is very convenient since it enables easy creation or deletion of a "synapse" or a "neuron" without altering the whole data structure. However, proper software layers are required which are discribed in the following section.

Another point relates to data addressing. As a distributed memory system is chosen, local address evaluation requires relatively few bits (the memory size to actually address is the size of the entire memory divided by the number of PEs). It can then be efficiently performed with low cost dedicated hardware. An implementation of such an architecture uses for instance a peak 1 Gops for sparse matrix support while delivering a sustained 100 MFlops without the cost becoming excessive. On the contrary, the efficiency of a CRAY2 falls to 10-30 MFlops (about one tenth of its linpack power) as it consumes power in addressing and in memory conflicts, despite the use of optimized procedures [ERH89].

These operations performed for sparse matrix support is implicit in the register model. The reference by an instruction of a vector register will automatically produce the parallel evaluation of the address of each component. Besides vector registers, FIFO registers are defined to consistently support data transfer and vector operation (hence extern I/O are directly supported within the processor model). FIFO registers also ease signal and image processing algorithms as they allow a simple implementation of delay lines (for convolutions) and are involved in Data Flow models [EDW87]. Finally, an implementation of FTT [COO65] is possible using the FIFO registers to perform a perfect shuffle [STO71] between cells.

The SMART machine which is under development at TIRF laboratory implements all the above architecture considerations [LAW90]. An overview of the complete system is given by Figure 7. The machine uses CMOS technology and an air-cooling system. It is built around a SPARC®

Integer Unit core which appeared to be the most scalable and open RISC standard. But the key part is the coprocessor controller which recognizes custom instructions and supports the register model.

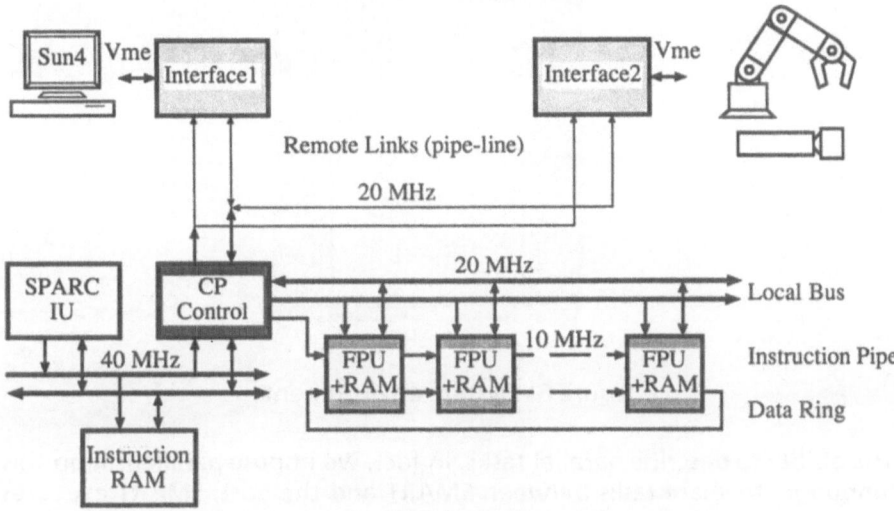

Figure 7   SMART coprocessor environment

Software layer. In order to achieve the best efficiency, any RISC machine needs an advanced compiler. In fact, the RISC model is endorsed to allow easy code optimization. That is obvious for a RISC vector machine like SMART which uses extensive pipe-lining, but SMART presents other features relying on software.

Clearly, the combination (workstation + SMART) forms a parallel system in which each machine is based on the same SPARC®processor model (see Figure 8). This point of view allows the definition of a simple model for the combination.

Then a new language programming the combination appears necessary to get the best implementation of ANN simulations on this system. This language should have certain features, such as:

The ability to implement algorithms other than the ANN ones, for instance signal processing, image processing, etc, as preprocessing or postprocessing tools (see "Some general ideas").

The language should have few instructions defining matrices and vectors and which perform all possible operations between them; sparse matrices will be treated as special types.

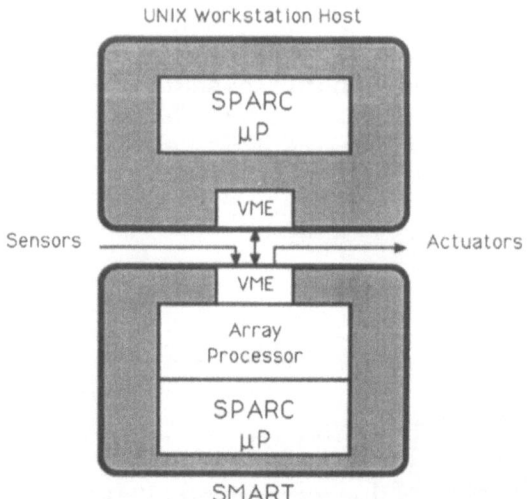

Figure 8   System environment

The ability to describe parallel tasks. In fact, we impose parallelism on this language to share tasks between SMART and the host; SMART executes all vectorial instructions concerning matrices and vectors and monitoring is performed by the host. Thus a data flow is established between the two machines and is controlled by a special protocol introduced into the language.The user may assign specific tasks for each machine in his program, therefore the synchronization is achieved using "rendez-vous" placed by either user or compiler.

A good choice is made for the language by extending C language vectorially. In this way, all available software on the host is reusable by adding a simple interface. This language will be the first step toward an implementation of a high level specification language for ANN on SMART.

An advanced compiler must then be developed for the extended C according to SMART architecture (obviously, the vector C language allows easy and efficient compilation for any UNIX® workstation if the program is to be executed without the SMART accelerator). This means, the compiler should handle the following points:

All classical obstacles caused by parallelism-communication, synchronization, etc.

Memory management for matrices, especially sparse ones. Matrices must be stored in memory in lists which provide information about matrix structures. Thus the compiler must deliver these lists for each matrix declared in a program, using some strategies for the best memory allocation and load balancing upon the cells [LAW90].

The various number of cells allowed by SMART. A recompilation is required for programs when a change in the number of cells takes place; the compiler must take into account different pipe-line models.

SOFTWARE ENVIRONMENT

Motivations and Needs

Having defined what a "Neurostation" core could be, the software tools may be divided into two classes, the executing environment and the programming environment:

- Executing environment - In most applications, neural algorithms are part of a complete processing chain (section "Environment for Artificial Neural Networks Development"); the neural network must be connected to its environment. It receives data from one or more sources and drives data outside the network. The nature of input and output data may be different from one another (for example, in a hierarchical architecture). Information types computed in neural networks are only activation states at the neural level and transmission efficiencies at the synaptic level. This implies data transformations before and after neuromimetic processing; at the beginning or end of the complete processing. "Real" or symbolic data must be "compiled" to be suitable for neural net representation.

- Programming environment - An ANN has to be programmed, debugged and controlled. This supposes the existence of a set of tools to interface the ANN with the programmer. Programming environment needs and proposed solutions are detailed in the following sections.

Overview of programming environment

An ANN programmer needs the following:

-- a powerful language to easily describe networks;
-- the possibility of using this same language to program different machines (either parallel or sequential);
-- man-machine interface graphical tools to provide clear representations of program behaviours and results.

To satisfy these requirements, a possible programming environment structure may be represented by the following schema (Figure 9):

The programmer has a mental representation of the nets to be coded and requires a High Level Description Language (HLDL) able to take into

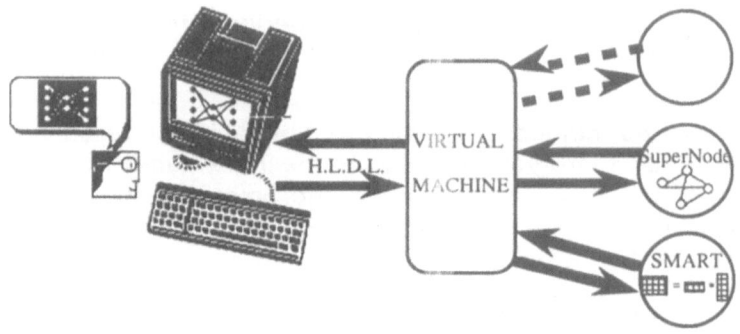

Figure 9   Programming environment overview

account the characteristics of this mental representation in order to have
a concise and easy expression of these nets. Given the requirement of
several target machines for the HLDL, compiling the HLDL program for a
virtual machine and then compiling the virtual machine code into one of
the target languages is necessary. Two target machines were selected:
SMART and Supernode. Supernode is based on CSP (Communicating
Sequential Processes programming model [HOA85]) and SMART is based
on sparse matrix operations. During the execution, data go backward
from target machine to virtual machine and from virtual machine to
graphical interface to provide easy to understand representations of
program behaviour and results. For this last step, an HLDL is also of critical
importance, allowing the design of man-machine interfaces
independently of target machines.

The introduction of a virtual machine has several advantages:

-- the definition of the virtual machine instructions ensures clarification
   of the fundamental concepts and their semantic properties,
-- the programming language for ANN is independent of the actual
   target machines,
-- only the compiler between the virtual machine and the target
   machine needs to be written when a new target machine is
   considered.

Virtual machine

The synthesis between the different parallel programming paradigms is
based on the following concepts introduced for the description of any
ANN at the virtual machine level:

-- cell,
-- aggregate and connector .

As a first approximation, a network could be represented graphically by using circles for cells, sets of cells for aggregates, and diamonds for connectors (Figure 10).

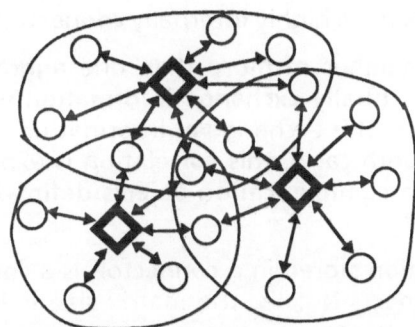

Figure 10   Graphic representation of a network

Cell.   A Cell is a process executing a program as soon as its inputs are available. Some output is computed and may be broadcasted through the net to be used as input to other cells.

Consequently, there is no way to explicitly control the execution of cell's programs since they are triggered by the availability of their inputs from other cells. As cell behaviour is only a function of their interaction and their descriptive program, control structures of the language should reflect the type of synchronizations attached to any interaction. This function will be devoted to objects called connectors. The only constraints imposed on a cell are on its interface with the connectors, there are none on the internal program. A synchronous cell should accept a scalar as its input and return a scalar as its output. An asynchronous cell, should also return a Broadcasting-time (which may be superior to the local time to allow modeling of delays) and a Next-wake-up-time (which tells the connector when to wake up the cell).

When a cell is shared by several aggregates, it is divided into as many processes as aggregates -one process per aggregate. To enable cell-sharing, a way of coping with two (or more) simultaneous calls to the one cell by different aggregates must be implemented. Thus a global variable Minimum-next-wake-up-time (the minimum of the next-wake-up-times of all the processes in the cell) is implemented. This variable is used by connector protocols. Some other global variables are also implemented and in some cases code can be shared between processes too.

A cell belonging to several aggregates may be activated independently by each aggregate (for instance, a cell belonging both to a layer and to a cluster in competitive learning [RUM86]) or all aggregates may be forced to synchronize when activating this cell (for instance, a cell belonging to several aggregates in a Boltzmann machine [ACK88]).

Aggregate and connectors. An aggregate is a set of potentially fully connected cells, whose activities are temporarily correlated; most of the time it corresponds to an highly internally connected part of the net.

A cell can be a member of more than one aggregate. Any cell in an aggregate can potentially exchange information with any other cell in the same aggregate. This exchange of information can be synchronous or asynchronous. In both cases, this correlation is expressed by a protocol. The connectors are communication objects defined to implement those protocols.

The main information stored in a connector, is a connection matrix **W**. If an aggregate connects N cells, the matrix size is N^2. In most cases the connection matrix is sparse, and its structure reflects the topological structure of the aggregate.

In fact, the connection matrix encodes more than communication capabilities, it contains entries (real or integer) usually called synaptic weights. These weights are used as parameters for a composition function Π and an integration function Σ which are used to transform information exchanged by the cells. The broadcast information should go through the same function $\Sigma o \Pi$ before being transmitted to the cells of an aggregate. Typically, Π is a multiplication function and Σ is a summation.

A basic operation of the virtual machine is **"activate"** -activating an aggregate. Roughly speaking, this activation consists in broadcasting the information delivered by every cell of the aggregate to the others, through the transformation $\Sigma o \Pi$. This process implies the automatic execution of the programs associated with each cell of an aggregate, as soon as their inputs are available. In fact, the differences between synchronous and asynchronous models appear in this operation. The synchronous and asynchronous activation of aggregates are detailed in the next two paragraphs ("Activation of synchronous aggregate" "Activation of asynchronous aggregate").

Another basic operation of the virtual machine is "learn" -executing learning in a given aggregate. A learning function L is associated with each aggregate. For any given connection matrix and some parameters reflecting the past activity of the network, L produces a new connection matrix.

Activation of synchronous aggregate. If we suppose that each cell delivers an output, then each synchronous aggregate has a vector V representing the output values for all cells; then v_j (j^{th} component of V) gives the output of the jth cell in the aggregate.

The composition function Π can be applied for each couple (w_{ij}, v_j) to obtain c_{ij}, then the integration function Σ produces for all N-tuples of c_{ij} the resultant vector R. In the case where Π is multiplication and Σ is addition, the whole operation can be written using a matrix operation R = W*V. Each r_i can be an input for the jth cell, whenever the inputs are available, the jth cell then re-executes its own program automatically to produce the new v_j, for another iteration.

In the synchronous case, the role of a connector is very simple; that of transmitting data between cells from a static output set to a static input set of cells.

Activation of asynchronous aggregate. The asynchronous model has "time" as extra information to be exchanged between cells. Three values are delivered by an asynchronous cell after execution of its program:

-- the output,
-- the time when the values are delivered,
-- the expected time for the next exchange of information between the connector and the cell.

The temporal information is controlled by the connector which wakes up the cells at the right time. Asynchronous models use the data structures described previously for synchronous models, with the additional items:

-- a vector B containing the arrival time of the last message coming from each cell (b_i for the ith cell);
-- a vector Nmin containing the next activation time for each cell;
-- a counter t representing the last activation time for a cell in the same aggregate.

After the initialization phase of W, V, B, Nmin, and t (t = 0), the activation of an aggregate is described by the following algorithm:

-- search the next cell(s) to wake up and update t,
-- send to all such cells the vector V after application of the composition function Π,
-- send to these cells the vector B,
-- all cells execute the function Σ and their own programs as soon as V and B are available,
-- each cell i sends to the connector its output values, the present time (b_i), and its next activation time (n_i).

Let us note that the synchronous mode can be emulated in the asynchronous mode by waking up the cells at each clock step.

The role of the connector is more complicated than in the synchronous case. In particular the ?-set of cells (cells that have to be consulted) and the !-set of cells (cells that have to be woken) are no longer statically specified and must be computed by the connector at each step. The !-set is determined by comparing t to the Nmin of all cells in the aggregate and the ?-set is then determined by computing all cells of the aggregate which are connected to those of the !-set.

We expect through the proposed approach to come out with a tool usable for a wide range of machines and offering full program portability. Using the proper aggregate activation mode will allow the best performance either using a synchronous machine or an asynchronous one. Obviously, the synchronous machine is usually more efficient, though the asynchronous machine is more general.

CONCLUSION

The concept of a "Neurostation" is becoming more and more important. On the one hand "Neurostations" are involved in the simulation of ANN and in connectionist applications with industrial constraints (data size, speed, ...). On the other hand, a VLSI integration of a particular ANN emphasizes technological constraints. These generally imply a new implementation of the algorithm; for instance, the communication inside a chip may be reduced. For numerical implementations, data bus width must be trimmed to the required accuracy. For algorithms working around these constraints, problems as stability, convergence, scalability... must be carefully analysed. All these studies are carried out using a "Neurostation".

A brief analysis of classical applications using ANN shows that a "Neurostation" must be computationally powerful and well balanced in processing, input-output transfers and graphics managements.

According to ANN architectures, a "Neurostation" is designed for huge networks (thousands of connections), as well as heterogeneous and convoluted networks composed of functionnally different pools of "neurons". Huge networks require suitable memory systems (large memory, sparse matrix handling...) and complex networks require suitable schemes to retrieve data in a convoluted data structure.

Research in Computer Science leads to some techniques for increasing frequency and coupling multiple processors. For a "Neurostation" based on a parallel coprocessor, some of these techniques are used. The systolic approach is promising for neural network implementations as matrix by vector produtcs are basic operations in connectionist computations. The problem lies in the architecture scalability. Nevertheless, the approch is strengthened as partitioning techniques are developed for sparse matrix handling.

Further, a "Neurostation" must be flexible for many reasons. First, the arithmetic core must support both common vector operations (matrix products, scalar products...) and more general ones (parallel conditional operations...). Second, monitoring the ANN requires extra data communication which must not halt processing for a large amount of time. For this, multiple data paths must be introduced. Thus a "Neurostation" open toward a wide range of algorithms (generally algorithms which are translatable into matrix transforms) follows; e.g. image and signal processing algorithms.

To achieve efficient implementation, any "Neurostation" must be supported by a powerful software environment. The proposed environment is based on a virtual machine approach. This enables designs which are independent of target machine peculiarities. Two parallel machine models are addressed -i.e. synchronous machines and a asynchronous machines. This approach is very powerful since the best performance can be achieved using the proper translation for either of the models.

Obviously the synchronous machine is usually more efficient though the asynchronous machine is more general. The proposed synchronous processor, named SMART, matches all hardware and software considerations. Accordingly, SMART which is developed at TIRF Laboratory, supports sparse matrix processing as a general formalism to compute heterogenous neural architecture and networks with dynamic topologies.

The work described in this paper is part of the ESPRIT-BRA-3049 project named NERVES. (NEurocomputing Research and VLSI for Enhanced Systems), and a project in PRC-AMN (Programme de Recherches Coordonnées en Architectures de Machines Nouvelles) of the French CNRS (Centre National de Recherche Scientifique).

REFERENCES

[ACK88]   Ackley D.H, Hinton G.E. & Sejnowsky T.J., "A learning algorithm for Blotzmann machines" in Connectionist models and their implications edited by David Waltz & Jerome A., Feldman Ablex publishing corporation, Norwood, NJ (USA), 1988

[ANN87]   Annaratone M., Arnould E., Gross T., Kung H.T., Lam M., Menzilcioghu O. & Webb J.A., "The Warp computer: Architecture, Implementation, and Performance", IEEE Trans. Comp., Vol. C36 N°12, December 1987, pp.1523-1537

[BAR68]   Barnes G.H. et al., "The Illiac IV computer", IEEE Trans. Comp., Vol. C17, August 1968, pp.746-757

[BLA88]   Blayo F., Hurat P., "A VLSI systolic array dedicated to Hopfield neural network", in VLSI for Artificial Intelligence, Delgado-Frias and Moore Eds., Kluwer Academic Publishers, 1988.

[BLA90]   Blayo F., Lehmann C., "A Systolic implementation of the self-organizing algorithm", INNC90, Paris (France), July 9-13, 1990

[COO65]   Cooley J.W., Tukey J.W., "An algorithm for the machine calculation of complex Fourier series", Math. Comp., Vol. 19, p297

[CRU88]   Cruz C.A., "Neural network hardware performance assessment", INNS 88, Sept. 88

[CYP90]   Cypress Semiconductor Corporation, "RISC Family Users Guide", second ed. February 1990

[DAR88]   Darpa, Neural Network study, AFCEA Int. Press, 1988

[EDW87]   Edward A.L., Messerschmitt D.G., "Synchronous Data Flow", IEEE Proc., Vol. 35N9, September 1987, pp.1235-1245

[ERH89]   Erhel J., "Sparse matrix multiplication on vector computers", Rapport de Recherche N° 1101, INRIA (Roquencourt), Octobre 1989

[FAU89]   Faure B., Mazaret G., "A VLSI implementation of multi-layered neural networks", in VLSI for artificial intelligence, Kluwer Academic Publishers, 1989, pp.159-168

[FEN81]   Feng T., "A survey of interconnection networks", Computer, December 1981, pp.2-27

[GUE88]   Guerin A., Jutten C. & Herault J. "Neurocomputer 'CRASY' a cost-effective solution", in Neural Networks from models to applications, L. Personnaz & G. Dreyfus Eds, I.D.S.E.T., Paris (France), 1988, pp.756-765

[HIL85]   Hillis W.D., "The Connection Machine", Cambridge: The MIT Press, 1985

[HOA85]   Hoare C.A.R., "Communicating sequential process", Prentice Hall International, Englewood cliffs, NJ (USA), 1985

[INM89]   Inmos Limited, "The Transputer Data Book", Redwood Burn Ltd, Trowbridge (UK), second ed. 1989

[KAN88]   Kane G., "mips: RISC Architecture", Prentice Hall International, Englewood cliffs, NJ (USA), 1988

[KOR89]   Korb T., Zell A., "A declarative neural network description langage", Microprocessing and microprogramming, N° 27, 1989, pp. 181-188

[KUN82]   Kung S.Y., Arun K.S., Gal Ezer R.J. & Baskar Rao D.V., "Wavefront Array Processor: Language, Architecture and Applications", IEEE Trans. Comp., Vol C31 N° 11, November 1982, pp.1054-1065

[LAW90]   Lawson J.C., « SMART » in Architectures for Neurocomputers, ESPRIT-BRA 3049 NERVES - DR1/D1, J.C. Lawson, F. Blayo, C. Lehmann & T. Muntean Eds, LTIRF-INPG, Grenoble (France), June 1990, pp.23-56

[MOR89]   Morton S.G., "Digital 'Intelligent Memory Chips' build neural networks with fully programmable synaptic weights", Int. Jour. Neurocomputing, Vol.1 N° 2, 1989, pp.29-46

[MUN89]   Muntean T., "Supernode : architecture parallèle et dynamiquement reconfigurable de Transputers", 11èmes Journées francophones sur l'informatique, Nancy (France), Janvier 1989, EC2 Editeur, Paris (France)

[RAM89]   Ramacher U., "Systolic architecture for neurocomputing", IFIP Workshop on parallel architectures on silicon, Grenoble (France), December 11-13, 1989, pp.1-8

[RUM86]   Rumelhart D.E., Zisper D., "Feature discovery by competitive learning" in Parallel Distributed Processing, D.E. Rumelhart, J.L. Mc Clelland & the P.D.P. research group Eds, The M.I.T. press, Cambridge, 1986

[STO71]   Stone H.S., "Parallel processing with the perfect shuffle", IEEE Trans. Comp., Vol. C20, 1971, p.153

Index

A

accelerator 31, 229, 326

accuracy 4 – 7, 14 – 15, 49, 55, 109, 126, 131, 136, 148, 164 – 165, 171 – 176, 178, 205, 234, 245, 276, 332

activated components 157, 159, 165

activation function 103 – 106, 189 – 190, 193, 195 – 196, 201, 206, 209, 232 – 233, 237

adaptation rule 65

adaptive weight 20

adjustable synaptic weights 47, 61

algorithm

    biological learning algorithm 85

    Boltzmann machine 330

    Booth's algorithm 192

    Hamming distance 175, 178, 180

    Hebb-rule 40

    iterative algorithms 311

Algorithm Mapper 225

analog

    analog data 21

    Analog Encoding 129

    analog multiplex method 91

    analog neural networks 57, 65

    analog neuron 65

    analog sample-and-hold 91, 94

    analog shift register 20

    analog storage 19

    analog synapse 94, 111, 149

    analog technology 169

annealing 179 – 180

application specific processor 187

artificial intelligence 185, 334

ASIC 125 – 126, 134, 137, 149, 191, 247, 249, 252

    ASIC cell library 125

    ASIC processor 125

associative

    association 19, 40, 84, 122, 156, 162, 167

    associative linking 100

    Associative Matrix 155 – 156, 158 – 160, 163, 165 – 166, 168

    associative memory 153 – 155, 157, 167, 169, 171, 174, 176, 184

    auto-association 175

    autoassociative memory 155

asynchronous communications protocol 217

attractor 177 – 179

auto-adaptive 169

auto-adaptivity 179

axon 84 – 86, 128

    axon hillock 85 – 86